FROM A GEOMET

LOGIC, EPISTEMOLOGY, AND THE UNITY OF SCIENCE
VOLUME 14

Editors
Shahid Rahman, *University of Lille III, France*
John Symons, *University of Texas at El Paso, U.S.A.*

Editorial Board
Jean Paul van Bendegem, *Free University of Brussels, Belgium*
Johan van Benthem, *University of Amsterdam, the Netherlands*
Jacques Dubucs, *University of Paris I-Sorbonne, France*
Anne Fagot-Largeault, *Collège de France, France*
Bas van Fraassen, *Princeton University, U.S.A.*
Dov Gabbay, *King's College London, U.K.*
Jaakko Hintikka, *Boston University, U.S.A.*
Karel Lambert, *University of California, Irvine, U.S.A.*
Graham Priest, *University of Melbourne, Australia*
Gabriel Sandu, *University of Helsinki, Finland*
Heinrich Wansing, *Technical University Dresden, Germany*
Timothy Williamson, *Oxford University, U.K.*

Logic, Epistemology, and the Unity of Science aims to reconsider the question of the unity of science in light of recent developments in logic. At present, no single logical, semantical or methodological framework dominates the philosophy of science. However, the editors of this series believe that formal techniques like, for example, independence friendly logic, dialogical logics, multimodal logics, game theoretic semantics and linear logics, have the potential to cast new light on basic issues in the discussion of the unity of science.

This series provides a venue where philosophers and logicians can apply specific technical insights to fundamental philosophical problems. While the series is open to a wide variety of perspectives, including the study and analysis of argumentation and the critical discussion of the relationship between logic and the philosophy of science, the aim is to provide an integrated picture of the scientific enterprise in all its diversity.

For further volumes: http://www.springer.com/series/6936

From a Geometrical Point of View

A Study of the History and Philosophy of Category Theory

by

Jean-Pierre Marquis
University of Montreal, QC, Canada

Springer

Dr. Jean-Pierre Marquis
University of Montreal
Department of Philosophy
C.P. 6218, succ. Centre-ville
Montreal QC
Canada H3C 3J7
jean-pierre.marquis@umontreal.ca

Cover image: Adaptation of a Persian astrolabe (brass, 1712–13), from the collection of the Museum of the History of Science, Oxford. Reproduced by permission.

ISBN 978-1-4020-9383-8   e-ISBN 978-1-4020-9384-5

Library of Congress Control Number: 2008938359

© Springer Science+Business Media B.V. 2009
No part of this work may be reproduced, stored in a retrieval system, or transmitted
in any form or by any means, electronic, mechanical, photocopying, microfilming, recording
or otherwise, without written permission from the Publisher, with the exception
of any material supplied specifically for the purpose of being entered
and executed on a computer system, for exclusive use by the purchaser of the work.

Printed on acid-free paper

9 8 7 6 5 4 3 2 1

springer.com

*À Marie,
   pour tout,
      tout simplement.*

Acknowledgements

I thank Mike Barr, Marta Bunge, André Joyal, Jim Lambek, Michael Makkai, Gonzalo Reyes, Robert Seely who have, in one way or another, taught me about category theory, its content, significance and history. I want to acknowledge indispensable help from Steve Awodey, Luc Bélair, Peter Freyd, Daniel Kan, Bill Lawvere and Colin McLarty. I also want to thank an anonymous referee for many helpful suggestions. Needless to say, I am solely responsible for all mistakes and misrepresentations left in the book. I owe gratitude to Lawrence Deck who has read the entire manuscript and made numerous corrections and suggestions. Special thanks go to my student Mathieu Bélanger who has translated my manuscript into LaTeX, corrected numerous mistakes and helped me improve the whole work tremendously. The first draft of this book, a mere 100 pages, was written while I was a Humboldt Fellow at the University of Konstanz and the final version was completed while I was on sabbatical at Stanford University Mathematics Research Center. I thank all these institutions for their help and for providing me extraordinary working conditions. Finally, I thank the SSHRC and the FQRSC for their financial support.

Contents

Introduction 1

1 **Category Theory and Klein's Erlangen Program** 9
   1.1 Eilenberg and Mac Lane's Claim 9
   1.2 Klein's Program: Basic Aspects 12
       1.2.1 A Philosophical Fable 12
       1.2.2 Transformation Groups: Encoding Basic Geometric Facts .. 15
       1.2.3 Transfer of Structure: The Irrelevance of the Nature of the Elements of a Space 25
       1.2.4 Why a Transformation Group is not Quite Enough 28
       1.2.5 But Then Again, why a Group is Enough 29
       1.2.6 Classifying Geometries 31
   1.3 Logical Remarks 32
   1.4 Main Ontological and Epistemological Consequences of Klein's Program 34
   1.5 Groups and Geometries: Formal Supervenience and Reduction 36
   1.6 Summing Up 39

2 **Introducing Categories, Functors and Natural Transformations** 41
   2.1 From a Transformation Group to the Algebra of Mappings 44
   2.2 Foundations of Category Theory 51
   2.3 Philosophical Interlude: An Argument Against the Foundational Status of Category Theory 54
   2.4 At Last, Natural Transformations 60
   2.5 Extending Klein's Program in the Wrong Direction 64
   2.6 Category Theory: The First Phase 1945–1958 67

3 **Categories as Spaces, Functors as Transformations** 73
   3.1 Universal Morphisms 74
       3.1.1 Mac Lane: Doing Duality without Elements 77
       3.1.2 Universal Morphisms 86

ix

	3.2	Grothendieck and Abelian Categories 90
		3.2.1 Abelian Categories 92
		3.2.2 Representable Functors 102

4 Discovering Fundamental Categorical Transformations: Adjoint Functors 109
   4.1 The Background: Homotopy Theory and Category Theory 114
   4.2 Kan's Discovery 125
   4.3 Kan's 1958 Papers "Adjoint Functors" 132

5 Adjoint Functors: What They are, What They Mean 147
   5.1 Adjointness 148
   5.2 Equivalence of Categories Again 161
   5.3 Back to Klein 164
   5.4 From Groups to Groupoids 166
   5.5 The Foundations of Category Theory... Again 175

6 Invariants in Foundations: Algebraic Logic 191
   6.1 Lawvere's Thesis 194
   6.2 The Category of Categories as a Foundational Framework 197
   6.3 The Elementary Theory of the Category of Sets 208
   6.4 Categorical Logic: the Program 210
   6.5 An Adjoint Presentation of Propositional Logic 216
   6.6 Quantifiers as Adjoint Functors 220
   6.7 Graphical Syntax: Sketches 225
   6.8 Categorical Theories: Conceptual and Generic Structures 234
   6.9 Summing Up 246

7 Invariants in Foundations: Geometric Logic 247
   7.1 Grothendieck Toposes: Generalized Spaces 248
   7.2 Elementary Toposes 261
   7.3 Invariants Under Geometric Transformations 267
   7.4 Invariants Under Logical Transformations 271
   7.5 Invariant Foundational Frameworks 276
   7.6 Using Geometric and Logical Invariants 282
   7.7 Summing Up 283

Conclusion 285

References 291

Index 303

Introduction

Category theory is now part and parcel of contemporary mathematics, theoretical computer science and even mathematical physics. And it is here to stay, for good reasons. However, these reasons are not clear to everyone, even within the mathematical community. Category theory is hard to learn and master. It is abstract. It is general. Someone has even said that it is "general abstract nonsense" (but it was meant as a joke). What is the point? Why bother? The point is that not only is it *mathematically* rich, deep and profound, but it is more generally *conceptually* rich, deep and profound. What does it take to see this? Understanding a specific important case is a place to start.

Nowadays, *no one* would deny the importance of group theory in mathematics, physics, chemistry and science in general. Groups show up everywhere and whenever they do, their presence is significant, useful, and reveals deep and essential *conceptual* components of a situation. But it took almost a century for the community of scientists to recognize this fact. According to many, groups were considered too abstract, too conceptual. The latter reaction might reveal more about those who pronounced it than about the field itself. The fact is: groups capture basic structural facts of a situation and allow for computations of otherwise intractable properties. It turns out, although this certainly was not intended to be an important part of the definition of a category, that groups are special cases of categories. It might take a century for the scientific community to recognize the importance of category theory. Indeed, it even took some time for category theorists themselves to recognize that *categories*, their *properties* and *structures* were as significant as groups, their properties and structures.

Category theory arose in the brains of Samuel Eilenberg and Saunders Mac Lane in the early forties and was immediately applied to algebraic topology. In the fifties, it made possible the creation of a whole new field of mathematics, namely homological algebra. It was the language in which Alexandre Grothendieck cast his revolutionary approach to algebraic geometry in the sixties and although it took a while before the community of algebraic geometers adopted Grothendieck's style, contemporary algebraic geometry is unthinkable without category theory. And the same can be said about algebraic topology and homotopy theory, to mention just these two obvious cases. It was only at that time, approximately twenty years after its creation, that category theory became an autonomous research area and that mathematicians working in the field could claim to be category theorists. This was an important moment in itself, but more was to come. In the seventies, to the surprise of its originators (and despite the contrary opinions of the majority of logicians at the time), the realms of logic and the foundations of mathematics were shown to fall under its coverage via one of the key concepts introduced by Grothendieck, namely the concept of a Grothendieck topos. At that point, category theory was already covering a large spectrum of mathematical concepts: from logic and sets to topological spaces. And the developments and applications went on. Categorical ideas and methods were in the background of Faltings' proof of the Mordell conjectures and of Wiles' proof of Fermat's last theorem, although not explicitly in the proofs themselves. Computer scientists understood its usefulness in the eighties and it has been applied in various ways in that field ever since. It is now finding its way into mathematical physics, especially in the search for a theory of quantum gravity.

So the applications of category theory, at first somewhat limited to a few specific fields of mathematics where its role was obvious, are nowadays varied, wide and deep. This vast success, in itself, calls for an explanation. One striking feature of the theory is that it turns up whenever the *conceptual* foundations of a problem or a field need to be clarified. The fact that some of the key concepts of mathematics, at first seemingly unrelated, turn out to be special cases of general categorical concepts indicates that it captures key features of mathematical situations. The fact that seemingly different constructions are nothing but the same abstract categorical concept suggests that there is a deep unity revealed by categorical methods.

Mathematics has changed radically during the first half of the twentieth century and the use of the axiomatic method certainly played a key role in this change. Mathematics underwent further profound changes in the second half of the twentieth century, changes that are still going strong and I believe that category theory has played, is playing and will play a crucial role in these changes. A basic feature of this role is the fact that it is possible to use the axiomatic method entirely *within* the language of category theory: the language contains the resources to *define*, axiomatically or directly, concepts, systems and theories, to *construct* in a unified manner fundamental mathematical systems of various types and *prove*, with the help of categorical methods, numerous theorems. These possibilities were *not* seen when the theory was introduced and, as I have already mentioned, it took approximately twenty to thirty years before some mathematicians could understand that it was possible. One of the goals of this monograph is to present how these changes came to be: how mathematicians moved from a convenient and heuristically valuable linguistic turn (I am even tempted to use the word "dialect" at this stage), to a systematic and powerful way of thinking—not only a "language" but also a whole collection of concepts, methods and procedures—whose level of abstraction and generality is without any comparable predecessor.

Of course, the fact that one can do something does not entail that one has to do it. Learning a new language is not easy and requires time and effort. One has to convince oneself that the acquisition of this new language is worth the investment, both at the individual level and at the collective level. Here a subtle and delicate issue arises: why should a mathematician or a community of mathematicians adopt category theory? For some, perhaps a majority, the adoption of such a general, abstract and abstruse framework can only be justified by its power and efficacy in proofs; that is, it has to be clear that proofs of important, deep and difficult theorems can be achieved with it. Even better: that it is *only* by these means that they can be proved. Features of a different kind will convince others: simplicity, unifying conceptual power, indispensability with respect to some specific problems in a field (e.g. homotopy theory), or generality. Still others will claim that category theory cuts "mathematical nature" at its joints or that it constitutes the exact mathematical expression of a fundamental metaphysical fact. Needless to say, it is not our intention here to argue in favor or against one or another of these justifying claims. Nor is it our intention to convince anyone to adopt category theory as a general framework. It is not even our intention to collect explicit reasons that would justify its endorsement. These reasons will show up in various guises as the book develops. They will be part of the story.

My basic claim in this monograph is that category theory is a conceptual extension of Klein's program in elementary geometry. Or rather, with hindsight one can argue that Klein's program is one very special case of the power, richness and persuasiveness of categorical methods. It illustrates in a particular setting how basic categorical ideas, namely morphisms between certain kinds of objects, capture essential and fundamental aspects of a situation. Furthermore, there are many historical threads between category theory and Klein's program. First, the fathers of category theory, Eilenberg and Mac Lane, made a direct reference to Klein's program in their original paper on category theory published in 1945. Second, there is a filiation from Klein to Mac Lane, via Hilbert and Noether. Third, given Noether's influence on the development of algebraic topology in the thirties, one can also conjecture a link between Noether and Eilenberg, perhaps via Hurewicz, who introduced, among other things, the notion of short exact sequences. Be that as it may, the conceptual link stands by itself. As I will try to show, most of the basic elements implicitly and explicitly present in Klein's program transfer almost directly to category theory. They make clear the distinctive features of category theory, its role in mathematics and in the foundations of mathematics. The main and basic point is that category theory is thoroughly *geometric* and this simple fact, correctly understood, goes hand in hand with facts which allow us to see in what way category theory differs from set theory and in what sense it can play a genuine foundational role in mathematics.

I will now sketch how these theses will unfold in the chapters of the book.

Chapter 1 opens up with Klein's program. Geometry is usually thought of as the science of figures and shapes in space. The latter are usually given in a certain language, analytic or synthetic, and one proceeds from postulates and definitions to prove properties of geometric objects. In Klein's time, various notions of space were developed. Klein's insight was that a space could be studied as an object in itself, from a global and external point of view. One then associates to a given geometric system an algebraic structure that captures its basic and essential properties, namely its group of transformations.

Once the importance of the notion of transformation group is recognized, an elementary geometry can be *given* by a manifold together with a criterion of identity for geometric objects and a criterion of meaningfulness for geometric properties. The transformation group acting on the manifold provides these criteria.

Klein systematically used what he called a "principle of transference" which allows the transfer of a geometric structure from one manifold to another manifold, provided there is a bijective correspondence between the manifolds. This is, as far as we know, the first systematic instance of what is now called a "change of base" or a transfer of structure. Although complex projective geometry occupied a privileged place in Klein's work, it is clear that his approach provided a thoroughly "democratic platform in geometry, establishing the group of transformations as the ruling principle in any kind of geometry, and yielding equal rights of independent consideration to each and every such group." ([268], 28) This does not mean that any geometric structure is as good as any other one, but, rather, that one has the systematic means to understand how geometric structures are related to one another, how they can be identified, compared and classified.

It is the transformation group which is seen as fundamental, for not only does it provide, as we have already said, a criterion of identity and meaningfulness, but also because it shows what is essential and what is inessential in geometry. Thus the study of a geometry becomes the study of a transformation group. Moreover, the systematic use of groups provides also a criterion of identity between geometries and therefore yields a systematic classification of geometries, which was Klein's principal goal.

In Chapter 2, I look at Eilenberg and Mac Lane's original claim that category theory can be considered a continuation of Klein's program. Along the way, I present and examine the definitions of category, functor and natural transformation. I try to clarify the status of these "entities" in their applications, in particular in algebraic topology. However Eilenberg and Mac Lane did not and could not exploit the connection between Klein's program and category theory fully. One of the main reasons is that the notion of category was, rightly for their purposes, secondary. It was only introduced so that the notions of functor and natural transformation could be given a firm basis and so that they could be defined in their full generality. Moreover, they did not have nor did they need the general notion of a categorical property, that is, of a genuine property of a category. Thus they could not consider genuine categorical structures.

Hence, they did not extend Klein's program as such, although they clearly made an effort to extend a part of it. Categories are not seen as playing in mathematics a role similar to groups in geometry. The connection they make to Klein's program, although perfectly correct and legitimate, is limited to one methodological dimension of Klein's work.

But in fact, the main reason why Eilenberg and Mac Lane could not see how to extend Klein's program by using categories is that they simply did not have in their possession some of the key concepts of the theory. It can be argued, I believe, that although Eilenberg and Mac Lane introduced categories, functors and natural transformations, they did not introduce category *theory*. What was missing was the *algebraic structure* of categories. It took about fifteen years for the latter to emerge and another three to five years for the algebra to be exploited systematically. In a sense, it is still not fully exploited by the community of mathematicians. It is as if mathematicians had noticed the presence of groups in geometry but had failed to exploit group theoretical concepts in geometry, thinking that groups were merely useful for the classification and organization of geometry. In Chapter 3, I examine therefore the emergence of the notion of universal arrow as it appeared in Mac Lane's paper in 1950. I then move on to Grothendieck and his use of universal arrows in the definition of Abelian categories. Abelian categories constitute the first historical example of a foundational use of category theory that goes beyond the merely heuristic and organizational. While doing this work, Grothendieck introduced explicitly the correct criterion of identity for categories, namely the notion of an equivalence of categories. It is a surprise to most category theorists to learn that the latter notion was *not* in Eilenberg and Mac Lane's original paper. I then move on to another key notion in Grothendieck's work: representable functors. Although the latter is equivalent to the notion of universal morphism, it leads to a better understanding of the criterion of identity for categories as well as the notion of a categorical property.

In Chapter 4, I proceed to examine *the* key categorical concept: adjoint functors. The whole chapter is devoted to Kan's discover of the concept in 1956 and published in 1958. I consider certain claims made by Mac Lane concerning the discovery of adjoint functors and present my own analysis of this historical event. I speculate as to why Eilenberg and Mac Lane did *not* see the notion right from the beginning, and why Kan *did* see it. This takes us through some homotopy theory and Kan's early work in the field. There is a circle of ideas showing up at this stage between the notions of universal morphism, representable functor and adjoint functors. Once these notions have been seen and understood, category *theory* came to life. Genuine categorical concepts and properties were formulated and exploited in numerous situations. In the same way that group theoretical properties and structures can be used to solve genuine geometrical problems (and others, of course), categorical properties and structures can be used to solve mathematical problems. Moreover, again as in the group theoretical case, the categorical representation yields a better understanding of the situation and of its essential elements.

I therefore pursue the notion of adjoint functors further in Chapter 5. I try to provide an informal understanding of the notion and present a series of paradigmatic examples. At this point, I make contacts with logic by showing how propositional connectives arise naturally as adjoints to elementary and trivial functors. This is a key and central phenomenon: given an elementary functor, very often its adjoints, when they exist, turn out to be fundamental operations. This illustrates in a striking manner the importance of the concept of adjunction. I go back to the notion of equivalence of categories and illustrate how two seemingly different categories turn out to be equivalent, or essentially the same, from a categorical point of view, in the same way that two seemingly different geometric systems turn out to be essentially the same when looked at from the group theoretical point of view. I also return to the issue of what constitutes a categorical property, first by considering a natural generalization of the notion of group—namely the concept of groupoid—and second by presenting Freyd's characterization of categorical properties.

At this juncture, I open up a parenthesis on the issue of the foundations of category theory and categorical foundations of mathematics. In the sixties, while category *theory* was gaining in popularity, clarity and strength, there were reasons to reconsider the question of its foundations and there were indeed proposals to build mathematics on categorical footings. I take a look at Mac Lane's hesitations on the question of the foundations of category theory, Feferman's proposal regarding the same problem, and the accompanying criticisms launched by Kreisel against categorical foundations. I believe that Kreisel's attitude and criticisms have been extremely influential among logicians and mathematicians interested in the foundations of mathematics. I also believe that although Kreisel was justified in voicing these criticisms in the late sixties and early seventies, they quickly became unjustified. Furthermore, it seems clear that Kreisel himself was open to various alternatives in foundational studies and that he might in the end have seen the relevance of category theory and categorical logic in these studies.

I then turn, in Chapter 6, to the categorical program in the foundations of mathematics, launched by Bill Lawvere in the early sixties and mostly developed by him during the same decade. Lawvere's vision was entirely original and based on a deep understanding of abstract category theory and its main concepts. Lawvere's goal was to find a new and, according to him, more appropriate foundation for differential geometry. He was convinced that the standard set-theoretical foundations for analysis carried with them extraneous and useless elements and that an alternative based on categories would be more faithful to the nature of the continuum. However, Lawvere did not start with analysis, but rather presented a categorical foundation for universal algebra. But the latter was conceived right from the start in a larger framework in which the category of categories constituted the universe in which mathematical concepts had to be developed, including the concept of abstract set. Soon after, Lawvere not only understood how logical operations, including quantifiers, were representable as adjoints to elementary operations, but also how formal systems and their models could be captured in the categorical space. This exploration culminated in the discovery, made in collaboration with Myles Tierney, of the concept of elementary topos and its links with logic and algebraic geometry. Thus by the early seventies, the direction to be taken was clear to all category theorists interested in the program and it did not take long before the roots to grow deep in the foundational soil. In this chapter, I take a closer look at the program and its main philosophical implications. I examine Lawvere's approach to the category of categories, the category of sets, and then I move to logic and some of the main and relevant results obtained in the early seventies by various category theorists. It turns out that one of the main ideas in the field is precisely the idea of invariants under various transformations and how category theory captures these invariants.

Chapter 7 is solely devoted to topos theory. I look at both Grothendieck toposes and elementary toposes. Topos theory gives rise to a fascinating interplay between logic, set-theoretical thinking, and geometry. Indeed a topos can be thought of as a higher-order type theory or a local set theory, but it can also be thought of as a generalized space. Topos *theory*, in turn, can be thought of as a generalization of topology, as Grothendieck has emphasized from the beginning, and a proper framework for the foundations of mathematics. Our goal in this chapter is to give a brief and partial survey of the main features of toposes and topos theory, more specifically those features which are closely related to the geometric and foundational nature of the field. Needless to say, a whole book could have been written on that particular aspect of category theory.

Category theory is radically different from set theory in many ways, one of which is a consequence of the fact that its methods are fundamentally algebraic, whereas set theoretical methods are, well, set-theoretical (some would say, and I would concur: combinatorial). In particular, since groups are a special type of categories, many methods developed in the study of abstract groups, including Lie groups, have found and are finding a natural generalization in categories, in particular presentations and representations of categories. These methods naturally lead to categories of categories, which are categories with more structure, this latter additional structure being essentially geometric. However, the development of the geometric point of view in the foundations of mathematics also leads to categories of categories. At this stage, two problems show up: a proper characterization of the so-called weak n-categories, and once the latter problem has been solved, a proper axiomatization of a category of categories as a foundation for mathematics has to be provided. These are still open problems at the moment of writing. No matter what the technical details of these developments turn out to be, the main fact remains: category theory is essentially geometric.

Assuming that I have been successful, I will have shown how many important conceptual aspects of Klein's program and its extension transfer to category theory and categorical foundations of mathematics. In fact, the parallel between Klein's approach to geometry and category theory might be reasonably easy to explain: in the same way that transformation groups and Lie groups are basically algebraic structures which turn out to have a rich geometric content because they are deeply rooted in geometry, categories are basically algebraic structures which turn out to have a rich geometrical content, also because they are deeply rooted in geometry, in a loose sense of that latter expression. Categories, like transformation groups and Lie groups, are more or less the algebraic embodiment or coordination of our thinking about geometric structures and properties. (And, as I have already said, Klein's program is nothing but a very special case of applied category theory!) Thus, although it is possible to do the algebra without thinking about geometry, I believe that to make sense of these algebraic concepts and techniques, it is preferable to keep in mind what it is that is coordinated by these techniques of thinking and what it was all along: geometry. Otherwise, one might come to think that category theory is merely a collection of purely algebraic tricks useful in the organization and the resolution of a certain large but limited kinds of problems.

It is not impossible that my attempt will, in the end, disappoint everyone. The book contains some history of mathematics and logic, some mathematics (but not one single proof) and logic, and some philosophy of mathematics. Historians will probably find my historical contributions obvious, simple-minded and narrow-minded: a kind of conceptual history that is far from a real historical study where problems are put in a conceptual network, mathematicians in a social network, and mathematical communities in a larger cultural and political context. Mathematicians will probably find many mathematical mistakes and misunderstandings in my presentation and discussion of mathematical concepts and theorems, simplifications of important ideas and results and a lack of a truly global mathematical perspective. Last but not least, philosophers of mathematics will assuredly find my philosophical contribution shallow and irrelevant. Be this as it may, I hope that my book will nonetheless stimulate others to contribute to a field that I believe to be in dire need of new ideas and approaches.

Chapter 1
Category Theory and Klein's Erlangen Program

1.1 Eilenberg and Mac Lane's Claim

In the paper "General Theory of Natural Equivalences", in which Eilenberg and Mac Lane formally introduced categories, functors, and natural transformations, they made an explicit reference to Klein's Erlangen Program.[1] They claimed that category theory could be considered a continuation of Klein's program. Here is how they put it:

> This may be regarded as a continuation of the Klein *Erlanger Programm*, in the sense that a geometrical space with its group of transformations is generalized to a category with its algebra of mappings ([74], 237).

This passage suggests an "equivalence" that can be written thus:

$$\frac{\text{Space}}{\text{Transformation group}} \sim \frac{\text{Category}}{\text{Algebra of mappings}}$$

Eilenberg and Mac Lane present this analogy as a *generalization* of Klein's program. As such, the equivalence is far from being clear. Are they making a *genuine* generalization or is it merely an analogy?[2] Furthermore, it is well-known that it is possible to generalize concepts in varied and non-equivalent ways. Indeed, before 1945 (the year Eilenberg and Mac Lane's paper appeared) Klein's program had been generalized already in different ways. To mention but one important case, the French geometer Elie Cartan had developed in the twenties what is considered to be a genuine generalization of Klein's program by developing a framework in which what he called Kleinian geometries and Riemannian geometries appear as special cases of his generalized spaces. (See [39] or [249].) Cartan's generalization extends

---

[1] Eilenberg and Mac Lane had introduced functors and natural transformations in 1942 in their paper "Natural Isomorphisms in Group Theory". But the general theory only was published in 1945.

[2] We leave open the question of what constitutes a *genuine* generalization. It is a difficult and delicate question.

9

Klein's ideas to differential geometry. It is a bold and powerful generalization that stays in the realm of geometry and in that sense is a "technical" generalization. For Cartan's goal is to find a way to apply Klein's ideas, which are restricted to elementary geometry, to the whole of geometry as it was known at that time. By doing so, Cartan extended the very notion of geometry by introducing new kinds of spaces. (See, for instance, [4].) Eilenberg and Mac Lane's generalization is of a different nature.

As the title of their paper suggests, the key idea is the concept of natural equivalence—to which we will turn later. In the introduction, they claim that the "study of naturality is justified (...) both by its technical and by its conceptual advantages" ([74], 236). Let us leave aside the technical aspects for the moment and concentrate on the conceptual elements, since it is at this level that the connection with Klein's program is made. The essential passage reads as follows:

> The theory also emphasizes that, whenever new abstract objects are constructed in a specified way out of given ones, it is advisable to regard the construction of the corresponding induced mappings on these new objects as an integral part of their definition. The pursuit of this program entails a simultaneous consideration of objects and their mappings (in our terminology, this means the consideration not of individual objects but of categories). This emphasis on the specification of the type of mappings employed gives more insight onto the degree of invariance of the various concepts involved. For instance, we show in Chapter III, §16, that the concept of commutator subgroup of a group is in a sense a more invariant one than that of the center, which in its turn is more invariant than the concept of the automorphism group of a group, even though in the classical sense all three concepts are invariant ([74], 236–237).

The main ideas seem to be that: (1) Abstract objects ought to be considered simultaneously with their mappings; that is, one should look at the *category* of those objects instead of taking the objects separately; (2) Considering the mappings allows a better classification of concepts in terms of their invariance. It is clearly the latter idea that Eilenberg and Mac Lane had in mind when they made the connection with Klein's program. Indeed, Klein himself had presented his program in the following words:

> Given a manifold and a transformation group acting on it, to investigate those properties of figures on that manifold which are invariant under [all] transformations of that group. ([138])

Klein's fundamental idea was that to study a geometry, one had to look at its group of transformations and, furthermore, the geometric properties of that geometry are those which are invariant under the group of transformations. This seems to be the core of the generalization that Eilenberg and Mac Lane had in mind. But as we will see in Chapter 2, if this is all that they had in mind, then we could say that the generalization was a total failure! What *I* want to claim is that they were correct in making this connection and in ways that they could not foresee when they made it.[3]

---

[3] I personally asked Mac Lane in 1993 whether he would still claim that category theory, as it had developed in the last 50 years, could be considered a generalization of Klein's program. After a moment of thought, he answered unhesitatingly that it definitely was.

1.1 Eilenberg and Mac Lane's Claim

Eilenberg and Mac Lane's work strongly suggests that they saw category theory as a convenient tool or language for certain purposes—at first, mostly in algebraic topology. They were keenly aware that categories, functors and natural transformations were completely general and could be applied to any field of mathematics.

> In a metamathematical sense our theory provides general concepts applicable to all branches of abstract mathematics, and so contributes to the current trend towards uniform treatment of different mathematical disciplines. In particular, it provides opportunities for the comparison of constructions and of the isomorphisms occurring in different branches of mathematics; *in this way it may occasionally suggest new results by analogy* ([74], 236. My emphasis).

The last claim is prudent, to say the least. Thus, category theory might suggest, *by analogy*, new results and constructions in various fields of mathematics: this is certainly correct, but it falls far short of what the theory would actually make possible. They could not know at that time, although as Mac Lane was just about to discover, all the fundamental concepts of mathematics can be defined or constructed in a categorical framework. In the words of Mac Lane: "Eilenberg and Mac Lane initially thought—and stated—that this first paper would be the only needed paper on categories; these ideas would then appear just as a useful language" ([197], 131).

Eilenberg and Mac Lane's use of categories, functors and natural transformations certainly marks the birth of what can be called the first phase of category theory, a phase during which category theory was seen as providing the right mathematical framework to give a precise expression of certain problems which were previously only informally formulated. As the quotation from Mac Lane makes clear, category theory was, during this period, which extends roughly from 1945 up to 1955–57, considered a convenient language, or more generally, a convenient framework. We will come back to category theory itself and its first phase of development in the next two chapters.

Before we do so, we have to take a close look at Klein's program from a philosophical perspective. For I believe that as the history of category theory unfolds, the theory turns out to be a generalization of the conceptual dimensions of Klein's program in a deep sense. Indeed, I claim that not only does Klein's program lead us straight to fundamental methodological principles of category theory—namely the very definition of a category and functoriality of mathematical concepts—but moreover the program provides us with the inherent conception of mathematical objects at work in category theory. Once this is clarified, one can move to the epistemological dimension of the theory and reveal the basic epistemological components involved. Finally, it will be easier to get a better understanding of the position of category theory with respect to other mathematical theories, e.g., set theory or constructive mathematics.

I will therefore look at Klein's program from a definite perspective. My main objective is to bring to the fore three philosophically relevant aspects of the program. First, the fact that transformation groups are not merely useful algebraic tools in elementary geometry, but that they constitute in a precise sense the algebraic encoding of a criterion of identity for geometric objects, or to be more precise for geometric object-types. Second, the same transformation groups also encode a definite criterion of meaningfulness for geometric predicates, or, equivalently, a definite criterion for geometric properties. Third, the main goal of the program is to classify geometric spaces systematically and see precisely how they relate to one another. It is these three philosophical dimensions of the program that I will then generalize to category theory.

1.2 Klein's Program: Basic Aspects

> Given any group of transformations in space which includes the principal group as a sub-group, then the invariant theory of this group gives a definite kind of geometry, and every possible geometry can be obtained in this way. Thus each geometry is characterized by its group, which, therefore, assumes the leading place in our considerations.
>
> ([139], 133)

1.2.1 A Philosophical Fable

In order to illustrate what are, from a philosophical point of view, the fundamental aspects of Klein's program, we will start with a philosophical fable.[4]

Suppose three geometers, call them A, B and C, each developing a geometric theory, go to the International Congress of Geometry held in beautiful Manifoldland that particular year. They each present anxiously their results to the audience.

Geometer A works synthetically and develops what is now called inversive or Möbius geometry. Her geometry is based on three primitive notions: points P, Q, $R, \ldots$, circles $a, b, c, \ldots$ and an incidence relation satisfying the following axioms ([82], 268–269):

Axiom 1: Three different points are on one and only one circle.

Axiom 2: If P is a point on the circle a and if Q is not on a, there exists a unique circle b passing through P, Q and having only P in common with a.

> **Definition 1.1.** a and b satisfying axiom 2 are said to be *tangent* at P.

---

[4] Except for the axiomatic presentation, what we are about to describe can be found in more detail, but in a different order, in [32], Chapter 5.

1.2 Klein's Program: Basic Aspects

[Figure: two intersecting circles labeled a and b, with points P and Q at their intersection]

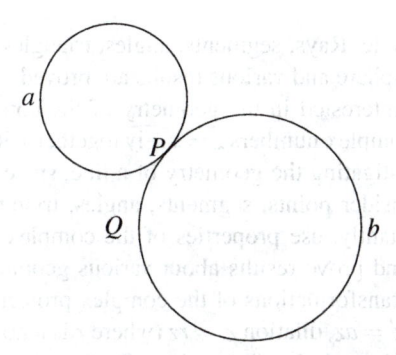

Axiom 3: On each circle there are at least three points. There exist a point and circle that are not incident.

Axiom 4: There exist four circles each two of which are tangent such that all six points of tangency are different and such that no further circle through one of these points is tangent to three of the circles.

Axiom 5: Let $P, P', Q, Q', R, R', S, S'$ be different points with the exception that P may coincide with P'. If $C(PQRS)$, that is the points P, Q, R and S lie on a common circle, similarly $C(QQ'RR')$, $C(RR'SS')$ and if PQQ', PSS' intersect in P, P', then $C(P'Q'R'S')$.

A also adds the usual primitive notions to write down the axioms for order and continuity of the (real) plane (with a point at infinity), and then proceeds to develop the geometric theory that follows. In particular, A introduces the fundamental transformations of her geometry: inversions.

An inversion is a transformation of the plane that generalizes reflection in a line. Roughly described, an inversion maps points inside a circle to points outside the circle, and vice versa. Given a circle C with center O and non-zero radius r and two points P and P' collinear with O such that $(OP)(OP') = r^2$, sending P to P' defines an inversion with the point O the *center of inversion* and C the *circle of inversion* with radius r. Notice that C is fixed by this transformation and that the center O is not in the domain or codomain of the transformation. Naturally, A explores the properties of these transformations and even the properties of the group of such transformations.

B studies a certain spherical geometry with analytic tools or in the analytic language. Thus she starts by defining the real numbers either from the standard axioms for a complete ordered field or begins with the natural numbers and proceeds as usual to the integers, the rationals and either Dedekind cuts or Cauchy sequences. Then B moves to the standard analytic construction of three-dimensional Euclidean space, namely the set $\mathbb{R}^3$ with the standard Euclidean metric. She identifies the equatorial plane $\{(x,y,z) \in \mathbb{R}^3 \mid z = 0\}$ with $\mathbb{R}^2$ and defines the standard unit sphere $S^2 = \{x \in \mathbb{R}^3 \mid |x| = 1\}$.

Using the ambient space $\mathbb{R}^3$ and algebra, she defines her geometric entities as being certain subsets of the sphere S^2. Thus a point is simply a point on the sphere. A line on S^2 is obtained by intersecting a plane through the origin of $\mathbb{R}^3$ and S^2: it is thus a great circle. Rays, segments, angles, triangles, circles, etc. can be defined similarly on the sphere and various results are proved.

Finally, C is interested in the geometry of the complex projective line $P^1(\mathbb{C})$. Points are thus complex numbers $z = x + iy$ together with a point at infinity ∞. Even though C is investigating the geometry of a line, since it is the complex projective line, she can consider points, segments, angles, triangles, circles, etc. on the line and, most importantly, use properties of the complex numbers like taking conjugates, to study and prove results about various geometric figures. In particular, C defines various transformations of the complex projective line. These are the standard translation $z' = az$, dilation $z' = rz$ (where r is a non-null real number), rotation $z' = az$ (where a is such that $\|a\| = 1$), reflection $z' = \bar{z}$, the complex conjugate of z, and inversion $z' = \frac{1}{\|z\|^2} z$. These transformations can be described succinctly as follows: they are all of the form

$$z' = \frac{az+b}{cz+d} \quad \text{or} \quad z' = \frac{a\bar{z}+b}{c\bar{z}+d}$$

where a, b, c, d are all complex numbers and $ad - bc = 0$.

The three geometers finally meet and discuss their respective results. They are all very interested in each other's work, but each finds that the motivation underlying her colleagues' work lacks something in comparison with her own. Then arrives K, also a geometer, who announces to A, B and C that they are all really doing the same thing after all, even though their geometric theories are different. They all readily agree to the differences: A works on the Möbius plane and her geometric entities are only circles, though sometimes with an infinite radius. This last possibility baffles B, since everything in her geometry is in a sense "finite" or at the very least "bounded". Furthermore, B uses analytic means whereas A restricts herself to synthetic methods. Even though C's methods seem closer to B's—for she also uses analytic and algebraic methods freely—C's geometry seems to be "closer" to A's. Indeed, C's complex projective line can be thought as a plane with a point at infinity, although it never occurred to her before to think of it that way. However, her work differs considerably from A's since to her, points are complex numbers and all geometric constructions are defined via properties of complex numbers.

K first convinces A and C that they are doing the same thing. K shows first that their underlying space is the same. Indeed, an analytical model of A's plane is constructed thus: taking the real plane and adding a formal point, the point at infinity. It is then possible to show that A's axioms are satisfied in this plane. K then shows that this new plane is equivalent to C's line. Inversions are then defined analytically directly on the plane. K defines the standard bijection between the real plane and the complex line and extends it by sending the point at infinity of the Möbius plane to the point at infinity of the complex projective line. Using the bijection, she shows how one can transfer A's results into C's language and vice-versa. Moreover, and this finally convinces A and C that they are indeed doing the same thing, she shows how A's inversion can be easily expressed in C's language, namely by complex linear transformations. In other words, she shows them that their geometries have one thing in common: the associated transformation group.

K then turns to B's case. Taking the north pole $p = (0, 0, 1)$ as the point of projection, K defines the standard stereographic π from S^2 minus the north pole to the equatorial plane and extends it to the map from the whole sphere to the Möbius plane by sending the north pole to ∞. (She notices that the projection *is* an inversion, but in space, not on the plane.) B realizes that her spherical geometry can be transported or transferred to the Möbius plane in a systematic manner and that the map π provides a dictionary between the two geometries: it sends lines to lines, circles to circles, and preserves angles. She realizes that certain lines on the sphere are sent to circles with an infinite radius, namely those lines passing through the north pole, and that her geometry, which seemed to be essentially bounded, is not intrinsically so. K then shows that transformations of the sphere become Möbius transformations via the map π. Again this shows that the groups of transformations are isomorphic. A, B and C are then convinced that K is correct: they are doing essentially the same thing. K then insists that what they are all doing should be called "inversive geometry" and should be described thus: it is the study of those properties of figures in the extended plane that are preserved by inversive transformations. A, B and C might still nonetheless want to develop their geometry with their own methods and in their original framework, although K urges them to study "the" transformation group, since this is really what their geometries are all about. In a sense, we have a case of what philosophers call "multiple realisability": one and the same basic structure is exemplified in many different ways that seem to be at first radically different.

K also points out that by fixing the point at infinity, one obtains Euclidean plane geometry and thus that inversive group contains the Euclidean group as a subgroup. She also notes that the inversive group does not contain the projective group of the plane, nor does it contain the affine group of the plane, although it also contains the group of similarities of the plane. Thus some of the relationships between these geometries are established. We will come back to this question later.

1.2.2 Transformation Groups: Encoding Basic Geometric Facts

Let us come back to Klein's program.[5] Although the fable is restricted to inversive geometry and that geometry was not primary in Klein's mind, the story illustrates an important aspect of the situation in Klein's time. Indeed, from the late 18th century until about the end of the 19th century, geometry exploded and grew into a complex and apparently disconnected tree. Projective geometry, done over the complex field, was slowly but surely moving towards center stage (before Klein conceived of his program, he thought, under the influence of Cayley, that projective geometry was central to the whole field). But it was not alone in the play: affine geometry,

---

[5] As was underlined by [244], the term "program" in the expression "Klein's program" does not refer to a program in the standard sense of that expression but is simply a consequence of the fact that Klein submitted his work as a "Programm zum Eintritt in die philosophische Fakultät".

Euclidean geometry, hyperbolic geometry, conformal geometry, descriptive geometry, Plücker's line geometry, Möbius' inversive geometry, *Analysis Situs*, etc., all had made or were making noteworthy appearances. The connections between these geometries were far from clear; especially unclear were the connections between projective geometry, which seemed to be incompatible with a metric, and other geometries that were based on the notion of a metric. What is more, there was a polemic over the "styles" in which a geometric theory had to be developed. Some mathematicians were fervent adherents of the analytic-algebraic method whereas others thought that only a synthetic approach could be faithful to the essence of geometric objects. Klein's main goal was to *unify* these various "characters" and "styles" under a "general principle"[6] and in such a way that the unification would be systematic and objective.

The key to the unification consists in finding some intrinsic properties of geometric spaces that can be encoded in a common framework, irrespective of particular, individual details of each one of them. These properties are encoded in the transformation group of a space and thus the group structure is used to shed a new light on the nature and organization of a large part of geometry. This is possible because the very notion of a transformation group is closely tied to basic features of a geometry. Indeed, a transformation group is no more nor less than a different representation of the basic properties of a geometry. Let us briefly recall how this is so.

Geometry is usually thought of as the study of figures in space, independently of the positions occupied by these figures and their orientations. We picture a geometric space by imagining points, lines, line-segments, circles, triangles, polygons, etc., and certain relations between them. It is also common to present the basic objects of a geometry via axioms in a certain language, be it synthetically or with the help of linear algebra, and then the various results are deduced from the axioms by logical or by algebraic means. A concrete example will clarify these general remarks.

To fix ideas, let us work in the real plane $\mathbb{R}^2$. In the real plane there is a well-known metric function defined by $d(\mathbf{x}, \mathbf{y}) = |\mathbf{x} - \mathbf{y}|$, where $\mathbf{x}$ and $\mathbf{y}$ are points of the plane, i.e., $\mathbf{x} = (x_1, x_2)$ and $\mathbf{y} = (y_1, y_2)$. Treating the plane $\mathbb{R}^2$ as a vector space over the real field $\mathbb{R}$, we can define various geometric objects by algebraic means. For instance, a line is defined in the following way. First, define a *direction* to be the set of all vectors proportional to a given nonzero vector. Thus, for a vector v, a *direction* is the set

$$[v] = \{tv \mid t \in \mathbb{R}\}.$$

A line l through a point P is then defined to be $l = P + [v]$, or equivalently,

$$l = \{x \mid x - P \in [v]\}.$$

It can be shown that this characterization is entirely equivalent to the purely analytic definition of a line in the plane. Recall that analytically, a line is given by the set

$$\{(x,y) \mid ax + by + c = 0\}, \text{ provided that } a^2 + b^2 \neq 0.$$

---

[6] This is Klein's own expression.

1.2 Klein's Program: Basic Aspects

We can obviously define other figures either with intersecting lines or directly by algebraic or analytic means, e.g. circles and conic sections in general. This way of setting up the stage for geometry is of course the algebraic approach. A purely logical description is also possible, and in fact it is possible to mix the logical structure together with the algebraic components in an axiomatic framework. (See for instance [25].)

So far, we have given the basic ingredients necessary to develop plane Euclidean geometry in an "internal" fashion, so to speak. It is entirely possible to develop the geometry with the standard algebraic methods. However, sooner or later, one realizes that some results are immediate consequences of certain inherent symmetries of the geometric figures. To use these symmetries, certain transformations of the plane that capture these symmetries are introduced. In the case of Euclidean geometry, these transformations are called "isometries".

An *isometry* is a bijective map $f \colon \mathbb{R} \to \mathbb{R}$ that preserves distances, i.e., $f(d(\mathbf{x},\mathbf{y})) = d(f(\mathbf{x}), f(\mathbf{y}))$. It can be shown that every isometry is either a translation along a line in $\mathbb{R}^2$ or a reflection in a line in $\mathbb{R}$ or a rotation about a point in $\mathbb{R}$ or a glide reflection in $\mathbb{R}$. The collection $S(\mathbb{R}^2)$ of all the isometries has a certain algebraic structure. Indeed, given two isometries $f, g \colon \mathbb{R}^2 \to \mathbb{R}^2$, their compositions $f \circ g$ and $g \circ f$ are well defined and it can be immediately verified that they too are isometries. Thus, the set $S(\mathbb{R}^2)$ comes equipped with a binary operation. It is easily seen that this operation is associative. Furthermore, there is an isometry $e \colon \mathbb{R}^2 \to \mathbb{R}^2$, namely a rotation through an angle that is a multiple of 2π, which acts as the identity with respect to the binary operation; i.e., for any isometry f in $S(\mathbb{R}^2)$, $f \circ e = e \circ f = f$. Finally, it can also be seen that for any isometry f, there exists an isometry f^{-1}, called the inverse of f, such that $f^{-1} \circ f = f \circ f^{-1} = e$. In other words, $S(\mathbb{R}^2)$ is a group, the group of transformations of the Euclidean plane.

Isometries can be represented algebraically with the help of matrices. An isometry $t \colon \mathbb{R}^2 \to \mathbb{R}^2$ is a function of the form

$$t(\mathbf{x}) = U(\mathbf{x}) + a$$

where U is an orthogonal 2×2 matrix and a is a vector in $\mathbb{R}^2$. Recall that an orthogonal matrix is a matrix such that its inverse is equal to its transpose; i.e., $U^{-1} = U^t$, which is equivalent to the claim that its columns are orthonormal.

The group of isometries is not only useful in deriving proofs in Euclidean geometry. Besides this heuristic role, two key features have to be underlined. First, the group of isometries encodes basic features of Euclidean geometry. Indeed, on the one hand, the notion of congruence of figures can be defined on the basis of this group: two figures F_1 and F_2 are said to be *(Euclidean-)congruent* if there is an isometry from F_1 to F_2. On the other hand (but this is in fact a direct consequence of the previous remark), the notion of what it is to be a meaningful property of a Euclidean figure is determined by the group of isometries. For a property P of a figure F is a meaningful Euclidean property of F if it is a property of F as a rigid body, that is, if the property does not change as F is moved around the plane. More formally, P is a meaningful Euclidean property of F if and only if for any isometry f, the figure

$f(F)$ has the property P. It can easily be seen that these properties include distance, angle, collinearity of points and concurrence of lines, among other things. Second, since it is possible to associate to any plane geometry its group of transformations and since these groups are structures in their own right which can be related to one another, it is possible to determine on that basis how different geometries are related to one another.

Consider, for instance, affine plane geometry. We take as the underlying space the same space as above, that is $\mathbb{R}^2$. In Euclidean geometry, the notion of distance is fundamental: circles with different radii are different. There are, however, geometric properties that are independent of metric information. Circles have properties as circles, over and above their properties as circles of a certain radius. The same is true of all other geometric figures. Developing geometry with these requirements in mind lead to affine geometry. The relevant transformations of the plane are in this case collineations: a *collineation* is a bijection $f: \mathbb{R}^2 \to \mathbb{R}^2$ satisfying the condition that for all triples P, Q and R of distinct points, P, Q and R are collinear if and only if $f(P), f(Q)$ and $f(R)$ are collinear. Let us immediately give the corresponding set of algebraic transformations, the so-called affine transformations on the space $\mathbb{R}^2$. An affine transformation of $\mathbb{R}^2$ is a linear map, that is, expressed in terms of matrices, a function $t: \mathbb{R}^2 \to \mathbb{R}^2$ of the form

$$t(\mathbf{x}) = A(\mathbf{x}) + b,$$

where A is an invertible 2×2 matrix and b is a vector of $\mathbb{R}^2$. Affine transformations map straight lines to straight lines, parallel straight lines to parallel straight lines and preserve ratios of lengths along a given straight line. As in the case of Euclidean geometry, the set of affine transformations forms a group and that group encodes the same type of information as above. Indeed, we can now determine which figures are affine-congruent. It can be shown that:

1. All triangles are affine-congruent;
2. Ellipses are affine-congruent to circles, more precisely, every ellipse is affine-congruent to the unit circle; in other words, all ellipses are affine-congruent to each other;
3. Every hyperbola is affine-congruent to the so-called rectangular hyperbola with equation $xy = 1$; in other words all hyperbolas are affine-congruent to each other;
4. Every parabola is affine-congruent to the parabola with equation $y^2 = x$; in other words all parabolas are affine-congruent to each other;
5. But no non-degenerate conic is affine-congruent to one of a different type.

Thus, all triangles share the same (affine) properties. The same holds for all ellipses, hyperbola and parabola. These facts are very different from Euclidean properties. It seems intuitively clear that affine geometry is more general than Euclidean geometry, but it is hard to express this fact using a purely geometric language.

However, the matrix presentation of the transformations allows us to see immediately, or, if you will, directly, the relationship between Euclidean transformations, i.e., isometries and affine transformations. Since every orthogonal matrix is invertible, it follows that every Euclidean transformation is also an affine transformation, but the converse is false. Hence, the group of Euclidean transformations is a *subgroup* of the group of affine transformations. In this sense, Euclidean geometry is a subgeometry of affine geometry. This is the relationship between the geometries we were seeking.

Klein's basic insight, although not expressed in terms of matrices at that time, was that the foregoing situation covers all cases of elementary geometry, including non-Euclidean geometry and projective geometry. Needless to say, the latter cases require more work, e.g. the projective plane is certainly not $\mathbb{R}^2$. Be that as it may, the fact that transformation groups play a key role in the program is no coincidence, nor should it be considered to be a purely heuristic classificatory device. Transformation groups occupy the forefront of the unification because they are intimately linked to fundamental aspects of geometry.

As in any systematic theory about a class of objects, we need a notion of equality for the different objects of our theory. Thus a criterion of identity for geometric objects is required. In other words, we need to know in some sense what the theory is about or what it is to be an object referred to by the expressions of the theory. This is a basic ontological requirement and transformation groups are intimately related to this requirement. Indeed, as I have already mentioned, a given transformation group is the algebraic encoding of the criterion of identity for the types of geometric objects that are admissible in a given theory. Here is how Elie Cartan presented this fact in a lecture on geometry and groups:

> If indeed one tries to clarify the notion of *equality*, which is introduced right at the beginning of Geometry, one is led to say that two figures are equal when one can go from one to the other by a specific geometric operation, called a motion. This is only a change of words; but the axiom according to which two figures equal to a third are equal to one another, subjects those operations called motions to a certain law; that is, that an operation which is the result of two successive motions is itself a motion. It is this law that mathematicians express by saying that motions form a group. Elementary Geometry can then be defined by the study of properties of figures which do not change under the operations of the group of motions ([42], 15–16. My translation[7]).

Let us unpack this claim, for it contains much of what I want to underline. First, what is usually taken as a logical notion, namely equality of objects, is captured in geometry by motions, or transformations of the given group. What the group clarifies is the notion of superposition of figures, which was already used informally in geometry by Euclid. Thus, given two figures, F_1 and F_2, we say that $F_1 = F_2$

---

[7] Si en effet on cherche à préciser la notion d'*égalité* qui s'introduit dès le début de la Géométrie, on est amené à dire que deux figures sont égales quand on peut passer de l'une à l'autre par une certaine opération géométrique, appelée déplacement. Cela n'est évidemment qu'un changement de mot ; mais l'axiome d'après lequel deux figures égales à une troisième sont égales entre elles assujettit les opérations appelées déplacements à une certaine loi, à savoir que l'opération résultant de deux déplacements successifs soit encore un déplacement. C'est cette loi que les mathématiciens expriment en disant que les déplacements forment un groupe. La Géométrie élémentaire peut alors être définie comme l'étude des propriétés des figures qui ne changent pas par les opérations du groupe des déplacements.

if there is a motion g in the given group G such that $g(F_1) = F_2$. One can even interpret the axioms of group theory as being the algebraic expression of the formal conditions imposed on the relation of identity. Generalizing the foregoing discussion about group of transformations, we see that a group G is given by the following data: a collection of elements together with a binary operation $\times$, a unary operation $^{-1}$ and a constant e such that:

1. $\forall x (x \times e = e \times x = x)$;
2. $\forall x (x \times x^{-1} = x^{-1} \times x = e)$;
3. $\forall x \forall y \forall z ((x \times y) \times z = x \times (y \times z))$.

The first axiom, looked at from the point of view of the identity of geometric objects, asserts that an object is identical to itself. The second axiom amounts to the symmetry of the identity relation: if F_1 is equal to F_2, then F_2 is equal to F_1. Finally, the third axiom stipulates that if an object F_1 is equal to an object F_2 and if F_2 is equal to an object F_3, then F_1 is equal to F_3. In other words, reflexivity, symmetry and transitivity are captured by the axioms of group theory. Thus, associating a transformation group to an elementary geometry is to fix a criterion of identity for geometric objects in this geometry.

One aspect of this criterion of identity has to be emphasized immediately: what we are characterizing with its help are *types* of geometric figures, not *tokens* of these figures. Since a generalization of this point will recur again and again in the chapters that follow, we will immediately try to give a first approximation of the distinction between types and tokens that is at work in the context of geometry.[8] To illustrate the distinction, suppose we are again in the Euclidean plane $\mathbb{R}^2$. A specific geometric figure in the Euclidean plane can be thought as a subset of the plane. For instance, the unit circle S^1 centered at the origin $(0,0)$, defined as the subset $\{(x,y) \mid x^2 + y^2 = 1\}$, is a *particular* figure in that geometry. It is, as we will say more generally, a *particular*. As such we will say that it is a *token* of a *type*, namely, in this case, the type "circle with radius $r = 1$". It is a specific instance of the type. But there are clearly infinitely many others like it in the plane. However, they are all, from the point of view of Euclidean geometry, the same. The group of isometries guarantees this. Any other circle S' with radius one in the plane can be mapped to the unit circle S^1 and vice versa. Thus, in Euclidean geometry, for each k, there is a type "circle with radius $r = k$" and for $k_1 \neq k_2$ the associated types are also different. Notice, however, that one can consider the collection of all circles in the Euclidean plane or the collection of all types of circles. These collections are different. For, the elements of the first are all particular circles, whereas elements of the second are all types. Of course, in this case, one could move from the first collection to the second by taking equivalence classes of circles, where two circles of the plane are considered to be equivalent when they have the same radius (from a group-theoretical point of view, circles in the same equivalence class belong to

---

[8] Philosophers will recognize the familiar philosophical type/token distinction introduced by C. S. Peirce.

1.2 Klein's Program: Basic Aspects

the same orbit). One could then represent a type by its equivalence class.[9] Notice that when we move to the affine plane, there is only *one* type "circle", for, as we have already pointed out, any two circles can be mapped to one another in this case. Thus, if one were to consider the *collection* of all types of circle in the affine plane, then one would end up with a singleton set, namely the set of the type "circle". That type, however, has infinitely many tokens, namely all circles in the affine plane. Geometry is not, in general, about particular figures, it is about *types* of figures.[10] In order to think about these types, we have to have the resources to define the types as well as the resources to define tokens of these types. This point is fundamental both ontologically and epistemologically and we will come back to it once we are in the categorical setting.

Thus, a transformation group specifies the types that are admissible in a geometric space, it determines what there "is" or what can be in a space in an essential way. As we have seen, circles and ellipses are distinguished in Euclidean geometry, whereas the distinction does not make sense in affine geometry or in projective geometry. Indeed, circles can be transformed into ellipses and vice-versa in both affine and projective geometry, but no such transformation exists in Euclidean geometry. In Euclidean geometry, circles of different radii constitute different types and ellipses with different eccentricities constitute different types, whereas in affine and projective geometry circles and ellipses are one and the same type, which might be called "circular figure".

Working with the transformations amounts to working with types instead of working with tokens. Notice, though, that the transformations are applied to tokens of these types and clearly the existence of the latter depends directly on the existence, or should we say the presence, of the former. Thus, a transformation group indicates the presence of geometric types whose existence depends on the existence of geometric tokens. Although this might seem to be a trivial point for elementary geometry, it will turn out to be crucial when we get to categories.

This ontological dimension is closely associated to an epistemological facet of mathematical knowledge. A transformation group is a way to *abstract* types from specific tokens. For although one might work with a specific concrete representation of a geometric figure in the course of a given proof, the transformations determine in a certain sense the domain of validity of the proof by circumscribing the type to which the proof ultimately refers. It should also be noted that the transformations, in the case of elementary geometry, are global; they act on the whole space. Thus, when one works on a proof, one usually concentrates on properties of a figure or a configuration in the space, in other words on local attributes. Moving to the transformation group constitutes a shift in attention, from the local to the global.

Furthermore, as Cartan also briefly indicates in the quoted passage—and as I have also mentioned in the two specific cases presented—the underlying group also determines what is a meaningful geometric property. The idea is simple enough:

---

[9] Although it is very tempting to *identify* a type with an equivalence class, we believe that it is in general a mistake, because we believe that types are not classes.

[10] Of course, traditionally, geometry was very much about particular figures and the computation of some of their properties, for instance areas or angles, etc.

a property P of a figure F in the space S should be considered to be *geometric* if and only if it does not depend on the choice of a coordinate system or, more roughly, F's position in the space S. This means that a genuine geometric property P is invariant with respect to relevant transformations. More precisely, given a space S and a group of transformations G associated to S, a property P of a geometric figure F_1 is said to be a *geometric* property of F_1 if and only if P is a property of F_1 and for any other figure F_2 equal of F_1, P is a property of F_2. Equivalently: for all transformations g in G, $g(F_1)$ has the property P. A property Q that is not invariant under the relevant transformations is not a genuine geometric property of that geometry. For instance, in Euclidean geometry the radius r of a circle C is a geometric property of C, whereas it is not a geometric property of a circle C in affine geometry nor is it a geometric property of a circle C in projective geometry. For in affine geometry, circles with different radii are transformed into one another, and in projective geometry, circles are sent to conic figures in general.

As we have seen, the group axioms are the algebraic formulation of the criterion of identity for geometric figures. More can be said about the translation of logical aspects of geometry into the algebraic structure of the transformation groups. Indeed, a given group of transformations literally contains all the logic of a geometry and since it provides a criterion of identity of objects and their properties, it can be said to provide, in a certain sense which will have to be clarified, a foundation of the geometry. In the words of Elie Cartan,

> (...) it is the whole *logical structure* of elementary Geometry which is contained in the group of motions and even, in a more precise manner, in the law according to which operations of that group compose with each other, *independently of the nature of the objects on which these operations act*. This law constitutes what we call the group *structure* ([42], 17. My translation.[11])

This is an extremely delicate passage. It contains two claims, both of which are fundamental. First, Cartan asserts that the law of composition of a group contains all the logical structure of the geometry. The reason is that the law determines all the subgroups of a given group and it is basically these subgroups that constitute the group structure. In fact, from the point of view of representation theory, which is Cartan's point of view in this instance, the law of composition can take many guises, depending on the choice of the "space" used to represent the group. But in each case, to know the group is to know all its subgroups in all these representations. (See, for instance, [41], Chapter VI. This is also related to what Cartan called the "structural equation" of a Lie Group, which contains, so to speak, all the geometry in a certain sense see, for instance, [249], Chapter 3, §3). Second, from the point of view of the group of transformations, the nature of the underlying objects of the space somehow "disappears" or becomes in a certain sense irrelevant or at least secondary as far as the geometry is concerned. The structure of the geometry is independent of the nature of the objects on which these operations act. This is *not*

---

[11] (...) c'est toute la *structure logique* de la Géométrie élémentaire qui est contenue dans le groupe des déplacements et même, d'une manière plus précise, dans la loi suivant laquelle se composent entre elles les opérations de ce groupe, *abstraction faite de la nature des objets sur lesquels s'exercent ces opérations*. Cette loi constitue ce qu'on appelle la *structure* du groupe.

1.2 Klein's Program: Basic Aspects

to say that objects are dispensable in geometry. What is claimed is that the specific *nature* of the objects used is irrelevant. To use the terminology already introduced and which will constitute our framework, the objects are used as token of certain types and it is only those features they possess as tokens of those types that are relevant. Their specific nature, the attributes they have as objects and not as tokens, is irrelevant. Klein himself was entirely aware of this fact and it even constitutes a crucial, if not the central, philosophical aspect of his program. In his own words:

> Instead of the points of a line, plane, space, or any manifold under investigation, we may use instead any figure contained within the manifold: a group of points, curve, surface, etc. As there is nothing at all determined at the outset about the number of arbitrary parameters upon which these figures should depend, the number of dimensions of the line, plane, space, etc. is likewise arbitrary and depends only on the choice of space element. *But so long as we base our geometrical investigation on the same group of transformations, the geometrical content remains unchanged.* That is, every theorem resulting from one choice of space element will also be a theorem under any other choice; only the arrangement and correlation of the theorems will be changed. *The essential thing is thus the group of transformations; the number of dimensions to be assigned to a manifold is only of secondary importance* ([138]).

This is an astonishing claim. The *group of transformations captures the geometrical content*. It is clear that for Klein, the group of transformations contains the whole geometric structure all at once. The choice of specific geometric "elements"—a notion to which we will come back in a short while—will determine the order of presentation of the theorems of the theory, but not the content of the theory itself. It could be said that the group of transformations encodes the *objective* content of the theory whereas any specific choice of element as a token of the geometric type yield a presentation of the theory, a presentation which contains an arbitrary component.

It should be stressed that this point of view was forced upon Klein by striking geometrical results of the time. Indeed, he was familiar with Plücker's idea that lines could be taken as the fundamental elements of a geometry. Probably more significant was Lie's introduction of his line-to-sphere "translation" which showed how one could systematically transfer a line geometry to a circle geometry and in such a way that the transformations of one geometry are also transferred systematically to the other. Here is how the French geometer Darboux described the situation:

> Nothing resembles a sphere less than a straight line and yet (...) Lie found a transformation which makes spheres correspond to straight lines and hence makes it possible to derive theorems about spheres from those about lines and vice versa (Darboux, quoted in [144], 487).

Again, this is striking. How can one derive theorems about spheres from theorems about lines and vice versa? What the line-to-sphere translation reveals is that both geometries have essentially the same group of transformations. In other words, both geometries have the same structure, it is only the concrete "visual" representations which change from one context to the other. From there, it is just but one step to conclude that what really matters in geometry is the group of transformations, not the geometric elements one might adopt as fundamental or primitive.

This brings us to an additional epistemological element. No one will deny that there are concepts which are properly geometrical, e.g. circle, line, triangle, sphere, polyhedron, etc. With these concepts one usually associates visual representations, or as some philosophers would have said not so long ago, "intuitions". According to a certain philosophical tradition, in order to have geometric knowledge, concepts and intuitions have to meet via a process of schematization. The introduction of groups of transformations forces us to modify the picture in many different ways. Lie's result and its extensions in the hands of Klein indicate that:

1. Geometric concepts have to be taken as a whole and in a specified *context*; it does not make much sense to talk about circles alone and in absolute terms;
2. Intuitions for one and the same system of concepts can vary considerably; as Darboux put it "nothing resembles a sphere less than a straight line and yet (...)"; these various intuitions are all adequate in the sense that they exemplify correctly the given geometry but as was already mentioned, their specific nature is irrelevant;
3. Nonetheless, the intuitions are still required in different ways and for different purposes; this becomes especially clear when we consider various aspects of representation theory;
4. According to certain interpretations of Kant's philosophy of geometry, e.g. [95], it is argued that the process of schematization is given by the axioms of Euclidean geometry, the axioms being interpreted as giving rules for the construction of certain geometric objects and relations. One could argue that the choice of a generating element is another possibility, although certainly not what motivated Kant himself; more generally, it seems legitimate to inquire whether it is not the whole status of axioms in geometry that becomes transformed;[12]
5. A group of transformations is an algebraic structure and it is well known that algebra in general has a problematic epistemological status. Algebra is usually considered to be a collection of tools used in the resolution of problems. To conceive of a group of transformations merely as a tool is to miss the foregoing facts related to the criteria of identity and meaningfulness.

The emphasis on transformation groups automatically led to an abstract point of view of geometry, something that Klein was well aware of, although my presentation is in a contemporary setting. From an *abstract* point of view, a geometry is conceived as a (connected) manifold S, e.g. a space, together with a transformation group G *acting* on S. Even though I have not mentioned the action of a group on a space so far, it is now time to underline its importance. It is not only the group G of transformations that counts but also its action on S. Here is the precise formal definition: given a Lie group G and a connected manifold S, G is said *to act on S* (from the left) is there is a (smooth) map $\alpha: G \times S \to S$ such that:

1. $\alpha(e,s) = s$ for all $s \in S$ and e is the identity of G;
2. $\alpha(gh,s) = \alpha(g,(h,s))$ for all $g,h \in G$ and $s \in S$.

The action $\alpha(g,s)$ is usually simply written as $g \cdot s$. Moreover, all groups considered by Klein act *transitively*:

---

[12] "Typically, Klein conferred on axioms evidence and exactitude, but not self-evident truth" ([98], 622).

1.2 Klein's Program: Basic Aspects

3. For all $s, t \in S$, there is a $g \in G$ such that $g \cdot s = t$

and are *effective*:

4. For all $g \in G$, for all $s \in S$, if $g \cdot s = e$, then $g = e$.

Notice that a group action can also be presented as a homomorphism of groups, or what is called a *representation* of G, $\alpha \colon G \to \text{Aut}(S)$, from G to the set of automorphisms of S. Indeed, given an action $\alpha \colon G \times S \to S$, for each $g \in G$, $\alpha_g \colon S \to S$ is easily seen to be a permutation of S (condition 2 and the fact that G is a group are essential here). Conversely, given a representation $\alpha \colon G \to \text{Aut}(S)$, define $\alpha(g,s) = \alpha(g)(s)$, which is easily seen to be an action. This way of thinking about a group action will also play an important role in what follows.

It is the group G and the action α that determines the geometry, for the geometry is neither more nor less than the invariant properties of the figures under the group action. As we have seen, Euclidean geometry is characterized by the group of rigid motions (isometries)—that is, the length-preserving transformations—whereas affine geometry is characterized by collineations—that is, mappings preserving collinearity, parallelism and ratios of division of line segments, etc. Again, to know the structure of the group (and the action) is to know the geometry. To mention but one simple example: the group of rigid motions of Euclidean geometry has as a subgroup the group of all translations and it is in fact a normal subgroup. In hyperbolic geometry, which is very "close" to Euclidean geometry since only the axiom of parallels of Euclidean geometry is false therein, there is no such subgroup. In Euclidean geometry, composing translations yields translations, whereas in hyperbolic geometry, composing translations can yield a rotation.

1.2.3 Transfer of Structure: The Irrelevance of the Nature of the Elements of a Space

> If, then, one takes away from the mathematical theory that which appears merely as an accident, namely its matter, only what is essential will remain, namely its form; and this form, which constitutes so to speak the solid skeleton of the theory, will be the structure of the group.
>
> ([240], 264. My translation)

The real and deep significance of the abstract approach emerges dramatically in Klein's application of what is called the "principle of transference". In fact, it constitutes the cornerstone of Klein's "comparative review", the expression found in the title of his program. The principle, whose name he borrowed from Hesse, can be formulated in its most general form as follows: given two spaces S and S' and a bijection $\phi \colon S \to S'$, it is possible to transfer a group action defined on S to a group action on S', of course for a fixed group G. Indeed, given a group action on S, define the group action $G \times S' \to S'$ by:

$$g \cdot s' = \phi\left(g\phi^{-1}(s')\right);$$

it is easily verified that this is a group action on S'. In terms of a group action as a representation on the set of permutations of the space, the transfer is defined as follows: given a representation of G on S, i.e., $\rho \colon G \to \mathrm{Aut}(S)$, a representation $\rho' \colon G \to \mathrm{Aut}(S')$ of G on S' is given by

$$\rho'(g)(s') = \phi\rho'(g)\phi^{-1}(s');$$

which is again easily verified to be well-defined.

Notice immediately that the foregoing principle is very general. We can think of a group action on S as being but one structure on S, namely a geometric structure on S. This is only one structure of a certain type, or, to use another expression, a structure belonging to a certain species of structures. Thus given a topology Γ on S and a bijection $\phi \colon S \to S'$, the topology can be transferred to S' simply by taking the image $\phi(U)$ in S' of each open set U of Γ. The transfer clearly yields a topology on S'. Or, given a binary operation $\odot \colon S \times S \to S$ and a bijection $\phi \colon S \to S'$, we can transfer the operation $\odot \colon S' \times S' \to S'$ on S' thus:

$$s' \odot t' = \phi\left(\phi^{-1}(s') \odot \phi^{-1}(t')\right).$$

And this clearly is a binary operation on S'.

Klein used the principle of transference to *identify* geometric spaces: two geometries are the *same* or *equivalent* if there is a bijective correspondence between the underlying spaces and the same transformation group acts on both of them. It is of course natural to include isomorphic groups as being the same. Indeed, if the structure defined on S is transferred to S', then S and S' are equivalent (and it is easily seen to be an equivalence relation). Note that the bijection $\phi \colon S \to S'$ literally provides a dictionary between the different geometries. It shows how to translate notions defined in S into notions of S'. The important point is that S can be a certain space in which certain geometrical concepts are taken as fundamental and from which all the others are derived, whereas S' can be constructed from entirely different principles. Thus they might "appear" considerably different, in the sense that the language used are different and the representations of the objects are also different, whereas they are fundamentally, structurally, the same. The dictionary helps us see how they are really the same and in what sense. What matters in geometry is not that one refers to points, lines, planes, etc., in a specific order and with certain notions taken as primitive or fundamental but rather that one studies a certain structure which is captured by a transformation group.[13]

---

[13] It is not clear whether these facts influenced what was to become the common attitude regarding the meaning of axioms in geometry, most notably represented by Hilbert. Recall his infamous remark: "It must always be possible to substitute 'table', 'chair' and 'beer mug' for 'point', 'line' and 'plane' in a system of geometrical axioms." (Hilbert, in [29]) In the axiomatic framework, the principle of transference becomes the principle of isomorphism, also expressed by Hilbert: "(...) any theory can always be applied to infinitely many systems of basic elements. One only needs to apply a reversible one-one transformation and lay it down that the axioms shall be correspondingly

1.2 Klein's Program: Basic Aspects

Klein uses the principle of transference in his paper to show that various geometries that appear different are, from the point of view of their groups, indistinguishable. Many of the arguments in his paper have the following form. The framework is complex projective geometry, for instance $P^3(\mathbb{C})$, the complex three-dimensional projective space. Then with a suitably chosen quadric one obtains a 2nd-degree surface S in that space, for instance a sphere. Then via the standard stereographic projection from a point P of S one obtains a bijection between this surface and a subspace of the complex projective space, usually the complex projective plane, $P^2(\mathbb{C})$. It is then possible to show that the group of linear transformations of $P^3(\mathbb{C})$ which leave the surface S and the point P invariant can be transferred, via the stereographic projection, to the group of transformations of the plane. Hence these geometries are essentially equivalent. But the argument is completely general and can easily be lifted to higher dimensional spaces, as Klein does. In this manner, Klein presents many examples of different geometries that are essentially the same. Here are a few examples mentioned by Klein himself.

1. Elementary geometry of the plane and projective study of a quadric with a fixed point are identical;
2. The theory of binary forms and projective plane geometry with a fundamental conic are equivalent.

This is just one special case. The philosophical fable in Section 1.2.1 provides another example of equivalent geometric theories. I should point out that the fable could have included others geometers working in non-Euclidean geometry. For one thing, the inversive group contains the isometries of the elliptic plane and of the hyperbolic plane and in fact of three-dimensional hyperbolic space. Furthermore, it is also connected to the Lorentz group and has applications in potential theory and crystallography and thus is connected to physics in various ways. (See [269] for references.) Thus there might be yet another geometer working, for instance, in three-dimensional hyperbolic space. A model of hyperbolic space can be constructed in $\mathbb{R}^3$ with an additional point, the point at infinity ∞. The domain of the model is the set of points $\{(x,y,z) \in \mathbb{R}^3 \mid z > 0\} \cup \{\infty\}$. Points, lines, angles, circles, figures, spheres, solids, etc. can be defined in that space. From our present perspective, the (x,y)-plane together with the formal point ∞, the so-called "sphere at infinity" of hyperbolic space, is particularly significant. Indeed, it can be shown that any inversion in a circle in the (x,y)-plane—and the latter can still be thought of as the complex projective line—extends to an inversion in a hemisphere orthogonal to the (x,y)-plane, which is an isometry of hyperbolic space. Conversely, any isometry of

---

the same for the transformed things. This circumstance is in fact frequently made use of, e.g. in the principle of duality, etc. (...) [sic]" (Hilbert, in [87], 40) But notice that the principle works differently in the axiomatic framework. For it is not so much a transfer of structure but simply that two systems satisfy the same axioms. Thus they are two examples, or in the technical jargon two models, of the same theory. Klein's principle is somewhat different: we have two different geometric theories that are at a different level identical. It is in a loose sense the dual of the principle of isomorphism. For whereas the principle of isomorphism starts with one axiomatic theory and allows us to see how the models are basically the same, the principle of transference starts in a way with different theories and allows us to see how the theories are fundamentally the same.

hyperbolic space induces a linear fractional mapping of the sphere at infinity. Thus, a *three*-dimensional geometry is essentially the same as a *two*-dimensional one! (See [54], [269] for more.) Many other examples of this kind can be given. For instance, the above results can be generalized to higher dimensions, e.g. n-dimensional inversive geometry can be shown to be essentially the same as $n+1$-dimensional hyperbolic geometry or complex *three*-dimensional projective geometry can be shown to be the same as *five*-dimensional complex hyperbolic geometry. (See also [58], Appendix II and III, or [40], Chapter III.)

I should finally note that it would be easy to develop a case similar to Benacerraf's famous argument about natural numbers. Simply add irrelevant properties, like purely set-theoretical properties of the sphere and similarly purely set-theoretical properties of the Möbius plane, and a geometric version of Benacerraf's problem shows up. (See [24].) However in this case, Benacerraf's problem is immediately resolved for we know precisely in what sense different geometries are nonetheless the same: they share the "same" transformation group and the latter settles all questions of identity and meaningfulness with respect to geometric matters. It is therefore possible to say precisely what it is to be a geometry of a certain sort and thus settle all possible disagreements between A, B and C about what is the case in each geometry. For instance, suppose A were to adopt K's model of the Möbius plane pick a certain set-theoretical property underlying her geometry. She then claims that this property is what her geometry is about. B and C could simply point out that the property is not a geometric property, since it is not invariant under the transformation group. A would have to agree since she agrees that to be a geometric property is to be preserved by transformations of the underlying group. Thus there is a fact of the matter to settle disputes as to what is to be a geometric statement and a geometry.

1.2.4 Why a Transformation Group is not Quite Enough

As we have seen, from Klein's point of view, a geometry is given by a space, or domain of action S and a group of transformations G acting on that space. The space is thought as being "matter" without "form". It is the group action that determines the form, the geometry. But we have to be somewhat more careful. In fact, one has to choose what can be called a *generating element* of the geometry, that is, a basic geometric figure out of which all the others will be constructed. Thus one can choose a point, or a line, or an oriented line, or a circle of fixed or arbitrary radius, or some even more exotic figure. The chosen figure constitutes the "atom" of the geometry. One is enough since the group of transformations will scatter the chosen element around. Thus, for instance, if the group G is the group of isometries of the Euclidean plane, then one can choose a point, or a line, or a circle of fixed radius, but not a circle of arbitrary radius, since circles with different radii are different in this geometry. Thus, a geometry is given by three components: (1) an underlying space, (2) a group acting on that space, and (3) a generating element. Epistemologically, these three components correspond to: (1) a given support on which the various

1.2 Klein's Program: Basic Aspects

geometric figures can be represented, (2) a structure imposed on this support which gives it a shape, and (3) a choice of a specific "intuition", or concrete representation from which all the others are obtained via the global structure.

Once a generating element γ has been chosen, one of the subgroups of G has a privileged status: the *stabilizer subgroup* of γ, that is the collection of all transformations g such that $g(\gamma) = \gamma$. It is easily verified that it is a subgroup of G and since G is a topological group, it can be shown that it is a closed subgroup of G. Its privileged status is explained by the fact that two generating elements γ_1 and γ_2 with the same stabilizer subgroup yield the same geometry. For instance, if the domain of action is a plane and the group of transformations is the group of isometries of the Euclidean plane, then one can choose a point γ or a circle σ with fixed radius r as generating elements and the resulting geometries will be the same. This can be seen informally as follows. First, it is clear that the stabilizer subgroup of γ will be identical with the stabilizer subgroup of the circle σ with center γ. Lines in the geometry generated by the circle are strips between parallel lines filled with circles of radius r; distance between two circles is computed by measuring the distance between their respective centers, etc. In fact, all notions of line geometry have an equivalent in the geometry generated by the circle. Notice, however, that once again, the geometries are different at a certain level, if only because the fundamental objects of the first one are points whereas, in the second, this role is played by circles and all the remaining notions are defined from those. Hence, the choice of the generating element will have important consequences at the analytic level, that is, the analytic description of the geometry. But one can also see how a transfer can be effected: there is an obvious bijection between points and circles, namely the one that sends each circle to its center and conversely sends each point to the circle of radius r with that point as center. The dictionary and equivalence follow. Thus, the geometries are fundamentally the same although they might appear different.

1.2.5 But Then Again, why a Group is Enough

The previous remarks lead us to shift from the underlying space to the transformation group and even to abandon the space in a certain sense. To see this, we first observe that given a manifold S, a generating element γ and a group G acting on S, we can construct the quotient G/H, where H is the stabilizer subgroup of γ. It is then possible to define a bijection between G/H and S and thus transfer the action of G on S to G/H. In the latter space, we do not choose a generating element of S, but a closed subgroup H of G and construct the space out of it. It therefore appears that what really matters is the closed subgroup H of G, for by choosing one such subgroup it is possible to define various spaces upon which G can act transitively.

To illustrate this procedure, let G be the group of matrices of order $n+1$, that is matrices M such that $M^{n+1} = 1$, of the following form: $\begin{pmatrix} A & \xi \\ 0 & 1 \end{pmatrix}$, where A is an orthogonal $(n \times n)$-matrix and ξ is a column vector of length n. The closed subgroup in this case can be chosen to be the group H consisting of matrices of the form $\begin{pmatrix} A & 0 \\ 0 & 1 \end{pmatrix}$.

The quotient space G/H is the collection of right cosets with respect to H. It can be verified that in each such coset there is exactly one matrix of the form $\begin{pmatrix} E & \xi \\ 0 & 1 \end{pmatrix}$, where E is the unit matrix. There is thus a bijection between G/H and the n-dimensional space of the columns ξ. We have thus recovered our space. Furthermore, it follows immediately that given an element ξ of the space and an element $g = \begin{pmatrix} E & \eta \\ 0 & 1 \end{pmatrix}$, of the group G, the action of g on ξ has the form $\xi \mapsto A\xi + \eta$. It can be shown that there is a unique (Riemannian) metric on the space S that is invariant with respect to the action and that the metric is, in this case, Euclidean. We have thus recovered Euclidean geometry.

Affine geometry can be characterized in a similar manner. In this case, G is the group of matrices of the form $\begin{pmatrix} A & \xi \\ 0 & 1 \end{pmatrix}$ where $\det A \neq 0$ and ξ is as above. H is the same subgroup as above of all matrices of the form $\begin{pmatrix} A & 0 \\ 0 & 1 \end{pmatrix}$ and we take once again the quotient G/H. We can construct the same bijection between G/H and the space of columns on which the action is defined in exactly the same way. However, in this case, no metric can be defined, although a connection that yields the affine structure can be defined. Hence, we have affine geometry.

Projective geometry, conformal geometry, etc. can be constructed by essentially the same means. The construction described at the beginning of this section leads to the classification of all possible two-dimensional "Kleinian" geometries: there are 23 of them. (See, for instance, [233], 12–21 or [249], Chapter 4. See also [261] and [88].)

We can interpret the foregoing constructions and classification result as follows. Even though the given transformation group G provides information about the geometry from "the outside", by simply showing how the space transforms onto itself, this knowledge is nonetheless enough to reconstruct the elements or components of the space S itself and the resulting geometry. Thus we can have knowledge of the "inside" also. Notice that we are not thereby reducing all the internal information to the external information. The foregoing construction shows that we can access facts about a manifold if we look at the abstract group of transformations in the right way. The quotient G/H as to be seen as a space—it is not a group in general anyway, for a stabilizer subgroup need not be a normal subgroup. Notice that there are important constraints to this construction, e.g., the group G has to act transitively.

We have now arrived at a fully abstract approach to elementary geometry. This is far from Klein's work itself. Indeed, Klein's groups are simply sets of functions that are closed under composition and the groups are given by parameters. In the foregoing construction, groups could be given in a purely abstract manner and geometries are then constructed from certain closed subgroups of the given groups. Matter, form and generating element are now all part of the transformation group. We have left behind almost all intuition. The space is constructed out of algebraic elements and procedures. This progression from "concrete" representations to abstract structures will accompany us throughout, for it is a key feature of the whole of 20th century mathematics.

1.2.6 Classifying Geometries

> It is when unsystematic classification gives place to systematic classification that we can begin to make sense of talking of general criteria of identity not just for things that belong to kinds, but for the kinds themselves.
>
> ([257], 27)

As we have seen in the foregoing examples, the principle of transference allows us to see that geometries that were thought to be unrelated could in fact turn out to be the same in some abstract sense. By looking at the structure of the groups involved, it is possible to determine when a given geometry Γ on a manifold S is a subgeometry of another geometry Γ^* on the same manifold S. Indeed, Γ is a *subgeometry* of Γ^* if G_Γ, the group of transformations of Γ, is a subgroup of G_{Γ^*}, the group of transformations of Γ^*.[14] This is a consequence of the fact that the group encodes the identity criterion and, derivatively, the criterion of meaningfulness. Indeed, every property of a figure in Γ^* makes sense in Γ, since there are fewer properties in Γ^* than in Γ. In other words, all distinctions feasible in Γ^* are also feasible in Γ. However, some distinctions made in Γ cannot be made in Γ^*. Furthermore, all theorems valid in Γ^* are valid in Γ. For instance, the transformation group of Euclidean space is a subgroup of the transformation group of projective space, so Euclidean geometry can be considered a subgeometry of projective geometry. Any property of a figure in projective geometry is a property of the corresponding figure in Euclidean geometry, and every theorem of projective geometry is a theorem of Euclidean geometry.

Informally, moving from one group G to a larger group G^* amounts to "relaxing" the criterion of identity of figures or geometric objects and geometric properties. What is different in the first setting becomes indiscernible in the second setting. Thus seen, a change of group is neither more nor less than a change in one's principle of classification and meaning, and the hierarchy of groups is a hierarchy of criteria of individuality and meaningfulness. Changing the criteria obviously amounts to a change of geometry. Hence the different geometries are related to one another systematically by some of their properties, namely their transformation groups, and geometries can be identified and classified in terms of transformation groups. Here is Klein's classification (taken from [253]):[15]

---

[14] There is some disagreement in the literature concerning the terminology. One could as well reverse the relationship between the geometries and claim that Γ^* is a subgeometry of Γ if G_Γ, the group of transformations of Γ, is a subgroup of G_{Γ^*}, the group of transformations of Γ^*. We will simply follow the order imposed by the groups.

[15] This is not the classification offered by [140], 919. In particular, Kline introduces affine geometry in the classification whereas Klein did not mention affine geometry in his work. Thus, although Kline claims that he presents the main geometries, which is certainly true, it is *not* the historical classification.

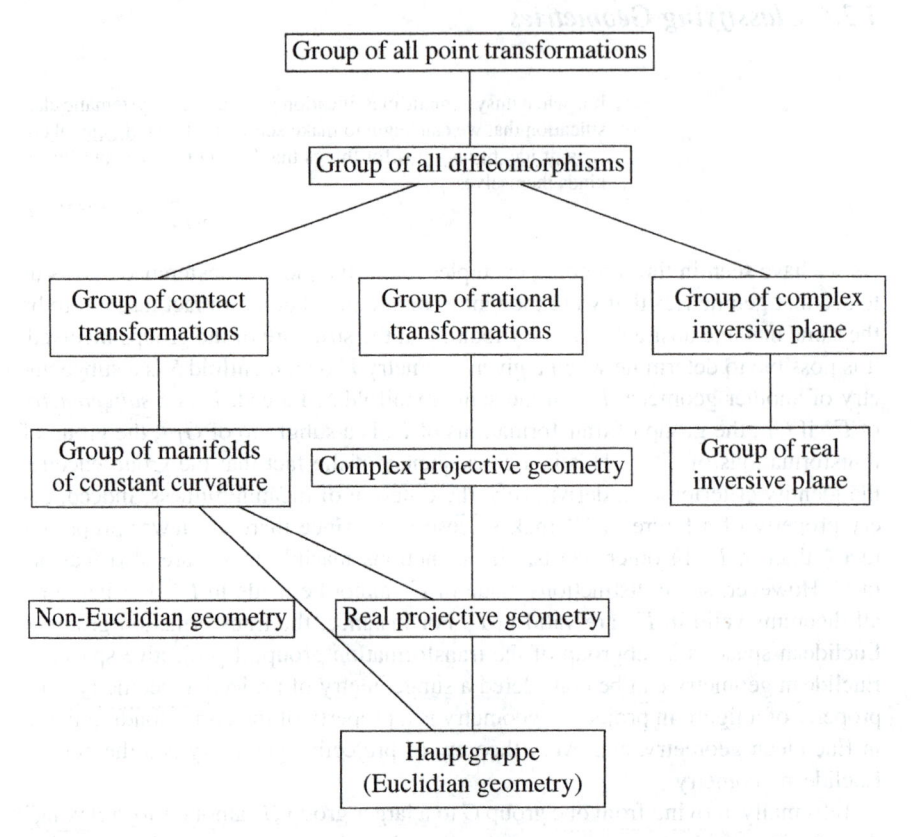

```
                    ┌─────────────────────────────────┐
                    │ Group of all point transformations │
                    └─────────────────────────────────┘
                                    │
                    ┌─────────────────────────────────┐
                    │  Group of all diffeomorphisms    │
                    └─────────────────────────────────┘
```

```
┌──────────────┐      ┌──────────────┐      ┌──────────────┐
│ Group of     │      │ Group of     │      │ Group of     │
│ contact      │      │ rational     │      │ complex      │
│ transformations │   │ transformations │   │ inversive plane │
└──────────────┘      └──────────────┘      └──────────────┘
```

```
┌──────────────┐      ┌──────────────────────┐      ┌──────────────┐
│ Group of manifolds │ │ Complex projective geometry │ │ Group of real │
│ of constant curvature │                              │ inversive plane │
└──────────────┘      └──────────────────────┘      └──────────────┘
```

```
┌──────────────────────┐   ┌──────────────────────┐
│ Non-Euclidean geometry │  │ Real projective geometry │
└──────────────────────┘   └──────────────────────┘
```

```
              ┌──────────────────────┐
              │      Hauptgruppe      │
              │  (Euclidian geometry) │
              └──────────────────────┘
```

Thus, transformation groups encode two different criteria of identity. On the one hand, a given transformation group G yields an *internal* criterion of equality for geometric objects in the space on which the group acts, namely for the equality of figures. On the other hand, from an *external* point of view, a given transformation group G yields an abstract criterion of identity for a geometry as a whole, namely that two spaces are identical as *geometric* spaces if they have isomorphic transformation groups. This is no coincidence: it is precisely because a transformation group encodes the internal criterion of identity for geometric objects that it constitutes at the same time the criterion of identity for the geometric structure as a whole. As we will see, this double aspect of transformation group will extend to categories.

1.3 Logical Remarks

Geometrical objects have an ambiguous status. When a geometer works on a theorem about triangles, she works with a triangle-type, not a specific triangle or a specific class of triangles. This suggests that a type-theory would be intuitively appealing for the development of geometry. Thus, one would start with a collection

1.3 Logical Remarks

of types, say points, circles, triangles, etc., each instance or representative of which would be a particular point, a particular circle, a particular triangle, etc. Note four points:

1. It simply does not make sense to consider whether two elements belonging to different types are identical to one another; the fact that they belong to different types settles the matter immediately. It simply does not make sense to ask whether this circle is identical with this triangle, even as geometric figures. But it does make sense to ask whether two triangles are in fact the same as triangle. It is entirely possible that each type determines its own identity criterion. This suggests that there is no universal criterion of identity for objects but rather a whole collection of identity criteria, one for each type.
2. For each type T, there is an identity criterion for objects of that type, namely the transformation group associated with the geometry we are describing. Thus, in this framework, the identity relation is a geometric relation presented in an algebraic dressing. In this case, it is one and the same group that is associated to each type. An obvious generalization would be to consider different groups for different types. How this ought to be done is not at first obvious. How are these groups to be related to one another?
3. Tokens of a type are not related to that type in the same way that elements of a set are related to that set. The key difference here has to do with the properties possessed both by the tokens and by the type of which they are tokens. Whereas elements of a set do not have to have properties in common with the set they belong to, tokens of a type necessarily have properties in common with the type they are tokens of.
4. Frege thought that concepts were related to objects in a simple and straightforward manner: to each concept C, it is possible to associate an object O_C, namely the extension of C, which can be thought of as the collection of objects *falling under C*. This leads immediately to Frege's infamous Axiom V that gives identity conditions on extensions. We apparently owe to Peano the idea that objects are *elements* or *belong to* a class determined by a concept (a "condition" in Peano's terminology). (See [163], 48.) But as my previous remark indicates, for types, a comprehension principle is clearly inadequate. Although given a type T, the class of its tokens can always be considered, doing so might direct our attention in the wrong direction. We use a token of a type in reasoning and we use it as a token of that type, that is, we concentrate on its properties as a token of the type and not on its properties as a particular. We seldom have to consider the collection of tokens as a collection of tokens of that type. It is certainly true that the tokens share crucial properties, but they share these properties with the *type*, not with the *collection* of objects having a certain property. Thus, we *might* want to collect the tokens of a given type, but this is not a fundamental conceptual operation. We might rather want to *declare* that a particular token is a token of a given type, or exhibit a token of a given type.

1.4 Main Ontological and Epistemological Consequences of Klein's Program

Of course, Klein's approach is restricted to a narrow class of group actions. It does not work for all geometries. One has to adopt a different standpoint to treat other possible geometries from a group-theoretical point of view, a standpoint which was elaborated by Elie Cartan and Hermann Weyl some fifty years later.[16] But many points need nonetheless to be emphasized, since they will have a parallel in the context of category theory.

1. Even though Klein himself always had geometry in mind and never fully considered abstract groups and their representations, he opened the door to that point of view by putting the group at the forefront. Given the importance played by groups, it is reasonable to wonder whether geometry could be expressed in purely group-theoretical terms, and this in two ways. First, from a purely abstract point of view, that is, from the point of view of abstract group theory, it is reasonable to ask whether one could determine abstractly which abstract groups are transformation groups of *geometries*. This amounts to the problem of classification of Lie groups and by now we know that it is a problem of representation theory: the goal is to provide a classification of all continuous groups acting on a n-dimensional manifold, for any n. (This program was initiated by Lie and Killing see [111].) Second, from an axiomatic point of view, it is also reasonable to ponder over the possibility of presenting a geometry axiomatically by using transformations as the primitive concept, and indeed this can be done. Whether it ought to be done is another matter, one that more often than not involves extra-mathematical arguments that are related to the way *we* do and learn geometry. (For a group-theoretical approach to geometry from the "ground-up" so to speak, see for instance [9] and [107]. In both cases, the group concept is considered fundamental.)
2. As Rowe puts it:

   > (...) what Klein accomplished was to provide geometers with a clear and (until about 1920) complete conception of what their subject was all about. He never intended, however, that the answer be abstract, axiomatic, or shorn of the rich material that motivated it, but rather that it should differentiate between what was *essential* and *inessential* when one considered a space and a certain group of transformations acting on it ([244], 47).

---

[16] It is clear that in his generalization Cartan kept the idea that groups provide a criterion of identity in geometry. In his method of moving frames, transformations still provide a criterion of identity, but this time for moving frames instead of figures. But it basically amounts to the same thing. Cartan's idea is to stipulate that a moving frame, instead of being given a priori is in a sense given by the transformation group and the latter also provides us with the criterion of identity for these frames. Thus, moving frames "are configurations in E [a homogeneous space] such that there exists exactly one transformation of G [the Lie group acting on E] carrying one such configuration into another" ([47], 239). In a Euclidean space, they can be taken to be standard tetrahedra, but in a general space, they are a family of "figures", or "configurations" in Chern and Chevalley's terminology, which are taken onto one another by the chosen group. Hence, as in Klein's program, it is the group that is fundamental. See, for instance, [39], 1279–1313.

1.4 Main Ontological and Epistemological Consequences of Klein's Program

Of course, what determines what is essential and what is inessential is precisely the group and the action on the space. As we have seen, the particular choice of generating element is, up to a technical constraint which we have ignored, inessential. The particular choice of axioms and the order of their development is, from the point of view of the structure of the geometry, inessential. This does not mean that the choice of axioms and their development are not important, either for pedagogical reasons or even for some "foundational" purpose, e.g. to prove the independence of some axiom from the others, etc., but from the point of view of the overall or global structure of the geometry, its "form" so to speak or what makes it what it is and not something else, it is the knowledge of the group which matters. Thus, one could claim that the knowledge of the group or rather its role in the proofs of theorems provides a better understanding of these results. It could even be claimed, as Cartan does at various points in his books and papers, that in some cases it provides an *explanation* for certain results.

As we will see, in the same way that groups allow to differentiate what is essential and inessential in geometry, more generally categories make it possible to differentiate what is essential and inessential in mathematical "spaces" in general. Moreover, it does so by looking at these fields in a geometric manner. However, in contrast with Klein's desire to avoid the abstract and axiomatic, category theory lives by it.

3. The lesson most mathematicians learned and remembered from Klein was this: one way to know a mathematical object is to look at its group of automorphisms. It is usually a good and effective tool, but not always, for there are many mathematical objects with a unique automorphism, namely the identity automorphism. Probably the best-known example of this phenomenon is the case of the field $\mathbb{R}$ of real numbers: the only automorphism of the field $\mathbb{R}$ is the identity morphism. Even in geometry, some Riemann spaces only support the trivial identity automorphism! Considering the group of automorphisms of an object is in a way a simple-minded generalization of Klein's program. Klein has shown that examining the group of automorphisms of a geometry is more than useful. Hence it seems to be a good idea to do the same thing for an arbitrary mathematical structure. But as I have just shown, it is not always revealing. For as I have tried to show above, in elementary geometry, a group of automorphisms has a very peculiar status. It does not merely reveal what might be called the "internal symmetries" of a mathematical system, but it encodes not only one but two different criteria of identity. *This*, I believe, is very unusual and is specific to its status in geometry.

4. Let us draw a few ontological morals. First, Klein's program offers a clear and unambiguous framework in which elementary geometry is cast: transformation groups. This is the unifying framework. It may not lead to new and deep results, although this is debatable, but it clearly delineates what it is to be an elementary geometry, or more generally a geometry on a homogeneous space.

   Second, even though the perspective offered by Klein's program does not determine which underlying manifolds (or sets) exist, it certainly yields definite results as to the possible existence of geometric structures on these sets, or more simply on the existence of geometric structures period. Thus, it provides the means required to determine whether a specific geometrical structure exists. Furthermore, the same means lead to a principled organization of geometry, an intrinsic classification based on structural properties. In this sense, group theory plays a crucial role in our understanding of the ontology of geometry.

Furthermore, it is clear that the existence of these geometric structures will *depend*, in one way or another, on the existence of something else, e.g. whatever collection of objects, abstract or concrete, which can be related in the specified geometric manner, although representation theory teaches us that the matter is more subtle than it might first appear. Klein was well aware of the alternative: it is possible to put the transformation group first and define a geometry afterwards *or* start from a geometry given in one way or another and then consider its transformation group. In the latter case, the relationship between geometric properties and transformation groups can be put in terms of a relationship which has attracted the attention of philosophers over the last twenty-five years, although in fields which have nothing to do with mathematics and logic. Indeed, I claim that one can say that group-theoretical properties *formally supervene* upon geometric properties. Since I believe that this claim stands on its own and will also help to clarify a certain view about categories in mathematics in general, we will now take a close look at it.

1.5 Groups and Geometries: Formal Supervenience and Reduction

What Klein and Lie have discovered and emphasized is the way group-theoretical concepts could be used fruitfully in geometry or for geometrical purposes. The usage of group-theoretical ideas is particularly effective for two reasons:

1. Fundamental geometric properties are closely tied to properties of transformations of the space, i.e., to elements of the associated group; that is, properties of points, lines, circles, etc., are closely related to properties of the transformations of the space;
2. The latter are in some sense independent of the former, since they are insensitive to some changes in the geometry, e.g. the choice of the fundamental concepts used in the presentation of a geometry, choice of coordinates, even the logical organization of the theory. There is thus a relationship of dependent-variation between transformation groups and geometries, a relationship, which at the informal level is very close to what philosophers call "supervenience". I want to borrow the latter *terminology* for my own purposes here. I am not claiming that the philosophical analysis of the concept of supervenience clarifies the mathematical situation, for the latter is clear enough, whereas the philosophical analysis of the concept of supervenience is, on the contrary, far from being clear. In fact, the clarity of the mathematical situation might lead to a better

1.5 Groups and Geometries: Formal Supervenience and Reduction

understanding of the notion of supervenience itself. For now, it helps me give a name to a situation which was identified by Klein and which I believe transfers to the case of categories.

In order to justify my choice of terminology, I will first show how the informal notion of supervenience in the philosophical literature can be directly applied to the case at hand.

The notion of supervenience is used in various places in philosophy, for instance in philosophy of mind, ethics and aesthetics. Various definitions of supervenience—weak, strong, global and modal—have been discussed in the literature. (See, for instance [137], [226], [254].) The concept of supervenience seems to be particularly apt here since it is usually presented as a relationship between sets of properties of some kind. Informally, the relationship of supervenience is a relation of dependent-variation. This, of course, is far from enough, for the mathematical concept of function captures a relation of dependent-variation, though no one would say that every functional relationship is a case of supervenience. The fact that we are dealing with sets of *properties* of a different type, and not just any set, is useful but not in itself enough either, for there are numerous functions between sets of properties too. As we will see briefly later, the language of category theory is helpful here.

Here is how the "core idea" of supervenience is presented in the literature: supervenience

> has to do with the fact that sameness in certain respects can exclude the possibility of difference in certain other respects. Suppose that there could be no difference of sort A without some difference of sort B. Then, difference in any A-respect requires difference in some B-respect. Thus, if there is no difference in any B-respect, then there can be no difference in any A-respect. Exact similarity in B-respects excludes the possibility of difference in A-respects. Moreover, if exact similarity in B-respects excludes the possibility of difference in A-respects, then there could be no difference in A-respects without some difference in B-respects. Thus, A-respects supervene on B-respects if and only if exact similarity in B-respects excludes the possibility of difference in A-respects ([226], 17).

We can reformulate in an abstract setting the basic situation underlying Klein's program and it will immediately allow us to see why we can say that group-theoretical properties supervene upon geometric properties. We will assume that we are working with the language of linear algebra and that a geometry is a structure defined on a vector space over a field. Since Klein, like almost everyone at that time, worked over the complex field, we can consider finite dimensional complex vector spaces. In this language, we can define the notion of point, line, and other geometric figures as usual. Notice that there are some choices that can be made here: the choice of a basis being the most obvious. But one can restrict oneself to line geometry or circle geometry in an obvious manner. Thus we can construct different geometries in the language, if only by choosing various bilinear forms. One way to look at the relation between a space and its group of transformations is to associate to each space S its group of automorphisms $\text{Aut}(S)$, which is also called the general linear group $\text{GL}(S)$ of the space (over the complex numbers). The idea underlying the principle of transference is the fact that any linear isomorphism φ between two spaces S and T will induce a group isomorphism between $\text{GL}(S)$ and

GL(T): indeed, given any $g \in$ GL(S), the mapping $\varphi g \varphi^{-1}$ is easily seen to be an automorphism of T.

It should be clear from the foregoing discussion that it makes sense to say that a group, thought of as a system of *properties* of a space, *formally supervenes* on (elementary) geometrical properties, or a geometry.[17] For, *two geometrical spaces cannot have different transformation groups without at the same time being different as geometries*. Indeed, if they have different transformation groups, this means that one of the spaces admits certain non-trivial transformations that are not possible in the other. These additional transformations will not only yield a different criterion of identity but will also allow for the proof of some theorems in the space admitting them. As the principle of transference shows, the converse is false: *two geometries can be, as geometries, different, for instance have different dimensions, and at the same time have isomorphic transformation groups*. These geometries are abstractly the same even though they are different as geometric theories.

Furthermore, there is an obvious dependency relation between groups of transformation and geometries: in the preceding framework, the groups are obtained from the geometries or the geometric properties. This dependency is rather strong: indeed, there is a correspondence between a given group, its subgroups and supergroups with a corresponding geometry, its subgeometries and supergeometries.

Finally, it *cannot* be said that group theory is reducible to geometry. It is hard to see what the latter claim would mean, what kind of reducibility relation one would have to use to make sense of this claim. For one thing, group theory is much more than elementary geometry. The notion of a solvable group comes from Galois theory, and even though one can claim that groups in Galois theory capture the geometry of the situation (i.e., the geometry of the roots of an equation), it cannot be said that it is about a geometrical space. In fact, Klein's program suggests just the opposite: it seems that geometry could be reduced to group theory, that it would just be a chapter of group theory! One has to reverse the picture of the relationship between geometries and groups and switch to representation theory to try to defend the idea that there is a reduction of group theory to geometry.[18]

We are thus lead to recognize two important complementary facts. First, if geometric spaces are taken first, then groups, seen as systems of *global* properties of homogeneous spaces, *formally supervene* upon geometric properties, that is properties definable in the language of the space, for instance via linear algebra. This is just a fact. Moreover, not only does a transformation group depend on an underlying geometry in a definite sense, but also the *identification* and *classification* of geometries is a function of this dependence. Second, it is *possible* to reverse the pre-

---

[17] We cannot refrain from pointing out that, as usual in mathematics, properties are objectified, which greatly simplifies our thinking about them. Furthermore, in this case the properties constitute a group: this means that they are closed under the operation of "product", in this case composition and not a logical operation. Philosophers thinking about closure conditions on properties should reflect upon this and similar cases. One does not have to impose a Boolean structure on properties. Many other type of structures make perfect sense, it all depends how properties are represented.

[18] In fact, even better here would be to look at more recent work on incidence geometry of the geometry of groups. What is indisputable is that group theory provides essential concepts for geometry and, vice-versa, geometry provides essential tools for group theory.

vious dependence and instead consider groups as being fundamental and construct the spaces from them or, more generally, to look at their various representations (in fact, preferably, at all of their representations) in a structure of a certain kind (for instance finite-dimensional linear representations). This is a more abstract point of view motivated by the observation that what is essential to a geometry is its underlying global structure, which is provided or captured by the group of transformations. For classificatory purposes, one wants to look at all such representations. However, as we have already seen, to finally "get down" to geometry in the usual sense of that expression, one has to choose a basic element of representation, in other words, what the geometry will be a geometry of. These choices can vary and none is canonical. What matters, once again, is the structure of the group, namely its stabilizer subgroups. Furthermore, once these choices have been made, the relationship between geometric structures and group-theoretical structures will be as above: groups still supervene upon geometries. Thus, in both cases, groups have a crucial, even indispensable ontological role to play in (elementary) geometry.

1.6 Summing Up

I have looked at Klein's program from a definite perspective. I have emphasized various points that are directly relevant to my analysis of the nature of category theory itself. The development of Klein's point of view—if one can talk in those terms at all—exhibits a shift from the idea that groups are merely tools for geometrical purposes to the idea that they are at the very core of elementary geometry, and thus have a foundational status. It is hard at this point to give a precise meaning to this foundational status. I believe that the main reason is that we have a purely logical model of what it is to be a foundational framework and it can be argued that transformation groups cannot be taken as being *logically* fundamental in geometry, a perfectly reasonable claim.

But I believe that the model leads us astray to some extent. Not that the model is false, but rather that while it leads to useful analyses, the latter hide important aspects of a situation. As I will argue in the last chapter of the book, I believe that this model follows from a "geocentric" metaphor or a "*ground*ing" metaphor of what it is to provide a foundation for a conceptual domain. Once this is seen, a different metaphor can be used to develop a perfectly reasonable alternative of what it is to be a foundation for a conceptual domain. We will turn to this alternative once we have accumulated enough material from category theory and categorical logic.

Hence, there is already in geometry a foundational dimension that can be transferred almost directly to categories. The main element here is that the notion of "form" can be captured by different means. "Logical form" is but one aspect of a theory. Transformation groups can be said to capture the "algebraic form" of a geometry. The logical form is assumed to provide a foundationally meaningful analysis of a domain, because the primitive terms are taken to be not only logically primitive but also epistemically primitive. The algebraic form also provides a foundationally meaningful analysis because the overall algebraic structure is closely related to the *meaning* of the terms involved and the *identity* of the space itself. The latter is particularly clear when one sees how the algebraic form is used for classificatory purposes, e.g. Klein's program itself or, for a slightly different case, [234]. But there is also an epistemic component in the algebraic form. As we have already seen in the case of geometry, very often the algebraic form points towards an *explanation* of various results and a better *understanding* of these results. This partly explains why mathematicians use, develop and apply algebraic techniques and machinery. Some of the advantages of the algebraic form are the following:

1. The algebraic form, e.g., a transformation group, is fundamentally the result of an attempt (sometimes unsuccessful, e.g., Weil's foundations of algebraic geometry or Bourbaki's general notion of structure) to extract essential elements from a given situation while ignoring irrelevant details; this is of the essence of algebra and the axiomatic method in general.
2. This process of extraction is directed towards the search of a solution to a given problem or a family of problems; it is the search for a solution to the given problem that determines what is extracted and what is left out.
3. This process of extraction leads to the development of various artifacts that are first thought of as tools and used as such for the solution of specific problems.
4. These tools are sometimes developed for their own sake and turn into systematic contexts, from the most elementary notational system, e.g. elementary algebra, to the most sophisticated machinery, e.g. homology, homotopy or algebraic K-theory. (For more on the latter, see [211], [212], [214].)
5. It is only within a systematic context that explanations and understanding can be reached.

We will now go back to Eilenberg and Mac Lane's introduction of category theory into the mathematical landscape and see more precisely in what sense it can be claimed that category theory can be thought of as a extension of Klein's program.

Chapter 2
Introducing Categories, Functors and Natural Transformations

Category theory is an embodiment of Klein's dictum that it is the maps that count in mathematics. If the dictum is true, then it is the functors between categories that are important, not the categories. And such is the case. Indeed, the notion of category is best excused as that which is necessary in order to have the notion of functor. But the progression does not stop here. There are maps between functors, and they are called natural transformations.

([90], 114) *Freyd*

Symmetries of a geometric object are traditionally described by its automorphism group, which often is an object of the same geometric class (a topological space, an algebraic variety, etc.). Of course, such symmetries are only a particular type of morphisms, so that Klein's Erlanger program is, in principle, subsumed by the general categorical approach.

([209], 4) *Manin*

The notion of group has wide and important applications, which will give us valuable insight into the meaning of categories. Indeed when Sammy Eilenberg and Saunders Mac Lane first introduced categories in 1940 [sic], one of the motivations was to extend the Erlanger Programm.

([258], 187) *P. Taylor*

Category theory began with a certain problem in algebraic topology formulated by Borsuk and Eilenberg in 1937. The problem was this: given a solenoid Σ in the three-sphere S^3, how many homotopy classes of continuous mappings $f: S^3 - \Sigma \to S^2$ are there? The problem is to find properties of certain spaces using algebraic tools. In this case, Eilenberg had showed that the homotopy classes were in 1-1 correspondence with the elements of a certain group, $H^1(K, I)$,[1] called the one-dimensional homology group. But this did not show how to *compute* this group, i.e. to provide an answer to the question. Then, Steenrod found a way to compute various groups

---

[1] I am temporarily using Eilenberg and Mac Lane's original notation. Nowadays, homology groups are denoted with a subscript, e.g. H_n, and superscripts are used for cohomology groups, e.g. H^n. Eilenberg, in a paper published in 1947 adopts what is now the usual convention. I do not know whether this paper established the convention.

for a class of relevant spaces. However, the computations were difficult since the various groups were complicated. Then, Eilenberg and Mac Lane found in 1942 that Steenrod's groups were isomorphic to what are called extensions of groups, a purely algebraic theory, and that the latter were much easier to compute.[2] In the 1942 paper, we find the following claim:

> The thesis of this paper is that the theory of group extensions forms a natural and powerful tool in the study of homologies in infinite complexes and topological spaces. Even in the simple and familiar case of finite complexes the results obtained are finer than the existing ones ([71], 759).

Hence, the goal is to develop the theory of group extensions with a solution to the problem given by Borsuk and Eilenberg in mind. In the introduction of their paper, Eilenberg and Mac Lane introduce the topological group of all homomorphisms $\text{Hom}(H,G)$ of H into G. In order to establish what they call their main theorem, they have to investigate various properties of this construction. In particular, they establish in lemma 18.2 the following isomorphism:

$$\text{Hom}(A \otimes B, C) \equiv \text{Hom}(A, \text{Hom}(B,C)).$$

This says that to every homomorphism from the tensor product $A \otimes B$ to C, there corresponds a unique homomorphism from B to the homomorphisms from A to C.[3] This isomorphism has a key property: it varies "naturally" with the Abelian groups A, B and C. The problem is to clarify what this "naturality" amounts to in full generality. Various isomorphisms of that type, which are also natural, play a key role in Eilenberg and Mac Lane's solution to the original problem. Two sections of the 1942 paper, namely §12 and §38, are devoted to the naturality of these maps. In §12, Eilenberg and Mac Lane give a precise meaning to the expression "natural homomorphism"—to which we will turn in due course. They wrote a second paper that same year, explicitly tackling the notion of naturality. This paper opens up with the following sentence:

> Frequently in modern mathematics there occur phenomena of "naturality": a "natural" isomorphism between two groups or between two complexes, a "natural" homeomorphism of two spaces and the like. We here propose a precise definition of the "naturality" of such correspondences, as a basis for an appropriate general theory. In this preliminary report we

---

[2] Here is Mac Lane's recollection: "We now turn to 1941. That year, Saunders Mac Lane gave a series of lectures on group extensions at the University of Michigan. (...) Samuel Eilenberg, (...), could not attend the last lecture and asked for a private lecture. Eilenberg immediately noticed that the group $\mathbb{Z}[1/p]$ was dual to the topological p-adic solenoid group Σ, which Eilenberg had been studying, and that Mac Lane's algebraic answer $\text{Ext}(\mathbb{Z}[1/p], \mathbb{Z}) = \hat{\mathbb{Z}}/\mathbb{Z}$ coincided with the Steenrod homology groups $H_1(S^3 - \Sigma; \mathbb{Z})$ calculated in Steenrod's 1940 paper. After an all-night session, followed by several months of puzzling over this observation, they figured out how Ext plays a role in cohomology; the result was the paper [1942a]" ([266], 805).

[3] As we will see in Chapter 4, this example will come to play a fundamental role in the history of the subject. However, in 1942, 1943 and 1945, Eilenberg and Mac Lane see it rather as an exceptional case. It is in fact an adjoint situation. It was only in 1958 that the latter notion was published by Daniel Kan. Notice that there is a similar isomorphisms in sets, where it is sometimes written as $C^{A \times B} \sim (C^A)^B$.

restrict ourselves to the natural isomorphisms of group theory; with this limitation we can present the basic concepts of our theory without developing the axiomatic approach necessary for a general treatment applicable to various branches of mathematics ([72], 531).

The axiomatic approach necessary for a general treatment is, of course, category theory.

Thus Eilenberg and Mac Lane started out by elaborating a tool, namely the theory of group extensions, and ended up discovering that some aspects of this tool revealed what appeared to be a common phenomenon in many different fields of mathematics and that it needed to be clarified in all its generality. This is the main goal of the 1945 paper, and it is by developing the proper framework for this clarification—namely, functors and categories—that Eilenberg and Mac Lane made an explicit connection with Klein's program.

It is clear from their 1945 paper that Eilenberg and Mac Lane had two simple elements in mind when they claimed that category theory could be thought as a continuation of Klein's program and both were essentially *methodological*.

1. The first element directly yields the notion of a category.
2. The second tries to extend the idea that what is basic or essential in a field are its invariants.

Although the first step is very natural and well motivated, it is not clear at first that it is of any value in itself; that is, it is not clear that the very concept of category is necessary or useful. In fact, Eilenberg and Mac Lane themselves thought that it was secondary. Indeed, in §6 of their paper, which deals with foundational issues and to which we will return, they explicitly state that:

> It should be observed first that the whole concept of a category is essentially an auxiliary one; our basic concepts are essentially those of a *functor* and of a natural transformation (...) The idea of a category is required only by the precept that every function should have a definite class as domain and a definite class as range, for the categories are provided as the domains and ranges of functors. Thus one could drop the category concept altogether (...) ([74], 247).

Since functors are functions and functions have to have definite domains and codomains, categories were introduced. Thus, as such, a category was merely the proper conceptual context in which natural transformations, the notion they were trying to elucidate and use systematically, could be defined. In itself, a category was thought to constitute neither a useful tool nor an object of study. As to the second step, they did not see exactly how it had to be developed and their original attempt falls short of success. Let us see how this is so.

2.1 From a Transformation Group to the Algebra of Mappings

> A great deal of modern mathematics, by no means just algebraic topology, would quite literally be unthinkable without the language of categories, functors, and natural transformations introduced by Eilenberg and Mac Lane in their 1945 paper. It was perhaps inevitable that some such language would have appeared eventually. It was certainly not inevitable that such an early systematization would have proven so remarkably durable and appropriate; it is hard to imagine that this language will ever be supplanted.
>
> ([218], 666)

Eilenberg and Mac Lane first suggested that the situation between a space and its group of transformations could be generalized to a category and its algebra of mappings. This way of looking at a category was explicitly adopted once again by Mac Lane in 1950, five years after Eilenberg and himself had introduced categories:

> The notion of an abstract group arises by consideration of the formal properties of one-to-one transformations of a set onto itself. Similarly, the notion of a category is obtained from the formal properties of the class of all transformations $\psi: X \to Y$ of any one set into another, or of continuous transformations of one topological space into another, or of homomorphisms, of one group into another, and so on ([186], 495).

Eilenberg and Mac Lane's generalization can be analyzed in the following distinct steps.

First, Klein started with a geometry and looked at the group of transformations of that geometry. One possible generalization, as I have already mentioned in Chapter 1, is to replace the geometry by a different structure X or even an arbitrary set S and consider its algebra of automorphisms. Very often, the latter constitutes a mathematical structure, but, as I have also mentioned, in some cases it is completely degenerate. Second, one can consider instead the algebra of all endomorphisms $\text{End}(X)$ instead of the algebra of automorphisms. What are in general the formal properties of this algebra? In fact, we lose one property from the algebra of automorphisms, namely the existence of inverses. Thus, two properties remain:

1. There exists an identity endomorphism, call it 1_X, such that $1_X(x) = x$ for all x in X. This endomorphism has the following characteristic algebraic property: for every endomorphism f of X, $f \circ 1_X = f$ and $1_X \circ f = f$.
2. Composition of endomorphisms is associative: for all endomorphisms f, g, h of X, we have that $f \circ (g \circ h) = (f \circ g) \circ h$.

Thus $\text{End}(X)$ is in general a monoid, not a group.

But Eilenberg and Mac Lane were looking at a whole collection of objects of a certain kind, namely topological spaces, and they considered not only endomorphisms of these objects but also morphisms between them as being valuable. This is the next step in the generalization: instead of considering the algebra of mappings of one object into itself, consider the algebra of *all* mappings between two objects of the same kind. Thus, if we consider two groups X, Y, we consider the collection of all homomorphisms between X and Y. If X and Y are topological spaces, one considers the collection of all continuous maps between X and Y. Since in the case of algebraic structures, the mappings are all called "homomorphisms", we denote the collection of mappings between X and Y by $\text{Hom}(X,Y)$. If these Hom-sets, as they are called, have an algebraic structure, then we can say that we are considering, instead of the group of transformations of a space, the algebra of mappings of the spaces. Since X and Y are arbitrary spaces, we have to collect all the Hom-sets together and "pasted" to one another in a coherent manner: any mapping in $\text{Hom}(X,Y)$ should compose with any mapping in $\text{Hom}(Y,Z)$ to yield a mapping in $\text{Hom}(X,Z)$; and any mapping in $\text{Hom}(X,Y)$ should compose with the collections $\text{End}(X)$ and $\text{End}(Y)$, which are just special cases of the previous pasting rule, since $\text{End}(X)$ is just $\text{Hom}(X,X)$.

The formal properties of this pasting yield the notion of a category. Here is Eilenberg and Mac Lane's *original* definition (slightly paraphrased and with a different notation). A *category* $\mathscr{C}$ is an aggregate of abstract elements $\text{Ob}(\mathscr{C})$, called the *objects of the category*, and abstract elements $\text{Mor}(\mathscr{C})$, called the *morphisms of the category*. The objects are denoted by X, Y, Z, etc. and the morphisms are denoted by f, g, h, etc. Certain pairs of morphisms f, g uniquely determine a product morphism $h = g \circ f$ subject to the axioms C1, C2 and C3 below. Corresponding to each object X in $\text{Ob}(\mathscr{C})$ there is a unique morphism, denoted by 1_X subject to the axioms C4 and C5. The axioms are:

C1. The triple product $(f \circ g) \circ h$ is defined if and only if $f \circ (g \circ h)$ is defined. When either is defined, the associative law

$$(f \circ g) \circ h = f \circ (g \circ h)$$

holds. This triple product is denoted by $f \circ g \circ h$.

C2. The triple product $f \circ g \circ h$ is defined whenever both products $f \circ g$ and $g \circ h$ are defined.

Definition 2.1. A morphism 1 will be called an *identity* of $\mathscr{C}$ if and only if the existence of any product $1 \circ f$ or $g \circ 1$ implies that $1 \circ f = f$ and $g \circ 1 = g$.

C3. For each morphism $f \colon X \to Y$, there is at least one identity 1_X such that $f \circ 1_X$ is defined and at least one identity 1_Y such that $1_Y \circ f$ is defined.
C4. The mapping 1_X corresponding to each object X is an identity.
C5. For each identity 1 of $\mathscr{C}$ there is a unique object X of $\mathscr{C}$ such that $1_X = 1$.

This is one way to define a category $\mathscr{C}$, although it is not the definition one finds in contemporary textbooks on the subject. Notice Eilenberg and Mac Lane's choice of terminology. They speak of an *aggregate*—not a *set*—of *abstract* elements, that is, objects whose nature is left entirely unspecified. However, they do speak of *elements*, suggesting that they are still in a set-theoretical framework, and although they say that everything is abstract, they talk about *mappings* (and not *morphisms*, as I do). In fact, it is clear that Eilenberg and Mac Lane assumed that structures of a certain type are *given* and that they are pasted together in a natural way to obtain a new structure, namely a category. It is indeed a natural extension of Klein's approach. Instead of having a geometric space and pasting together its transformations into a group, we start with a space of structures and we paste together all admissible transformations of these structures into one another. This way of thinking about categories, as a pasting of structures of a certain type, immediately leads to the standard examples of categories. Given a type S of structures and a well-defined notion of homomorphism between them, one can consider the category $\mathscr{C}_S$ of structures of that type. Thus, we get:[4]

1. The category **Set** of all sets and functions between them;
2. The category **Grp** of all groups and group homomorphisms between them;
3. The category **AbGrp** of all Abelian groups and group homomorphisms between them;
4. The category **Top** of all topological spaces and continuous functions between them;
5. The category **TGrp** of all topological groups and continuous homomorphisms between them;
6. The category **Vect** of all vector spaces and linear mappings between them;
7. The category **Ban** of all Banach spaces and linear transformations of norm at most 1 between them;
8. The category **Rng** of all rings and ring homomorphisms between them;
9. The category **CRng** of commutative rings and ring homomorphisms between them;
10. The category **Preord** of preorders and monotone functions between them;
11. The category **Pos** of posets and monotone functions between them.

And the list could be extended considerably. Notice that in fact it is the collection of morphisms that determines the properties of the objects and the category. Indeed, if one considers, for instance, the category of all sets but with, say, *surjections* between them, then one obtains a different category than the category **Set**. We will see more clearly why later. But it is an indication that the algebra of mappings does indeed make a difference.

However, whereas in Klein's program the new structures obtained were clearly useful and even basic to the study of geometry, in the case of categories, it was not at first clear how categories *themselves* could be used to solve problems or that they captured important properties of the original structures.

Nonetheless, there are some prima facie considerations that suggest that the algebra of mappings could be useful. As I have already mentioned, the Hom-sets sometimes have a structure. If X and Y are, for instance, Abelian groups, then it is possible to define an operation "+" on the collection $\text{Hom}(X,Y)$. Indeed, for every group homomorphism $f, g: X \to Y$, define $(f+g)(x) = f(x) + g(x)$. It can be verified that, thus defined, the operation is associative, that the trivial homomorphism

---

[4] Again, we ignore questions of size. It should be clear that we simply cannot consider the category of all sets without some restrictions. Notice that in the context where we consider a category as being a pasting of given collections, questions of size are inevitable and important. It is not, however, entirely clear that a given solution to the problem of size developed in this context will immediately transfer to a context in which a category is presented differently. But this is sheer speculation for the time being.

2.1 From a Transformation Group to the Algebra of Mappings

$e(x) = e_Y$—the morphism that sends every element of X to the identity element of Y—is the identity element of the newly defined operation, and finally that for every $f: X \to Y$, there is an inverse, defined by $f^{-1}(x) = (f(x))^{-1}$. Hence, in the case of Abelian groups, the collection $\text{Hom}(X,Y)$ of homomorphisms is *itself* an Abelian group. We see that for objects of a certain kind, for each X, Y, $\text{Hom}(X,Y)$ is more than a collection. This immediately suggests that there may be different *kinds* of categories. Indeed, one could take the foregoing definition of a category and *require* that for all objects X, Y, $\text{Hom}(X,Y)$ be an Abelian group. Once this is done, the axioms have to be slightly modified to make sure that all the operations meet the new requirement. Although the hom-sets and their structure played a key role in the *applications* of category theory Eilenberg and Mac Lane had in mind from 1942 until 1945, this usage of the algebra of mappings had to wait for three years—it appears in a somewhat different guise in [185]—before it was explicitly recognized. The reason seems to be that Eilenberg and Mac Lane were thinking primarily about functors and natural transformations and not about categories as such.

The notion of isomorphism or of isomorphic objects can be defined directly from the definition of a category.[5] A morphism $f: X \to Y$ is an *isomorphism* if there is a (necessarily unique) morphism $g: Y \to X$ such that $f \circ g = 1_Y$ and $g \circ f = 1_X$. If there is an isomorphism $f: X \to Y$, then X and Y are said to be *isomorphic*. Thus the notion of isomorphism is *not* defined pointwise, i.e. on elements, and it is not defined as a bijection preserving a certain structure as it is usually done in a set theoretical framework. In the category **Set** of sets, isomorphisms *are* bijections, in the category **Grp**, isomorphisms are the usual group isomorphisms (one-to-one and onto homomorphisms), and in the category **Top**, the isomorphisms are homeomorphisms. But it is *not* always the case, as we will later see.

The algebra of all mappings *coming into* an object and *going out of* an object could presumably capture essential properties of that object.[6] The underlying idea is still fundamentally geometric, although Eilenberg and Mac Lane did not explicitly present it as such. Consider, by way of example, a solid in three-dimensional space. By projecting it along three orthogonal axes, one obtains three two-dimensional images of the solid. It is well known that it is in general possible to reconstruct the solid from the three projections and vice-versa. The moral is this: the group of automorphisms of an object X can be revealing, but in some cases, projecting the object X into known objects or projecting known objects into X constitute another way of probing it. Thus, one can know a geometric object by:

1. Projecting into it certain known geometric objects and studying the images, direct and inverse, of these projections, and/or

---

[5] This is indeed what Eilenberg and Mac Lane did: they give the definition immediately after the axioms of a category. Notice that they call an isomorphism an "equivalence". The terminology is still in used in algebraic topology. See, for instance, [243], 12.

[6] This could be turn into a general epistemological maxim: to know an object is to know how it transforms into different objects, of the same kind or not (functors are allowed here), and how other objects, of the same kind or not (functors are allowed here), transform into it. This maxim could be applied to many domains of the natural sciences where scientists use various machines to access properties of the objects and phenomenon studied. For more on this parallel, see [211–213].

2. Projecting the object itself into or onto known geometric objects and study the direct and inverse images.

Thus if one were to use morphisms coming into an object or morphisms going out of an object in some way, then we could claim that the algebra of mappings is used as a tool of analysis in a geometric fashion. But this is not how Eilenberg and Mac Lane used it. It did not take long, as we will see, before Mac Lane understood how morphisms could be used to characterize concepts and analyze objects.

Notice how close the axioms of a category are to the axioms of a group: a category with *one* object X *is* a monoid. Indeed, any monoid M can be presented as a one-object category. The *elements* of the monoid M become the *morphisms* of the category. *Multiplication* in M becomes *composition* of morphisms. It is immediately verified that in this way the axioms for a monoid translate into the axioms of a category and vice-versa. Consider, for instance, the monoid of natural numbers $\mathbb{N}$ with the operation of multiplication. The identity element is the number one. Viewed as a category, it has one object, let us also call it $\mathbb{N}$, and the morphisms can be denoted by $1, 2, 3, 4, \ldots$ Composition of two morphisms n, m is defined by $n \times m$. What is missing from the axioms of a category to get a group is the existence of inverses. But since the objects we add in an arbitrary category are of "arbitrary dimensions", so to speak, it is natural not to require the existence of inverses and consider all transformations between objects. A category with one object such that all morphisms are invertible is a group.

Notice also that if for every X and Y in $\mathrm{Ob}(\mathscr{C})$, the collection $\mathrm{Hom}_{\mathscr{C}}(X,Y)$ is at most a singleton, i.e. there is at most one mapping from X into Y, then $\mathscr{C}$ is just a different presentation of a partial order: write $X \leq Y$ if there is a map in $\mathrm{Hom}_{\mathscr{C}}(X,Y)$. Again, the axioms for a category guarantee that $\mathscr{C}$ is a partial order. This is still geometric, since in this case the mappings can be thought as exhibiting a part–whole relation, as in the case of the lattice of open sets of a topological space. Hence, the notion of a category can be thought of as being a generalization of the algebraic notion of a monoid and the order-theoretic notion of a partial order.

It is also interesting to note, as Eilenberg and Mac Lane immediately did in the appendix of their 1945 paper, that Cayley's representation theorem for groups can be readily extended to (small) categories. Recall that Cayley's representation theorem asserts that any group G is isomorphic to a group of permutations, i.e. it can be represented as a group acting on a set. To construct the required group of permutations corresponding to a given group G, one first forgets that G is a group and simply considers it as a set. Then one can reintroduce a group structure on the set G by considering the group $\mathrm{Aut}(G)$ of automorphisms of the set G. The required group of permutations will in general be a subgroup of $\mathrm{Aut}(G)$. Each fixed $g \in G$ defines a function $\tau_g \colon G \to G$ by $\tau_g(h) = gh$ for each $h \in G$. Let $C = \{\tau_g \mid \forall g \in G\}$. One can easily show that C is a group of transformations on the set G. The function $\varphi \colon G \to C$ defined by $\varphi(g) = \tau_g$, is then seen to be an isomorphism of groups and indeed it is a representation of G by a group of transformations.

In order to see how Cayley's theorem is generalized to categories, we need to introduce morphisms between categories. For in general a representation of a group is a specific type of homomorphism of groups. Thus, a representation of category

2.1 From a Transformation Group to the Algebra of Mappings 49

will be a specific type of homomorphism of categories. These are precisely what Eilenberg and Mac Lane called "functors".[7]

Informally, a functor F from a category $\mathscr{C}$ into a category $\mathscr{D}$, is a mapping from $\mathscr{C}$ into $\mathscr{D}$ that preserves the structure of $\mathscr{C}$. More precisely, a *functor F* can be defined as a pair of functions:

1. $F_o \colon \mathrm{Ob}(\mathscr{C}) \to \mathrm{Ob}(\mathscr{D})$;
2. $F_m \colon \mathrm{Mor}(\mathscr{C}) \to \mathrm{Mor}(\mathscr{D})$ such that $F_m(1_X) = 1_{F_o(X)}$ and $F_m(f \circ g) = F_m(f) \circ F_m(g)$.

Usually, the functions F_o and F_m are not distinguished since they are coherent with one another. A functor such as F is called a *covariant* functor. Whenever the direction of arrows is reversed by F, i.e. $F(f \circ g) = F(g) \circ F(f)$, then we say that F is a *contravariant* functor. Both types of functors are common and play a crucial role in category theory.

Notice that functors compose: given $F \colon \mathscr{C} \to \mathscr{D}$ and $G \colon \mathscr{D} \to \mathscr{C}$ define the composition of F and G as $GF(X) = G(F(X))$ for all objects X in $\mathscr{C}$ and $GF(f) = G(F(f))$ for all arrows $f \colon X \to Y$ in $\mathscr{C}$. It is easily verified that this yields a functor $GF \colon \mathscr{C} \to \mathscr{D}$. Furthermore, functors automatically preserve commutative diagrams and therefore, in particular, isomorphisms. For if $f \colon X \to Y$ is an isomorphism in $\mathscr{C}$, this means by definition that there is a $g \colon Y \to X$ such that $f \circ g = 1_Y$ and $g \circ f = 1_X$. Thus, applying a functor $F \colon \mathscr{C} \to \mathscr{D}$, we get $F(1_Y) = F(f \circ g) = F(f) \circ F(g) = 1_{F(Y)}$ and therefore $F(f)$ is also an isomorphism. This simple property of functors plays a key role in many applications of category theory.

In the appendix of their paper, Eilenberg and Mac Lane stipulated that a *representation* of a category $\mathscr{C}$ in **Set** is a covariant functor $F \colon \mathscr{C} \to \mathbf{Set}$. Moreover, a representation, and more generally a functor, is said to be *faithful* if for every morphisms f, g of $\mathscr{C}$, $F(f) = F(g)$ only if $f = g$. We can now see how Cayley's representation theorem for groups is generalized to categories. For each fixed object X of $\mathscr{C}$, let $\varphi_X = \{f \colon Y \to X \mid \text{for some object } Y \text{ in } \mathscr{C}\}$, that is the set of all morphisms f of $\mathscr{C}$ such that X is the target of f. For each morphism $g \colon X \to Y$, φ_g is a function from φ_X to φ_Y defined by $\varphi_g(f) = g \circ f$, where $f \colon Z \to X$. It is easily verified that $\varphi_{()} \colon \mathscr{C} \to \mathbf{Set}$ is a functor and moreover a faithful functor between $\mathscr{C}$ and a (sub)category of **Set**. Thus every (small) category $\mathscr{C}$ can be faithfully represented by a collection of sets and functions between them. It is easy to verify that when the category $\mathscr{C}$ is a group and the set representing the unique object of $\mathscr{C}$ is the set of permutations of that set, then the categorical representation of $\mathscr{C}$ is the standard Cayley representation.

Functors pervade mathematics. Let us look at some examples.

1. Given a category $\mathscr{C}$, the identity functor $1_\mathscr{C} \colon \mathscr{C} \to \mathscr{C}$ always exists and is defined in the obvious manner.

---

[7] "There was also some fun with the choice of terminology. Since the philosopher Kant had made ample use of general categories [sic], the term was borrowed from him for its present mathematical use, while Carnap, in his book on *Die Logische Syntax der Sprachen* had talked of functors in a different sense and made some corresponding mistakes. It seemed in order to take over that word for a better and less philosophical purpose" ([197], 131).

2. $\wp$: **Set** → **Set** defined by $\wp(X)$, the usual power set of X, and given $f\colon X \to Y$, $\wp(f)\colon \wp(X) \to \wp(Y)$ defined by $\wp(f)(A) = f[A]$, the image of A under f. It is easily seen that this is a functor.
3. $\wp°$: **Set** → **Set** defined on objects as in the previous case. But this functor is contravariant: given an arrow $f\colon X \to Y$, $\wp°(f)\colon \wp(Y) \to \wp(X)$ is defined by, for all $B \subseteq Y$, $\wp°(f)(B) = f^{-1}[B] = \{x \in X \mid f(x) \in B\}$, the usual inverse image of B under f. Again, it is easily verified to be a functor.
4. Let us briefly come back to the collections $\mathrm{Hom}(X,Y)$ of morphisms between X and Y. For a fixed object X in a category $\mathscr{C}$, the following two functors can be defined: first a covariant functor $\mathrm{Hom}(X,-)\colon \mathscr{C} \to$ **Set** and second a contravariant functor $\mathrm{Hom}(-,X)\colon \mathscr{C} \to$ **Set** by:

   a. For each object Y, $\mathrm{Hom}(X,Y)$ is the collection of all morphisms $f\colon X \to Y$ and $\mathrm{Hom}(Y,X)$ is the collection of all morphisms $g\colon Y \to X$;
   b. For a morphism $g\colon Y \to Z$, $\mathrm{Hom}(X,g)\colon \mathrm{Hom}(X,Y) \to \mathrm{Hom}(X,Z)$ is defined by $\mathrm{Hom}(X,g)(f) = g \circ f$, where $g\colon Y \to Z$;
   c. For a morphism $h\colon Y \to Z$, $\mathrm{Hom}(h,X)\colon \mathrm{Hom}(Z,X) \to \mathrm{Hom}(X,Y)$ is defined by $\mathrm{Hom}(h,X)(f) = f \circ h$.

   Notice that in the foregoing representation of a category $\mathscr{C}$, the functor φ_X was in fact $\mathrm{Hom}(-,X)$.
5. In their 1945 paper, Eilenberg and MacLane defined functors with two arguments. However it may come as a surprise that they did not first define the product of two categories in order to define functors. The notion of (Cartesian) product category appears only in §13, whereas functors in two arguments are defined in §3. Given two categories $\mathscr{C}$ and $\mathscr{D}$, the *product category* $\mathscr{C} \times \mathscr{D}$ has as objects pairs (X,Y) where X is an object of $\mathscr{C}$ and Y is an object of $\mathscr{D}$ and as morphisms pairs $(f,g)\colon (X,Y) \to (U,V)$ where $f\colon X \to U$ in $\mathscr{C}$ and $g\colon Y \to V$ in $\mathscr{D}$. It is now possible to define a functor in two arguments, covariant in the first variable and contravariant in the second variable, $T\colon \mathscr{C} \times \mathscr{D} \to \mathscr{E}$: for each pair of objects (X,Y) in $\mathscr{C} \times \mathscr{D}$, $T(X,Y)$ is an object of $\mathscr{E}$ and for each morphism (f,g) of $\mathscr{C} \times \mathscr{D}$ as above, $T(f,g)\colon (X,V) \to (Y,U)$ satisfying the foregoing conditions in the definition of a functor, e.g.:

   a. $T(1_X, 1_Y) = 1_{T(X,Y)}$;
   b. Whenever $g \circ f$ is a morphism of $\mathscr{C}$ and $k \circ l$ is a morphism in $\mathscr{D}$, $T(g \circ f, k \circ l) = T(g,l) \circ T(f,k)$.

   For instance, the functor $\mathrm{Hom}(-,-)\colon \mathscr{C} \times \mathscr{C} \to$ **Set** is a functor in two arguments. The main examples presented by Eilenberg and MacLane are those particular cases where the category $\mathscr{C}$ is a category of topological spaces, a category of Abelian groups or a category of Banach spaces, together with the obvious product functor.
6. The so-called "forgetful" functor $U\colon$ **Grp** → **Set** that forgets some or all the structure of an object. It assigns to each group G, the set $U(G)$ of its elements, thus forgetting the group structure, and to each group homomorphism $f\colon G \to H$, the function $U(f)\colon U(G) \to U(H)$ between the sets. Needless to

2.2 Foundations of Category Theory

say, this is but one example among many. For instance, there is a forgetful functor $U\colon \mathbf{Rng} \to \mathbf{Set}$ that assigns to each ring R, the additive Abelian group of R and to each homomorphism of rings $f\colon R \to S$, the same function now regarded as an homomorphism of Abelian groups.

7. If the categories $\mathscr{C}$ and $\mathscr{D}$ are preorders, then a functor $F\colon \mathscr{C} \to \mathscr{D}$ is a monotone function.
8. If the categories $\mathscr{C}$ and $\mathscr{D}$ are monoids (respectively groups), then $F\colon \mathscr{C} \to \mathscr{D}$ is a monoid (group) homomorphism.
9. Consider a group $\mathscr{G}$ as a category with one object and morphisms the elements of $\mathscr{G}$. A functor $F\colon \mathscr{G} \to \mathbf{Set}$ to the category of sets sends the unique object of $\mathscr{G}$ to a set S in $\mathbf{Set}$ and each morphism of $\mathscr{G}$ to an automorphism of S. Notice that F is forced to send morphisms of $\mathscr{G}$ to automorphisms of S by definition. F is precisely a representation of $\mathscr{G}$.
10. If $\mathscr{G}$ is a group, then a functor $F\colon \mathscr{G} \to \mathbf{Vect}$ is a linear representation of the group.
11. It is possible to see how to extend the concept of a group acting on a set to the concept of a category acting on a set. We can also think of a representation of a category $\mathscr{C}$ as an action of $\mathscr{C}$ on the underlying sets (notice the plural). For a functor $F\colon \mathscr{C} \to \mathbf{Set}$ is defined thus: for each object X of $\mathscr{C}$, $F(X)$ is a set, for each morphism $f\colon X \to Y$ in $\mathscr{C}$, $F(f)\colon F(X) \to F(Y)$ is a function between the respective sets. If a transformation group G is essentially a collection of transformations of a space onto itself, a category $\mathscr{C}$ can by extension be thought as a collection of transformations of a family of "spaces" into one another. We have seen that Cayley's representation theorem is a special case of these.

I have already mentioned that functors compose and that for each category $\mathscr{C}$, there is an identity functor $1_\mathscr{C}$. It is therefore tempting to consider the category $\mathbf{CAT}$ whose objects $\mathscr{C}, \mathscr{D}, \mathscr{E}, \ldots$ are categories and the collections $\mathrm{Hom}(\mathscr{C}, \mathscr{D})$ of morphisms are collection of functors $F, G, H, \ldots$ It is easily verified that functors satisfy the conditions of the definition of a category given above. Hence, the category of all categories is a category. However, it quickly appears that the very existence of $\mathbf{CAT}$ is problematic: since $\mathbf{CAT}$ is a category, it should be one of its objects, i.e. $\mathbf{CAT}$ should be in $\mathbf{CAT}$. Is this a problem? Although Eilenberg and Mac Lane did not envisage the category $\mathbf{CAT}$ in their original paper, they had to face the same problem. Indeed, consider the category $\mathbf{Set}$ of all sets. What is it? Is it a set? It cannot be, since, as is well known, there is no set of all sets. What is its status? Is it definable at all? Have we been considering illegitimate collections all along?

2.2 Foundations of Category Theory

I introduced categories as pastings of structures of a certain type by morphisms between these structures. The problem is that to form a category, it seems necessary to collect together all structured sets of a certain type, in particular all sets, and all morphisms between these structured sets and this is logically impossible. Eilenberg and Mac Lane recognized the problem immediately and devoted §6 of their paper to it. It is rather interesting to see how they reacted to the problem. They more or less decided to ignore the technical details involved, since according to their analysis there is essentially no new problem involved. Their diagnosis is that the problem is the standard problem of size, well known from set theory:[8]

> We remarked in §3 that such examples as the "category of *all* sets," the "category of *all* groups" are illegitimate. The difficulties and antinomies here involved are exactly those of ordinary intuitive *Mengenlehre*; no essentially new paradoxes are apparently involved. Any rigorous foundation capable of supporting the ordinary theory of classes would equally well support our theory. *Hence we have chosen to adopt the intuitive standpoint, leaving the reader free to insert whatever type of logical foundation (or absence thereof) he may prefer.* These ideas will now be illustrated, with particular reference to the category of groups ([74], 246).

In other words, the solutions to the problem can simply be the standard solutions to the paradoxes of set theory. This clearly indicates that: (1) Eilenberg and Mac Lane assumed that category theory could be developed in a set theoretical framework; (2) To a large extent, the development of category theory—inasmuch as they thought that category theory could be developed—appeared to them to be independent of the chosen foundation.[9]

They go on and consider more specific solutions. As was already pointed out, it is at this point that they indicate that the notion of category is after all secondary and could be dismissed altogether. Thus the quotation I have given in the first section of this chapter continues as follows (see p. 43):

> Thus one could drop the category concept altogether and adopt an even more intuitive standpoint, in which a functor such as "Hom" is not defined over the category of "all" groups, but for each particular pair of groups which may be given. The standpoint would suffice for the applications, inasmuch as none of our developments will involve elaborate constructions on the categories themselves ([74], 247).

This is their first solution. It clearly indicates that categories did not have a specific status in their mind. The algebra of mappings, namely the Hom-sets together with the morphisms between them, is what matters in applications.

Eilenberg and Mac Lane then move on to consider more technical solutions to the problem. They entertain the possibility of considering *a* category of groups (any such that it would be a legitimate collection from the set theoretical point of view) instead of *the* category of groups. We would in this case have a multiplicity of categories together with a multiplicity of functors for structures of a given type. Thus,

---

[8] This is already debatable for sets themselves! To see the antinomies as problems related to the size of sets is but one way to look at the problem. See, for instance, [109] and [163].

[9] This is still a common attitude. Indeed, here is a typical statement: "It seems that no book on category theory is considered complete without some remarks on its set-theoretic foundations. The well-known set theorist Andreas Blass gave a talk (...) on the interaction between category theory and set theory in which, among other things, he offered three set-theoretic foundations for category theory. One was the universes of Grothendieck (...) and another was systematic use of the reflection principle, which probably does provide a complete solution to the problem; but his first suggestion, and one that he clearly thought at least reasonable, was: None. This is the point of view we shall adopt" ([15], ix).

2.2 Foundations of Category Theory

there would be different Hom-functors, one for each category considered. This solution would raise practical problems since it would become difficult to define the composition of functors in general.

They then suggest that one could adopt the (unramified) theory of types as a foundation for the theory of classes. In this context, one would consider, for instance, the category **AbGrp**$_m$ of all Abelian groups of type m. A functor in two arguments, e.g. Hom$(-,-)$ would then have both arguments in **AbGrp**$_m$, but its values would be in the category of groups **Grp**$_{m+k}$ of higher type $m+k$. They immediately point out that the well-known ambiguity of the type theory would complicate the study of natural isomorphisms since one would have to consider isomorphisms between groups of different types.

Finally, standard set theoretical solutions are mentioned. They suggest that a set of axioms for classes as in the Frænkel-von Neumann-Bernays system (NBG) would be adequate.[10] In this framework, a category is any legitimate class of the system. An alternative is to restrict the size of the sets considered. Thus, one could take the category of all denumerable groups, or the category of all groups of cardinality at most that of the continuum, etc.

What is striking in Eilenberg and Mac Lane's response to the question of the foundations of category theory is that they do not *see* the potential problems that are raised by the theory. For the specific *applications* that they have in mind and that they can foresee, the aforementioned "tricks" seem entirely satisfactory. And as we have seen, categories themselves are not even considered to be indispensable. But their paper already contains two elements that raise serious issues about the foundations of category theory. I have already hinted at one of them by suggesting the possibility of the category of categories. This is but one case of a general situation: given certain categories, it often seems natural to construct new categories from them, as in the case of the product of two categories, and this irrespective of their size. The other problem, to which we will turn shortly, has to do with functor categories. Let us briefly examine the first problem.

Assume that we work in NBG, which is the simplest technical solution. Recall that in NBG, it is possible to distinguish *classes* and *sets*. Informally, sets are classes that can belong to other classes. A *proper class* is a class that cannot belong to another class. In other words, a proper class is a class that is not a set. Thus, in this context, the category of all sets is a proper class, not a set and so is the category of all groups, etc.

Armed with this distinction, we say that a *large* category is a category whose class of morphisms is a proper class of NBG. Otherwise, the category is said to be *small*. In general for each pair X, Y of objects of a category $\mathscr{C}$, the class Hom(X,Y) is assumed to be a set. Since the class of objects of a category is in one to one correspondence with a subclass of the morphism class of a category, namely each object X corresponds to the identity morphism $1_X: X \to X$, if a category is small, its class of objects is also a set. Most of the examples of categories that we have seen are large categories and they are definable in NBG. We have also seen many

---

[10] This is how they present the theory. In his book on category theory, Mac Lane will talk about the Gödel-Bernays axioms instead.

examples of small categories. Any monoid, group, pre-order, partial order, lattice, Heyting algebra, Boolean algebra, etc., seen as a category, is a small category. These are all sets and they are obviously definable in NBG.

Suppose that we have a number of large categories $\mathscr{C}, \mathscr{D}, \mathscr{E}, \ldots$ and that we want to consider the category whose objects are the given large categories and functors between them as morphisms. Since in NBG, proper classes cannot be elements of another class—although they can be subclasses of another class—we cannot collect the various large categories into one class, nor can we form the morphism class of such a category.

There is, however, a standard way to avoid the foregoing difficulty. One starts with *small* categories $\mathscr{C}, \mathscr{D}, \mathscr{E}, \ldots$ and constructs the *large* category whose objects are these small categories and functors between them. But this does not allow us to consider functors between large categories, something which seems to be entirely natural and even indispensable. As we will see, the problem will become even more acute when functor categories enter the scene in their full generality.

2.3 Philosophical Interlude: An Argument Against the Foundational Status of Category Theory

If we think of category theory as a geometric framework for studying the structure of a family of objects, as Eilenberg and Mac Lane did, then the axioms of the theory as presented above *hide* a crucial aspect of the theory. It is as if someone were to define a group as a group of transformations, e.g., as a set of points and a collection of mappings on these points that satisfy certain properties. Given this presentation, we might be led to believe that a group is merely a tool to study the properties of the given set of points, and that the sets of points are therefore primordial. Thus transformations groups are part of applied mathematics; they constitute an example of the application of mathematics to itself. When we move to the abstract notion of group, we can look directly at the group structure, defined or characterized by means of group theoretical concepts and then look at representations again, namely go back to groups as transformation groups. We are at this point in the same position with respect to categories. The foregoing axiomatization, which is found in many textbooks of category theory, presents categories as a *tool* used to examine a collection of structured objects, and one is tempted to conclude that the objects are primordial.[11] Thus, one could argue that, just as a transformation group G in geometry is abstracted from an underlying geometric structure, a category $\mathscr{C}$ is abstracted from a collection of structures and homomorphisms between them. Hence, one would be tempted to conclude that categories are fundamentally collections and thus rest upon set theory.

---

[11] Algebra is generally seen as a tool. To give but one example from the literature: "It is now taken for granted that the methodology of algebra is an essential tool in mathematics. On the other hand, in recent research one can observe a return to the challenge presented by fairly concrete problems, many of which require for their solution tools of considerable technical complexity" ([115], xi).

2.3 Philosophical Interlude

This remark can lead to what is taken to be a philosophical argument against the foundational status of category theory. It can be summarized as follows: in order to explain the concept of a category, one needs to explain the notion of a collection and the notion of a structure. Hence, the notion of a collection comes before that of a category and therefore, category theory cannot be taken as a foundation for mathematics. Feferman first presented the standard formulation of this argument:

> The point is simply that *when explaining* the general notion of structure and of particular kinds of structures such as groups, rings, categories, etc., we implicitly *presume as understood* the ideas of *operation* and *collection* ([85], 150).

Three points have to be made immediately. First, the argument rests on a certain order of *explanation* and thus of what appears to be cognitive dependence. Second, Feferman explicitly places categories alongside groups, rings and presumably other algebraic structures and treats categories merely as one kind of structure among others. This is certainly correct in a certain sense, since a category is an algebraic structure and the first order theory of categories is an algebraic theory. However, category theory is a general theory of structures and it is therefore in a sense a metastructure. And third, the argument has a family resemblance with Poincaré's argument against logicism. It is certainly different in its details, but it relies on a similar strategy. (See, for instance, [56].)

The argument clearly rests upon: (1) a purely logical conception of explanation, probably along the lines of the deductivo-nomological model of explanation; and (2) an assumption that *understanding* follows directly from logical explanations. Needless to say, both claims are highly debatable. It would be pointless to analyze the nature of explanations in general or even only in mathematics—assuming that there *are* explanations in mathematics. (See, for instance, [241], [108].) However, I cannot refrain from making remarks concerning the underlying conception of understanding.

The concept of understanding implicitly used by Feferman assumes that to understand a structure is to grasp, first, that we are dealing with a set of bare elements and, second, that certain operations satisfying specific conditions are definable on these elements and these are grasped afterward. Thus, Feferman assumes that understanding proceeds along the lines of model theory. However, in practice, mathematical structures almost always appear in a given context and are *abstracted from* that context. Thus, by looking at different definitions of homology groups and the fact that the different definitions nonetheless yielded isomorphic groups, Eilenberg and Mac Lane were led to the concepts of natural transformation, functor and category by abstracting from each specific context the essential features of the situation. It is therefore entirely possible that although some structures can be reconstructed logically from the general idea of collection and operation, they might be best *understood* first by way of examples. Our point, at this stage, is not that Feferman is wrong but that the argument presupposes a definite epistemology of understanding that has to be seen as such.

It is revealing to note that an argument very similar to Feferman's argument could be constructed against the role of group theory in the foundations of geometry: in order to explain the concept of transformation group, one has to explain the notion of a geometric structure. Hence transformation groups cannot be taken as a foundation for geometry.[12]

But wait. The geometric case points toward an important fact: the dependence involved is of a different *nature* than what we usually have in mind. Indeed, it is true that the concept of a transformation group depends upon the concept of a geometry. But at the same time, a transformation group is independent, at least in general, from any *specific* geometry. This is the whole point of Klein's transfer principle according to which a group of transformations can be transferred from a given geometric space to a different geometric space provided that a bijection between the geometric elements exists. In other words, one could start, say, with a certain spherical geometry, and investigate its group of transformation, or one could start with Möbius geometry, investigate its group of transformation and realize that both geometries have isomorphic transformation groups and are, in a sense, the same. Although one has to choose one geometric structure to exemplify a given transformation group G, various geometric spaces can be chosen in order to illustrate and explain what G is. Thus, G is in some respect independent from any particular geometry that exemplifies it without being completely independent from all these geometries. Moreover, as we have seen in the case of elementary geometries, one can start with a purely algebraic and abstract description of a group of transformations G and construct from G and some of its subgroups the various geometric structures that exemplify it.

As I have also pointed out, looking at elementary geometry from the point of view of transformation groups provides an insight into the nature of geometry and geometrical concepts. Thus it could be argued that the traditional approach to the foundations of elementary geometry did not reflect the essential features of geometry (i.e., what makes geometry what it is), and that what does reveal the essential features is precisely the nature, role and structure of transformation groups. I am not equating transformation groups with geometry. As Poincaré said in a conference commenting on Cartan's work:

> If we then strip the mathematical theory of what appears in it merely as an accident, that is of its matter, only the essential is left, that is its form; and this form, which constitutes, one might say, the solid skeleton of the theory, will be the structure of the group ([239], 264).

In other words, the transformation group is what makes the whole thing hangs together as a whole, as a geometry. Clearly, when we start with a given geometric theory and we move to its associated transformation group, we are abstracting from a certain description of a geometric space to its skeleton, its structure. Conversely, when we start from a transformation group, given algebraically, and move to its representations, we are putting some flesh around the skeleton, we are giving a concrete exemplification of the structure. As we have seen, there are usually many different ways of adding flesh to a skeleton.

---

[12] However, as we have seen, they can. See for instance, [9]. It could be argued that such a presentation is cognitively awkward, but then the argument shifts from the purely logical possibility of providing such a foundation to its cognitive value.

2.3 Philosophical Interlude

Let us come back to the argument from this point of view. It is clear that when we explain the notion of a transformation group, we implicitly take as understood certain notions of geometry. I can certainly grant this fact. What can we conclude from it? Can we conclude that transformation groups have no role to play in the foundations of geometry? I do not think that this is the proper conclusion. Here is a different claim: someone who does not understand the nature and role of transformation groups in elementary geometry does not understand what elementary geometry is about. On the other hand, someone who does not know how to move from a transformation group to one of its geometric representation does not understand geometry either. Thus, we are led to a situation where a foundational framework somehow has to include two dimensions: a purely "abstract" component that provides the overall structure of the theory, together with one or more singular representations or, to use a metaphor, "embodiments". Notice how close this sounds like what we find in logic where we have a theory and its models. As we will see, categorical logic provides a bridge between representation theorems and completeness theorems and it shows how they are in fact two faces of the same coin.

In the next few chapters I will argue that the foundational status of category theory in mathematics is analogous to the status of transformation groups in elementary geometry (and groupoids in differential geometry). In fact, the role played by categories in mathematics is much *stronger* than the role played by transformation groups in geometry. Notice immediately that the analogy presupposes that mathematical concepts in general have a lot in common with geometrical concepts. This should be compared with the set-theoretical context in which, I believe, mathematical concepts are taken to be very much like concepts of arithmetic, or more generally, combinatorics. (Even at its origin, set theory had a lot to do with counting, whereas category theory had more to do with the classification of forms via topological spaces.) Be that as it may, if the foregoing considerations concerning transformation groups are sound, then I claim that they can be generalized to categories, viewed in the right way, and that the argument presented by Feferman misses its target.

Let us now consider a second aspect of Feferman's argument. First, it assumes that categories are structured collections in general and thus are presented as classes or sets with a certain structure, and second, it assumes a definite conception of sets. As I will argue later, category theory relies on a different conception of sets and once this is understood, categories can be presented in a way that does not depend on the standard conception.

John Bell, inspired by Feferman, presented arguments against category theory in the foundations of mathematics that go in a somewhat different direction. First, Bell distinguishes two different senses in which category theory could serve as a foundation for mathematics, a strong sense and a weak sense. According to the strong sense, "*all* mathematical concepts, *including* those of the current logico-mathematical framework for mathematics" should be explicable in category-theoretic terms ([19], 353). In the weaker sense, "one only requires category theory to serve as a [possibly superior] substitute for axiomatic set theory in its present foundational role"([19], 353).

Bell asserts that it seems implausible to him that category theory could be foundationally adequate in the stronger sense. He gives two reasons. The first reason is that metamathematics has two irreducible aspects: the combinatorial and the semantical. The combinatorial "is concerned with the formal, finitely presented properties of the inscriptions of the ambient formal language, and the semantical (...) is concerned with the interpretation and truth of the expressions of that language" ([19], 353). Bell argues that neither of these aspects can be accounted for by category theory. For, according to Bell, combinatorial objects are generally intensional and categories are extensional; that is, Bell presents categories as *classes* of objects and morphisms satisfying the usual axioms, and the latter cannot "cover" the former. Once more, this argument presupposes that there is a unique conception of classes. As to the combinatorial aspects of metamathematics, I can certainly grant that *any* foundational framework has to have a combinatorial dimension. As I will show later, category theory and categorical logic have certain things to say about this aspect of mathematics. We will come back to this point in due course.

Let us now turn to the semantical aspects of metamathematics. This is where Bell explicitly falls back on Feferman's argument, but with a slightly different twist. Semantics deals with the notion of truth, and in particular logical truth. Bell claims that in order to grasp the concept of logical truth for sentences of classical first-order languages, one has to grasp the concept of a class. Thus, if category is to serve as a foundation for mathematics in the strong sense, it has to give a satisfactory account of the notion of class solely in its terms. Bell believes that this is not possible. Here is how he puts it:

> But this seems to me highly dubious, for it is surely the case that the unstructured notion of class is epistemically prior to any more highly structured notion such as category: in order to understand what a category is, you first have to know what a class is ([19], 352–354).

And we are back to Feferman's argument.

Notice that it seems that both Feferman and Bell conflate the logical and the epistemic. It might be perfectly correct to say that from an *epistemic* point of view a notion of class precedes the notion of a category. This does not entail, however, that from a *logical* point of view, a notion of class has to come before the notion of a category. Once again, transformation groups and elementary geometric structures allows us to see a parallel situation. It makes perfect sense to claim that geometric notions are epistemically prior to group theoretical notions. However, it is perfectly possible, from a purely logical point of view, to start from abstract transformation groups and move to geometric spaces.

Let us now turn to the weak sense of "foundation" discussed by Bell. In this sense, the question is whether category theory could serve as a substitute for axiomatic set theory in its foundational role. Bell's analysis is based on work done in the seventies in which a systematic translation between axiomatic set theory and topos theory was developed. On the basis of this translation, he concludes that

> it would be *technically* possible to give a purely category-theoretic account of all mathematical notions expressible within axiomatic set theory, and so formally possible for category theory to serve as a foundation for mathematics insofar as axiomatic set theory does ([19], 355).

2.3 Philosophical Interlude

He then claims that although possible, the actual translation is awkward and "unsuitable as a means of formalizing those mathematical notions which are normally expressed set-theoretically" ([19], 355). I agree with Bell on this line and I will not look at the technical aspects of the translation in this book. I would even go further and argue that the possibility of such a translation has no real foundational relevance. It was an interesting technical question to be able to compare topos theory as it was then developed to axiomatic set theory. But it certainly cannot serve to establish the possibility of topos theory and a fortiori of category theory as a foundation for mathematics.

Bell then turns to Lawvere's early axiomatization of the category of all small categories (in [167]). This is an attempt at developing a foundational framework in which mathematical objects are categories and the universe itself is a category. Unfortunately, Lawvere's original attempt suffered from a slight technical problem. (See Chapter 6.) However, Bell believes that fixing these technical difficulties would not do category theory any good. The problem here, according to Bell, is that, in order to define notions of "workaday" mathematics, one has to introduce discrete categories and discrete categories are nothing but sets. In a way, we get back to Feferman's argument. Here is how Bell puts it:

> But the question automatically arises as to exactly *why*, in introducing ordinary mathematical notions into the theory, one must make a detour through the somewhat opaque notion of discrete category. It is difficult to see how this can be explained except by appeal to the notion of 'unstructured' category, *i.e. set* ([19], 355).

Notice that the problem is not that the notion of set has to be defined independently of the given theory, for the concept of discrete category is defined within the theory, rather the problem is that discrete categories make the whole system appear "artificial" as a "foundation". Two remarks have to be made. First, it now appears that a categorical framework would be syntactically and semantically rather different from what Lawvere presented in the sixties. (I will give pointers to these differences as well as reference in the conclusion.) Second, introducing sets or collections as discrete categories does not merely amount to introducing sets as they are described in an axiomatic set theory, e.g., Zermelo-Frænkel (ZF). For sets as discrete categories certainly do *not* satisfy the axiom of extensionality nor do they satisfy a comprehension or a separation principle unless they are defined within a set theory, which is just what the new framework is attempting to avoid. Thus, although collections cannot be bypassed even in a categorical framework, this does not mean that there is not a coherent and adequate notion of collection that can be developed within that framework, a coherent notion that is compatible with a general notion of mathematical concept.

Let us now close this philosophical interlude and return to categories and their algebra of mappings. What is still not entirely clear is *how* the algebra of mappings captures properties of a structure and *how* it should be used to discover these properties. Eilenberg and Mac Lane had an idea about this usage and it is precisely at this point that category theory should be seen as a generalization of Klein's program. To see how this was done, we have to turn to the core of their work, namely natural transformations.

2.4 At Last, Natural Transformations

Informally, a natural transformation is a translation or a transformation of one functor into another. More pictorially, one can think of the situation as follows: given two parallel functors $F\colon \mathscr{C} \to \mathscr{D}$ and $G\colon \mathscr{C} \to \mathscr{D}$, a natural transformation is a way to transform globally or systematically the image of F in $\mathscr{D}$ into the image of G in $\mathscr{D}$. The formal definition is usually given thus: a *natural transformation* $\eta\colon F \to G$ is a function which assigns to each object X of $\mathscr{C}$ an arrow $\eta_X\colon F(X) \to G(X)$ in such a way that for every morphism $f\colon X \to Y$ of $\mathscr{C}$, the following diagram commutes:

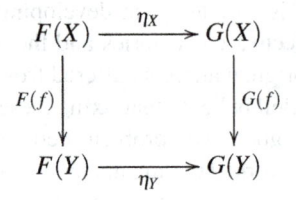

$$\begin{array}{ccc} F(X) & \xrightarrow{\eta_X} & G(X) \\ {\scriptstyle F(f)}\downarrow & & \downarrow{\scriptstyle G(f)} \\ F(Y) & \xrightarrow[\eta_Y]{} & G(Y) \end{array}$$

In particular, whenever η_X is an isomorphism for each X, then we say that η is a *natural isomorphism*. Here are some examples of natural transformations and natural isomorphisms.

1. Consider the functor $1_{\mathbf{Set}}\colon \mathbf{Set} \to \mathbf{Set}$ and the covariant functor $\wp\colon \mathbf{Set} \to \mathbf{Set}$ defined above. There is a natural transformation $\{-\}\colon 1_{\mathbf{Set}} \to \wp$, such that for each set X, the map $\{-\}_X\colon X \to \wp(X)$ is defined by $\{-\}_X(x) = \{x\}$ for all x in X. It is easy to verify that it is a natural transformation.
2. Here is the example given by Eilenberg and Mac Lane themselves in the opening paragraph of their original paper ([72]). Let V be a finite dimensional real vector space. As usual, let $V^* = \{f\colon V \to \mathbb{R} \mid f \text{ is real-valued fonction}\}$, the dual space of V. It is well known that V^* is itself a finite dimensional real vector space and that it is in fact isomorphic to V. However, in order to define an isomorphism between V and V^*, one has to choose a basis for V and the isomorphism will vary with different choices of basis vectors. It is also well known that the double dual V^{**}, i.e. the space of all linear functionals of linear functionals, is also isomorphic to V. However in this case the isomorphism does *not* depend on the choice of a set of basis vectors. The isomorphism can be defined for *all* finite-dimensional vector spaces simultaneously. This is why such an isomorphism was said to be "natural". It can easily verified that the operation of taking the dual space of a real finite-dimensional vector space is functorial, i.e. going from V to V^* is a functorial operation, more precisely a contravariant functor. It is then possible to show that there is a natural transformation, in fact a natural isomorphism, between the identity functor $1_{\mathbf{Vect}}$ and the functor $(-)^{**}$.
3. Let H be a fixed group. Then it is easy to verify that the map $H \times -\colon \mathbf{Grp} \to \mathbf{Grp}$ is a functor. Indeed, for any group G, $H \times G$ is the usual group product and for any homomorphism $f\colon G_1 \to G_2$, $(H \times -)(f)\colon H \times G_1 \to H \times G_2$ is the homomorphism $(1_H, f)$ defined pointwise. It is easy to see that this construction satisfies the conditions of the definition of a functor. Now, each homomorphism

2.4 At Last, Natural Transformations

$h\colon H \to K$ defines a natural transformation $H \times - \to K \times -$. This means that the following diagram commutes:

$$\begin{array}{ccc} H \times G_1 & \xrightarrow{\eta_{G_1}} & K \times G_1 \\ {\scriptstyle (1_\eta, f)}\Big\downarrow & & \Big\downarrow{\scriptstyle (1_K, f)} \\ H \times G_2 & \xrightarrow[\eta_{G_2}]{} & K \times G_2 \end{array}$$

4. In their 1945 paper, Eilenberg and Mac Lane gave a host of examples of natural transformations and also examples of mappings that are *not* natural. Here are the basic examples of natural transformations given. When standard constructions on sets, topological spaces, groups, Banach spaces are seen as functors, then well-known maps are shown to be natural transformations. For instance, for X, Y and Z sets, topological spaces or groups, the following isomorphisms are natural transformations (isomorphisms):

   a. $(X \times Y) \times Z \longrightarrow X \times (Y \times Z)$;
   b. $X \times Y \longrightarrow Y \times X$;
   c. $\mathrm{Hom}(Z, X) \times \mathrm{Hom}(Z, Y) \longrightarrow \mathrm{Hom}(Z, X \times Y)$;
   d. $\mathrm{Hom}(X \otimes Y, Z) \longrightarrow \mathrm{Hom}(X, \mathrm{Hom}(Y, Z))$, where in this case if X and Y are groups, then the product $X \otimes Y$ is the tensor product.

Once the concept of natural transformation is defined, it is easy to define a new kind of category, namely a category of functors or a functor category thus: given two categories $\mathscr{C}$ and $\mathscr{D}$, the category $\mathscr{D}^\mathscr{C}$ has as objects functors $F\colon \mathscr{C} \to \mathscr{D}$ and morphisms natural transformations $\eta\colon F \to G$. It is easy to verify that this is indeed a category (natural transformations compose in the obvious manner, the identity natural transformation exists for each functor F and satisfies the conditions of the definition). Since these categories play a key role in category theory and in our story, we will give some simple and hopefully illuminating examples of such categories.

1. Let us start with a trivial example. Let **1** denote the category with one object and one morphism, namely the identity morphism on the unique object. Then a functor $F\colon \mathbf{1} \to \mathscr{C}$ picks out an object X of $\mathscr{C}$ and the unique morphism is sent to the identity 1_X of X. In particular, when $\mathscr{C}$ is the category **Set**, then a functor picks a set X together with its identity map. Trivially in this case, the category **Set**^{1} is "the same"—although we still have not made precise under what conditions two categories should be considered to be "the same", an important issue to which we will turn shortly—as **Set**.
2. By modifying slightly the foregoing picture, we get a very different category. Instead of the category **1**, we take the category $\circlearrowright$ with one object and one nontrivial endomorphism. Then a functor $F\colon \circlearrowright \to \mathbf{Set}$ picks a set X together with an endofunction $X \to X$. The functor category **Set**$^\circlearrowright$ is the category whose objects are sets X *together* with an endofunction $f\colon X \to X$. A morphism of this

category is a natural transformation η with the property that it preserves endofunctions. It can be represented thus:

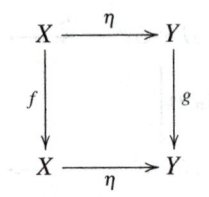

$$\begin{array}{ccc} X & \xrightarrow{\eta} & Y \\ f \downarrow & & \downarrow g \\ X & \xrightarrow{\eta} & Y \end{array}$$

As can be read directly from the diagram, preserving endofunctions means that $\eta \circ f = g \circ \eta$. This functor category can therefore be thought of as a category of sets with structure.

3. Let **2** denote the category with two different objects, denoted by 0 and 1, and one non-trivial morphism $0 \to 1$. A functor $F \colon \mathbf{2} \to \mathbf{Set}$ picks out two sets and a map between them, i.e. the functor category $\mathbf{Set}^{\mathbf{2}}$ is the category of morphisms of **Set**.

4. Let $\downdownarrows$ denote the category with two objects and two parallel non-trivial morphisms between them. A functor $F \colon \downdownarrows \to \mathbf{Set}$ picks two sets and two parallel morphisms $s, t \colon A \rightrightarrows D$. Such an object can be seen to be an irreflexive directed multigraph. The set A is the set of arrows of the graph and the set D is the set of dots of the graph. The map s sends an arrow to its source and the map t sends a arrow to its target. Thus the functor category $\mathbf{Set}^{\downdownarrows}$ is the category of irreflexive directed multigraph. A morphism of this category between objects $A_1 \rightrightarrows D_1$ and $A_2 \rightrightarrows D_2$ is given by a pair of functions $f_A \colon A_1 \to A_2$ and $f_D \colon D_1 \to D_2$ such that the diagram commutes:

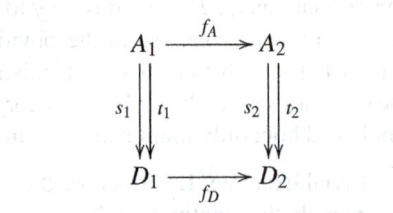

$$\begin{array}{ccc} A_1 & \xrightarrow{f_A} & A_2 \\ s_1 \downdownarrows t_1 & & s_2 \downdownarrows t_2 \\ D_1 & \xrightarrow{f_D} & D_2 \end{array}$$

that is, $s_2 \circ f_A = f_D \circ s_1$ and $t_2 \circ f_A = f_d \circ t_1$.

5. Let **3** denote the category with three different objects, denoted by 0, 1 and 2 and three non-trivial morphisms between them, namely $0 \to 1$, $1 \to 2$ and the composition $0 \to 2$. A functor $F \colon \mathbf{3} \to \mathbf{Set}$ picks three sets X, Y and Z and a commutative triangle between them. Thus, the category $\mathbf{Set}^{\mathbf{3}}$ is the category of commutative triangles of **Set**.

6. Let $\langle \mathbb{N}, \times, 1 \rangle$ denote the monoid of natural numbers with multiplication and one as the identity element, now seen as a one object category with denumerably many morphisms. A functor $F \colon \mathbb{N} \to \mathbf{Set}$ picks one object, thus a set X, together with an endofunction $f \colon X \to X$. More specifically, $F(1) = 1_X$ and $F(n) = f^n$,

2.4 At Last, Natural Transformations

i.e. f composed with itself n times. Thus, this is simply a different description of the category in the example 2 above.

7. Let now $\mathbb{N}$ be the set of natural numbers linearly ordered. A functor $F: \mathbb{N} \to \mathbf{Set}$ picks a sequence $X_0 \to X_1 \to \ldots$ of sets X_n and functions $X_n \to X_{n+1}$. Lawvere has suggested that such a sequence should be thought of as a set through time or as a *variable* set. A morphism between two such sequences is given by a sequence of functions thus:

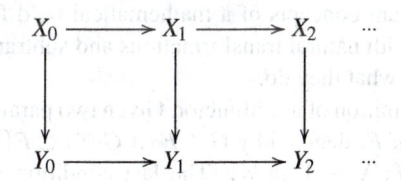

$$\begin{array}{ccccccc} X_0 & \to & X_1 & \to & X_2 & \cdots \\ \downarrow & & \downarrow & & \downarrow & \\ Y_0 & \to & Y_1 & \to & Y_2 & \cdots \end{array}$$

where every square is commutative.

8. Let G be a group. The functor category $\mathbf{Set}^G$, also denoted by $\mathbf{B}G$, is the category of all representations of a fixed group G, where a representation is defined as in the previous chapter, namely a representation of G is given by a set X together with an action $\alpha: X \times G \to X$ on X.

Although Eilenberg and Mac Lane did consider functor categories of a very special type in §8 of their paper, namely for functors with two arguments, such categories did not play a significant role in their work, nor did they notice that these categories raise another foundational problem. Indeed, they explicitly claim that "this category [namely a functor category] is useful chiefly in simplifying the statements and proofs of various facts about functors, as will appear subsequently" ([74], 250). Once again, a category is merely heuristically useful and does not, as such, play any significant role in the story.

As for the foundational problem, it is similar to the problem we have encountered earlier with the category of categories. Indeed, if $\mathscr{C}$ and $\mathscr{D}$ are large categories in the sense specified previously, then a single functor $F: \mathscr{C} \to \mathscr{D}$ is a proper class and therefore we cannot collect all functors from $\mathscr{C}$ to $\mathscr{D}$; that is, the functor category $\mathscr{D}^{\mathscr{C}}$ cannot be formed in NBG. However, if $\mathscr{C}$ and $\mathscr{D}$ are small categories, then their morphism classes are sets and the functor category $\mathscr{D}^{\mathscr{C}}$ is a set, since it is a subset of the class of all functions from $\mathscr{C}$ to $\mathscr{D}$, i.e. not necessarily structure-preserving. Thus, certain constraints have to be imposed on an operation that seems to be intuitively natural between categories.

Natural transformations and in particular natural isomorphisms were Eilenberg and Mac Lane primary targets in their 1945 paper. Once the general definition was given, they turned in Chapter III to functors in the category of groups and presented there the extension of Klein's program as they saw it. Let us turn to their attempt and try to see why it was inadequate.

2.5 Extending Klein's Program in the Wrong Direction

Eilenberg and Mac Lane's idea seems to have been the following: Klein's classification of geometric theories was based on the relationships between a group of transformations and its subgroups. Eilenberg and Mac Lane replaced the transformation group, let us denote it by $\mathrm{Aut}(X)$, by the algebra of mappings $\mathrm{Hom}(X,Y)$. Since they see the latter as a functor, it seems natural to consider its subfunctor and see what are the invariant concepts of a mathematical field for the functor and its subfunctors, together with natural transformations and subtransformations between them. This is precisely what they do.

We first need the definition of a subfunctor. Given two parallel functors $F, G\colon \mathscr{C} \to \mathscr{D}$, G is a *subfunctor* of F, denoted by $G \subset F$, if $G(X) \subset F(X)$ for all X in $\mathscr{C}$ and $G(f) \subset F(f)$ for all $f\colon X \to Y$ in $\mathscr{C}$. (The last condition means that $G(f)$ is a submapping of $F(f)$, i.e. $G(f)(x) = F(f)(x)$ for all x in $G(X)$.)

Given parallel functors $F_1, G_1\colon \mathscr{C} \to \mathscr{D}$ and $F_2, G_2\colon \mathscr{C} \to \mathscr{D}$, natural transformations $\eta_1\colon F_1 \to G_1$ and $\eta_2\colon F_2 \to G_2$, we say that η_1 is a *subtransformation* of η_2, denoted by $\eta_1 \subset \eta_2$, if $F_1 \subset F_2$ and $G_1 \subset G_2$ and if for all X in $\mathscr{C}$, η_{1X} is a submapping of η_{2X}.

Following a similar pattern, Eilenberg and Mac Lane then define in §15 quotient functors. Assuming now that $F, G\colon \mathbf{Grp} \to \mathbf{Grp}$, that $F \subset G$ and that F is a *normal* subgroup of G, then the quotient functor $Q = G/F$ is defined by $Q(X) = G(X)/F(X)$, where both $G(X)$ and $F(X)$ are groups now, and given a group homomorphism $f\colon X \to Y$, $Q(f)\colon Q(X) \to Q(Y)$ is defined for each coset by $Q(f)(x + F(X)) = [G(f)(x)] + F(Y)$. Given these definitions, Eilenberg and Mac Lane consider examples of subfunctors of the identity functor $1\colon \mathbf{Grp} \to \mathbf{Grp}$. This is where the classification of group-theoretical constructions is introduced. They mention that, for any group X, the commutator subgroup $C(X)$, namely the subgroup generated by elements of X of the form $xyx^{-1}y^{-1}$, x, y in X, is a normal subfunctor of the identity functor 1. Furthermore the quotient functor $(1/C)(X)$ is the factor commutator group of X.

Another important construction in group theory is the *center* $Z(X) = \{x \in X \mid xyx^{-1} = y \ (\forall y \in X)\}$ of the group X. (The center $Z(X)$ is also defined as the set of elements x that commute with all the elements of X, i.e. $\{x \in X \mid xy = yx \ (\forall y \in X)\}$, which is easily seen to be equivalent.) However, $Z(-)$ is *not* a functor and a fortiori a subfunctor of the identity functor on the category of groups. But if one restricts the algebra of mappings $\mathrm{Hom}(X,Y)$ to morphisms *onto*, then it *is* a functor. Similarly, the automorphism group $\mathrm{Aut}(X)$ of a group X is *not* a functor unless the Hom-sets are restricted to *isomorphisms* between groups. This is how the algebra of mappings enters the classification of the concepts or constructions of group theory and how, as we are told in the introduction of their paper.

> The invariant character of a mathematical discipline can be formulated in these terms. Thus, in group theory all the basic constructions can be regarded as the definitions of co- or contravariant functors, so we may formulate the dictum: The subject of group theory is essentially the study of those constructions of groups which behave in a covariant or contravariant manner under induced homomorphisms ([74], 237).

2.5 Extending Klein's Program in the Wrong Direction

In §16, Eilenberg and Mac Lane emphasize the *functorial* character of their classification. It is defined for all groups in the category of groups. This is the second aspect that seems to allow them to make a connection with Klein's program. For as a group of transformations apply to a whole space, a functor is defined over a category as a whole. In particular, one can consider a single group as a category (with one object, as we have seen) and in this case, one recovers a classification of subgroups that was already existent in 1945. Thus, Eilenberg and Mac Lane's approach is more general.

The category of groups, together with the category of Abelian groups, are the only cases considered in any detail in their paper. Eilenberg and Mac Lane did *not* present a general framework in which these ideas could have been developed. In the category of groups, it does make sense to consider a *normal* subfunctor and then quotients of a functor by a subfunctor, for these concepts are definable in this category. But they do not make sense in general. As far as I know, no one has referred to these sections of the paper since then. And it is no surprise, for Eilenberg and Mac Lane did *not* hit on the right way of thinking about invariance in this new context.[13]

Eilenberg and Mac Lane's approach implies that the concept of commutator subgroup is basic to group theory whereas the concept of center of a group is not as basic as the concept of commutator, though certainly useful and important in certain cases. Is this correct?

There seems to be something missing. Simply considering the (non)functoriality of the concept of center $Z(-)$ might lead us astray. Even though sending a group to its center is not a functorial construction, among all the subgroups of a group G, the center $Z(G)$ occupies a significant place: it is the *largest* subgroup of G having a certain property, i.e., that each of its elements commutes with all the elements of G. It is, in a sense, the best possible solution to the problem of finding a subgroup of G whose elements commute with all the elements of G. This fact can be translated in terms of the algebra of mappings as follows:

1. There is a monomorphism (in this case an injection) from $Z(G)$ into G and
2. *Any* group H which has an homomorphic image into G possessing the characteristic property of $Z(G)$ factors through $Z(G)$.

We can even use these facts to define the center $Z(G)$ with the help of the algebra of mappings. Something is at work here and it is not the "brute" fact of functoriality. In fact, what is missing is a basic categorical concept, some would rightly say *the* fundamental categorical concept: it is the concept of a universal arrow, or equivalently from the point of view of functors, the concepts of representable functor or, equivalently, of adjoint functors or adjointness.

It is important not to forget what Eilenberg and Mac Lane's goal was in their 1945 paper. Mathematicians knew informally that some mappings were "natural"

---

[13] Notice that determining the invariant properties under a family of transformations is more often than not a difficult task, even in geometry. Indeed, Klein's characterization of geometry as the study of the properties of figures that are invariant under all transformations of a group is in general hard to implement.

and others were not. They could not, however, say precisely or exactly what "being natural" amounted to. This is where Eilenberg and Mac Lane succeeded. The key observation Eilenberg and Mac Lane made was that "being natural" in the case of certain isomorphisms between groups could be defined for any group *as a group*, that is, without considering specific properties or elements of the group. One has to look at what they have done from an epistemological point of view. The notion of natural isomorphism was in the mathematical practice of the time. It had to be made exact. This is far from being a trivial process, for it is not obvious how such a notion should be made exact, and even more obvious that the set-theoretical framework is not adequate for the purpose. Indeed, "being a natural isomorphism" is not a property that is best defined in terms of the elements of the sets involved. Thus, Eilenberg and Mac Lane had to create the proper context in which the notion could be defined in full generality. Once the context is defined, one has to verify that it provides the means to give a general definition and then that the general exact definition *does* capture the informal notion adequately. One has to show that the examples mathematicians already considered as being natural *are* natural in the new, precise sense. Furthermore, one has to show that the examples of non-natural mappings *are* indeed non-natural in the new, exact sense. Lastly, one has to give new and convincing examples and applications of natural mappings from the exact definition in such a way that they are interesting, useful and coherent with the informal notion already used in practice. These were certainly the goals Eilenberg and Mac Lane had in mind when they wrote their paper and in these respects, they succeeded beautifully. The connection with Klein's program probably came as an afterthought and I for one am convinced that it was not one of their main motivations to generalize Klein's program, contrary to what Taylor claims in the foregoing quote.

But as often happens when a new general framework is introduced, this one opens the door to new possibilities: problems that had been vaguely conceived can at last be formulated precisely; sometimes these problems can be solved with the new tools provided by the framework; questions that were unthinkable become unavoidable; fields that were fragmented can be unified; new analogies can be discovered, new concepts can be introduced and results that had been proved in an obscure fashion can be proved and explained, sometimes in a way that appears to be trivial; all in the new framework. All this turned out to be in the future of category theory.

I will end this chapter by briefly looking at the status of category theory during what I call its first period of development, namely from 1945 until around 1957–1958.

2.6 Category Theory: The First Phase 1945–1958

> The work done by Eilenberg and Mac Lane on the role of the groups Ext(G,H) in homology led them almost immediately afterward to very general considerations on various aspects of groups theory that would ultimately bring new points of view in many parts of mathematics and exert a deep influence on subsequent work in algebra, algebraic topology, and algebraic geometry in particular.
> ([59], 96)

It is worth looking at the status of category theory from its inception in 1942 until about 1957–1958, two years that mark an important turning point in its development. It is probably fair to say that category theory was in this period first and foremost a useful framework for algebraic topology and homological algebra. I will therefore concentrate on two influential books published in that period, namely Eilenberg and Steenrod's *Foundations of Algebraic Topology*, published in 1952 but announced already in 1945 ([79]) and Cartan and Eilenberg's *Homological Algebra*, published in 1956 (but based on seminars given in 1950/51).

Eilenberg and Steenrod's main goal was to give the essential ingredients, via an axiomatic theory, of "the part of algebraic topology called homology theory" ([80], vii). In the thirties and forties, many apparently different and complicated homology theories were defined and used by various authors and as Eilenberg and Steenrod identify them in the preface, they are the singular homology groups of Veblen, Alexander and Lefschetz, the relative homology groups of Lefschetz, the Vietoris homology groups, the Čech homology groups, the Alexander cohomology groups, etc. Even though the experts knew implicitly what homology theory was, no one had written down explicit conditions. This is what Eilenberg and Steenrod decided to do by providing an *axiomatic* characterization of homology theory. One has to emphasize the originality of Eilenberg and Steenrod's contribution. What they axiomatized is not a domain of entities of a certain kind, e.g., groups or topological spaces, but *connections* between topological spaces and groups, namely *functors* between a category of topological spaces and a category of groups and this just after the concepts of category, functor and natural transformation had just been introduced.

> The various homology and cohomology theories appear as complicated machines, the end product of which is an assignment of a graded group to a topological space, through a series of processes which look so arbitrary that one wonders why they succeed at all. In a remarkable book (...) Eilenberg and Steenrod endeavored to break through this maze of unpleasant mathematics by adopting a totally different viewpoint, concentrating on *properties* of these end products rather than on the various methods devised to get them. This is the *axiomatic theory of homology (and cohomology)* ([59], 107).

Homology groups are constructed in an intricate manner and in various ways from topological spaces. In the words of Eilenberg and Steenrod, "the construction of homology groups is a long and diverse story, with a fairly obscure motivation" ([80], ix). Instead of focusing on the details of the various procedures to construct the diverse homology groups, Eilenberg and Steenrod concentrated on their common properties. Thus, not only does one end up with a better picture of the various essential components involved in all the constructions, but one presumably has a better understanding of their success. To present a homology theory as a family of *functors* between categories that satisfying certain specific properties is to define and clarify the *context*, the *framework* of homology theory and, more generally, algebraic topology.

Eilenberg and Steenrod define a homology theory as a family of functors $\{H_i : \mathscr{C} \to \mathscr{D}\}$, i an integer, from a certain category of topological spaces into a category of algebraic structures, e.g. Abelian groups or modules over a ring R, together with a family of natural transformations $\partial_i : H_i \to H_{i-1}$, called *boundary operators*, satisfying specific axioms which essentially stipulate how functors ought to behave with respect to certain obvious topological constraints, for instance subspace, homotopy equivalence and dimension. Moreover, and this detail turns out to be capital for the ensuing development of category theory, these axioms also define a cohomology theory, since the latter is simply the *dual* of a homology theory, i.e. it suffices to reverse all the arrows in the axioms. Using the categorical terminology, a homology theory is given by a family of covariant functors, whereas a cohomology theory is given by a family of contravariant functors. Four elements have to be underlined:

1. A homology theory is basically given by two families of *transformations*:

    a. Functorial transformations of topological spaces into algebraic structures; these provide an algebraic encoding of topological properties;
    b. Transformations of these functors into one another via natural transformations; these transformations are "natural translations" of one algebraic encoding into another;

2. These transformations have important properties themselves:

    a. The functors H_i, like any functor, preserve isomorphisms between topological spaces. Thus, if for two spaces X and Y, there is a functor H_i with $H_i(X)$ *not* isomorphic to $H_i(Y)$ for some i, then X and Y *cannot* be homeomorphic, for if they were, then there would be an isomorphism in the category of topological spaces and a functor H_i would take it to an isomorphism in the category of groups, which was shown to be impossible;
    b. Two homology theories are said to be *equivalent*, that is essentially the same, if there is a family of natural isomorphisms between them. Notice that as a matter of fact, homology theories can be very different with respect to other properties, e.g. facility of computation. It is the fact that homology theories can be systematically related to one another via natural transformations that allowed their classification. One can compare systematically, that is with the help of the morphisms in the target category, different homology theories and determine how and where they differ with respect to their role, that is capturing fundamental properties of the spaces studied. Chapter 3 of Eilenberg and Steenrod's book culminates in the so-called "uniqueness theorem" which stipulates under what conditions two homology theories are essentially the same, i.e. naturally isomorphic.

2.6 Category Theory: The First Phase 1945–1958

3. One can think of a homology theory as establishing a relation of formal supervenience between properties of mathematical objects. In this context, the given objects are topological spaces. In the category of topological spaces, some of their most important properties are captured by the existence and properties of continuous maps. The homology groups represent properties of topological spaces. In the words of Spanier:

> The functor H_n measures the number of 'n-dimensional holes' in the space (or simplicial complex), in the sense that the n-sphere S^n has exactly one n-dimensional hole and no m-dimensional holes if $m < n$. A 0-dimensional hole is a pair of points in different path components, and so H_0 measures path connectedness. The functors H_n measure higher dimensional connectedness, and some of the applications of homology are to prove higher dimensional analogues of results obtainable in low dimensions by using connectedness considerations ([252], 155–156).

Supervenience is simply a different way of saying that the homology theories are functorial: two identical, i.e. homeomorphic, spaces cannot have different homology groups, whereas a family of homology groups can be "exemplified" by two different topological spaces.

4. We can see more precisely what is the role and status of categories in this context. As we have already mentioned, the central objects of study in Eilenberg and Steenrod's work are functors and natural transformations between functors, namely homology and cohomology theories and their links. In turn, homology and cohomology theories are fundamentally and essentially complicated constructions whose purpose is to capture or measure basic properties of topological spaces. In Eilenberg and Steenrod's own words:

> Speaking roughly, a homology theory assigns groups to topological spaces and homomorphisms to continuous maps of one space into another. To each array of spaces and maps is assigned an array of groups and homomorphisms. In this way, a homology theory is an *algebraic image* of topology. The *domain* of a homology theory is the topologist's field of study. Its *range* is the field of study of the algebraist. Topological problems are converted into algebraic problems ([80], vii).

Hence a homology theory, or in fact most of algebraic topology, is comparable to a translation device. Seen this way, a category is merely the framework in which the "data" has to be organized for the translation to be effected. For the basic objects of study are still the topological spaces (and the mappings between them) and in order to apply the translation device, they have to be "prepared" in a certain way, that is they have to be presented to the device in the form of a category. Hence a category has absolutely *no* ontological relevance in this context. It is part of the required "preparation" of the data for the translation to take place. If a homology theory is a machine, to use Dieudonné's expression, then the categories involved are simply parts of the machine.

However, the language of diagrams *was* seen as being a new and useful language.

> Successful axiomatizations in the past have led invariably to new techniques of proof and a corresponding new language. The present system is no exception. The reader will observe the presence of numerous diagrams in the text. (…) Certain diagrams occur repeatedly

in whole or as parts of others. Once the abstract properties of such a diagram have been established, they apply each time it recurs.

The diagrams incorporate a large amount of information. Their use provides extensive savings in space and in mental effort. In the case of many theorems, the setting up of the correct diagram is the major part of the proof. We therefore urge that the reader stop at the end of each theorem and attempt to construct for himself the relevant diagram before examining the one which is given in the text. Once this is done, the subsequent demonstration can be followed more readily; in fact, the reader can usually supply it himself ([80], xi).

These remarks hold true immediately in the first chapter of the book where the axioms are stated and general results proved, but one might think that they are not indispensable. It is however difficult to see how Chapter 3 on the homology of simplicial complexes and especially the proofs of the main isomorphism, its properties and the uniqueness theorem could be written without the diagrams one finds there. There is no doubt that the diagrammatic language, derived from the categorical framework, contributed to the success of the whole enterprise.

> Eilenberg and Steenrod's book effected a revolution in mathematical notation. Perhaps not since Descartes's *La géométrie* has a book influenced how we write Mathematics. One [sic] knew they were looking at mathematics before 1600 because of the geometric diagrams with vertices and sides labeled by alphabetic letters. *La géométrie* in 1637 gave us nearly modern forms of equations, especially the notation of the exponent, i.e. a^3. The diagrams of Eilenberg-Steenrod not only made algebraic topology intelligible, but eventually swept out to other parts of mathematics, providing an efficient way to express complex, functorial relationships and giving us powerful methods of proofs by means of diagram chasing ([18], 733).

The categorical language provided "powerful methods of proofs". Hence, the categorical language is certainly useful, but categories themselves and their properties are, once again, secondary. At best, categories provide a convenient organization of mathematical data, they allow one to state certain theorems precisely and in ways that were not hitherto possible. But it certainly does not, so the argument goes, reflect an intrinsic mathematical structure. It does not capture or constitutes what mathematics is fundamentally about. Thus, we have the category of topological spaces, the category of Abelian groups, the category of groups, the category of rings, the category of modules over a ring R, the category of sets and so on and so forth, with various functors between them and natural transformations between these functors. As such, although it is an organization of mathematics, it is not a genuine classification, for in this case we have not stipulated when or under what conditions these different "categories" are different, or equivalently, identical. For instance, how do we know that the category of topological spaces is different, as a category, from the category of Abelian groups? Is it possible that they could be equivalent as categories? What are the identity criteria at work here? Does it matter at all? As we will see in a short while, it is possible—and in fact necessary—to give a precise answer to this question. It is probably this type of organization that mathematicians have in mind when they claim that category theory provides an organization of mathematics. But this point of view misses categories altogether as objects of mathematical interest themselves. In the same way that there was an ontological shift in geometry which went from the notion of a transformation group

2.6 Category Theory: The First Phase 1945–1958

as a methodological device to the notion of a transformation group as the basic geometrical structure, a similar ontological shift had to occur within category theory too. But we are getting ahead of ourselves.

Eilenberg and Steenrod's work showed what is essential for a family of functors to be a homology theory. It specified what homology theory is about. The details of the specific constructions and computations, which are chosen for some purposes and can be extremely complicated, are left to the applications of actual, specific theories. Categories remain in the shadows and are not seen as being indispensable or even useful.

Cartan and Eilenberg's work is in the same spirit as Eilenberg and Steenrod. Here is how they present the goal of their book in the opening sentences of the preface:

> During the last decade the methods of algebraic topology have invaded extensively the domain of pure algebra, and initiated a number of internal revolutions. The purpose of this book is to present a unified account of these developments and to lay the foundations of a full-fledged theory.
>
> The invasion of algebra has occurred on three fronts through the construction of cohomology theories for groups, Lie algebras, and associative algebras. The three subjects have been given independent but parallel developments. We present herein a single cohomology (and also a homology) theory which embodies all three: each is obtained from it by a suitable specialization ([43], v).

Thus, as in the case of Eilenberg and Mac Lane and Eilenberg and Steenrod, we are dealing with a unification accomplished once again with the help of categories, functors and natural transformations. In a nutshell, mathematicians progressively realized that certain homology and cohomology groups could be defined in terms of certain purely algebraic structures (e.g. groups, algebras and Lie algebras). Thus, methods developed for topological spaces could be transposed directly to algebraic structures. This is what Cartan and Eilenberg report in their preface. They coined the term "homological algebra" to circumscribe that part of algebra whose methods were based on homology (and cohomology) theory. Their book is filled with innovations: projective module, left exact and right exact functors, projective and injective resolutions, derived functors, to mention the most obvious. As they mention immediately in their preface,

> This unification possesses all the usual advantages. One proof replaces three. In addition an interplay takes place among the three specializations; each enriches the other two.
>
> The unified theory also enjoys a broader sweep. It applies to situations not covered by the specializations. An important example is Hilbert's theorem concerning chains of syzygies in a polynomial ring of n variables. We obtain his result (and various analogous new theorems) as a theorem of homology theory ([43], v).

In the words of Weibel, Cartan and Eilenberg's book "revolutionized the subject" ([266], 812). It would probably be better to say that the book *created* the subject altogether, since there was nothing to overturn.

There is, however, one striking fact about Cartan and Eilenberg's book. Although the book is mainly about certain functors and their properties, categories are *entirely* assumed! Indeed, in Chapter 2, where functors are explicitly introduced, we find the casual statement: "We consider functors (in the sense of [74]), defined for

Λ-modules and whose values are in the category of Γ-modules, where Λ and Γ are two given rings" ([43], 18). Functors are then defined directly on modules and homomorphism, without any mention of categories. Only in the Appendix written by Buchsbaum do we find categories explicitly mentioned and used. We will come back to the appendix in the next chapter.

However these applications do *not* use categorical concepts as such in any essential way. Both in Eilenberg and Steenrod's and in Cartan and Eilenberg's pioneering works in the foundations of algebraic topology and homological algebra, there is no categorical concept at work or which can be said to capture essential ingredients of the situation. As Eilenberg and Steenrod themselves declare:

> These [the concepts of category, functor and related notions] are needed in the subsequent chapters to facilitate the statements of uniqueness and existence theorems. Only as much of the subject is included as is used in the sequel. (...)
>
> The ideas of category and functor inspired in part the axiomatic treatment of homology theory given in this book. In addition, the point of view that these ideas engender has controlled its development at every stage ([80], 108).

Thus, as we are told, categories are: (1) heuristically useful to facilitate the *statement* of certain theorems, and (2) methodologically useful, for they provide certain *constraints* that help in finding proper definitions and the correct context for certain ideas and proofs. It is in this sense that category theory is merely a language while group theory is not. In the latter case, group-theoretical concepts—e.g., orbit, stabilizer subgroup, quotient group, conjugacy class, kernel, center, etc.—play an important role: they correspond to important geometrical facts. Investigating group-theoretical facts by group theoretical means is more than significant: very often this is what really matters to a geometric problem. For category theory to play a similar role in mathematics, genuine categorical concepts have to be found, developed and applied. One has to *see* that there are such concepts and that categories themselves encode basic mathematical facts. How do categories provide such an encoding? What are the concepts required? There is something missing in the categorical landscape.

Chapter 3
Categories as Spaces, Functors as Transformations

> In all these applications [homological algebra, algebraic topology, theory of schemes] it was soon realized that it was necessary to make frequent use of properties and constructions applicable to *all* categories, which had not been mentioned by Eilenberg and Mac Lane.
>
> [59], 149
>
> They are discoveries that *significantly different* and *mathematically important* subject-areas share a common structure (...) they suggest a formalism which implies that there are forms or structures which are common to significant and significantly different particular subject-areas. (...) The more global unification (...) authorize not only the transfer of knowledge from one mathematical subject-area to *another* [sic], and not only the transfer of knowledge from one mathematically significant subject-area to another, but, indeed, the transfer of knowledge from one mathematical subject-area to *many* other subject areas, both inside and outside of mathematics.
>
> [57], 291–292

As we have seen in the previous chapter, Eilenberg and Mac Lane's main goal was to give a precise definition of an informal notion. In order to provide a fully rigorous concept, they needed to introduce categories. The latter concept was not seen as being of real importance. Functors and natural transformations certainly were important and continued to be central in the work of Eilenberg and Steenrod and Cartan and Eilenberg. These two books are representative of the research based on categorical ideas in the period from 1945 until about 1957–58. Then, in 1957, Grothendieck published a paper on homological algebra in which he defined *Abelian categories*, then *used* the latter to unify and expand various methods and prove new and powerful theorems and, in 1958, Kan introduced the concept of adjoint functors, undoubtedly the concept that allowed a general and unified treatment of many diverse mathematical concepts, including logical and foundational concepts in general. After the publication of these two papers and until about 1970, research in category theory can be divided in three related areas:

73

1. The extension and development of work done in the first period, e.g., applications to algebraic topology, to homological algebra (which really took off as a subject of research after the publication of Cartan and Eilenberg's book), and most notably Grothendieck's project of rewriting the foundations of algebraic geometry and proving the Weil conjectures using categorical notions;
2. The study, development, applications and extensions of Abelian categories in general;
3. The search for and study of adjoint situations and their applications, in particular the associated monads or triples, in many fields of mathematics, including foundational research.

I will concentrate on the last two threads, since they contain the elements I want to put forward first — I will briefly come back to the impact of Grothendieck's work in algebraic geometry in later chapters. In particular, I want to focus on three interrelated notions, namely universal mappings, representable functors and adjoint functors, which constitute key conceptual elements in the categorical landscape. Furthermore, I claim that they are at the very heart of the analogy between Klein's program and category theory. If a category is to be thought of as a space, or as an algebraic encoding of a space — and as we will see, this way of thinking turns out to be much more than an analogy — and we look for the equivalent of transformation groups, we should consider functors, and more particularly adjoint functors, as the key notion. I will therefore look at universal mappings, since they were introduced first historically, then at representable functors. These two notions are intimately linked to Abelian categories in the history, and I will maintain this parallel in my exposition. Finally, I will turn to adjoint functors as Kan introduced them, and consider some examples, including some in logic and foundations.

3.1 Universal Morphisms

In the same way that mathematicians knew about natural isomorphisms in the early forties, they also knew implicitly about universal mappings: various cases of free structures, e.g., free groups or free topological groups, various extensions, completions and compactification procedures, etc., were known to be similar. Let us consider two typical cases to illustrate the situation.

Recall that a metric space (X,d) is said to be *complete* if every Cauchy sequence in X converges. For instance, the standard Euclidean space is complete in either of its usual metrics, i.e., the Euclidean metric or the square metric. It is well known that not all metric spaces are complete. For instance, the metric space $(\mathbb{Q},d)$, where $d(x,y) = |x-y|$, is not complete. The sequence

$$1.4, 1.41, 1.414, 1.4142, 1.41421, \ldots$$

converges in $\mathbb{R}$ and it is a Cauchy sequence in $\mathbb{Q}$ that does not converge in $\mathbb{Q}$. Given these facts, it is then natural to consider whether any metric space can be imbedded

3.1 Universal Morphisms

isometrically in a complete metric space, that is in such a way that the distance function is preserved by the imbedding. And indeed this can be done: given a metric space (X,d), there is a complete metric space (Y,D) together with an isometric imbedding $h\colon X \to Y$ which satisfies the following key property:

Property (Uniqueness of the completion). Let $h\colon X \to Y$ and $h'\colon X' \to Y'$ be isometric embeddings of the metric space (X,d) in the complete metric spaces (Y,D) and (Y',D'). Then there is an isometry between the embeddings.

The metric (sub)space $(\overline{h(X)},D)$, that is the closure of the image of X under h together with the ambient metric D restricted to it, is complete and is called the *completion* of X. For instance, the completion of $(\mathbb{Q},d)$ is the Euclidean space $(\mathbb{R},d)$.[1] The notion of complete metric space and the operation of completion play an important role in functional analysis.

Let us now consider what appears to be a slightly different problem. Recall that an *open cover* of a topological space X is a collection $\mathfrak{U}$ of open subsets of X such that the union of the elements of $\mathfrak{U}$ is equal to X. A space X is said to be *compact* if every open covering $\mathfrak{U}$ of X contains a finite subcollection that also covers X. Compactness is a very useful property in general, and in particular in analysis. It is usually hard to determine whether a given space is compact or not. The real line $\mathbb{R}$ with the standard topology is not compact, but the closed unit interval $[0,1]$ is compact. Again it seems natural to inquire whether topological spaces can be compactified, and if so, under what conditions and in what ways. For, there are many ways to compactify a topological space X: the simplest way is certainly the one-point compactification. If the space X satisfies certain natural conditions, i.e., if it is a locally compact Hausdorff space, then X can be imbedded in a compact Hausdorff space by adjoining a point to X and defining the appropriate topology on the new space. For instance, the one-point compactification of the real line $\mathbb{R}$ with the standard topology is homeomorphic to the unit circle and the one-point compactification of the real plane $\mathbb{R}^2$ is homeomorphic to the sphere. In fact, we have already seen these identifications, for if we look at $\mathbb{R}^2$ as the complex space $\mathbb{C}$, then $\mathbb{C} \cup \infty$ is the Riemann sphere that we encountered in chapter 1. This compactification is in a way the "smallest" possible way to compactify a space X. Is there a "most general" way to do it? The answer is provided by the Stone-Čech compactification: every completely regular space X can be imbedded in a compact Hausdorff space βX containing X such that X is dense in βX, i.e., the closure of X is βX. The Stone-Čech compactification has the following extension property: every bounded real-valued function on X can be uniquely extended to a continuous real-valued function on βX. In turn, it can be shown that given a completely regular space X, two compactifications Y and Y' which satisfy the previous extension property can be shown to be homeomorphic. In other words, the Stone-Čech compactification is essentially unique.

---

[1] In fact the construction of the complete metric space in which the given metric space is imbedded is a generalization of the standard construction of the real numbers via equivalence classes of Cauchy sequences.

The similarity between the metric case and the topological case is obvious. It would be easy to extend the list of examples and with examples coming from different fields. There seems to be a general pattern at work. The question is whether one ought to look for this pattern and if so, what would be the gain. If one is convinced that there is some sort of unity underlying mathematical knowledge, a belief which goes back at least explicitly to Klein and Hilbert and which was and still is certainly held by many mathematicians, then it seems natural to try to describe this pattern in its most general form.

One way to state the general pattern is as follows: given a set E with a structure of a certain kind, say of kind S, some sets F with a structure of a different kind, say T extending S, the problem consists in finding a T-structure F_0 together with an S-mapping $\varphi_0 : E \to F_0$ such that for every F and S-mapping $\varphi : E \to F$ there is a unique T-mapping $\varphi' : F_0 \to F$ such that $\varphi = \varphi' \circ \varphi_0$. The mapping $\varphi_0 : E \to F_0$ is said to be *universal*, for it satisfies a universal property: *every S-map $\varphi : E \to F$ factors uniquely through φ_0 by a T-map*. This situation can be represented by the following commutative diagram:

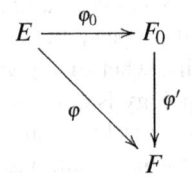

$$E \xrightarrow{\varphi_0} F_0$$
$$\varphi \searrow \quad \downarrow \varphi'$$
$$F$$

Thus, if this is adequate, one has to give a general definition of the notions required to state and solve this problem, which was called the "problem of universal mappings". It seems that, in particular, one has to clarify the notions of structure, structure-preserving mapping, isomorphism of structure, i.e., identity of structure, and composition of mappings.

In 1948, Pierre Samuel, a member of the Bourbaki group, presented the first attempt to axiomatize such a general theory. Instead of giving definitions of structure, kind of structure, isomorphism, etc., Samuel refers the reader to Bourbaki's book *Théorie des ensembles*, published in 1939.[2] ([31])

He then proceeds to give axioms for mappings of a certain kind. These axioms stipulate that mappings between structures of a kind T, that is T-mappings, compose and the result is a T-mapping, that T-isomorphisms are T-mappings and a one-to-one and onto T-mapping f is a T-isomorphism if and only if its inverse is also a T-mapping. Then, certain axioms are given for substructures, the "closure" of a subset of a set, a limitation on the cardinal number of that "closure" and axioms for Cartesian products. Clearly, these axioms, especially the first three on mappings, do not yield the definition of a category: nothing is said about associativity of composition, nor about identities. In fact, categories and functors are not mentioned in

---

[2] That little *fascicule* was reviewed by Samuel Eilenberg. The sentence before last of his review reads as follows: "The last section outlines an interesting method of treating structures, such as order, topology, group, ring, etc., on a general basis and having concepts like isomorphism defined quite generally." [67]

3.1 Universal Morphisms

Samuel's paper. Samuel does formulate the universal mapping problem in the general way given above, mentions that all the examples he listed earlier in his paper fall under the general scheme (except the case of the field of quotients of an integral domain, since this case does not satisfy one of his axioms on products), and shows explicitly how the completion of a uniform space is also a special case of the universal mapping problem. The two remaining sections of the paper deal with imbedding problems and free topological groups, seen as applications of the same problem.

But Samuel's axiomatization never caught on, and, as far as I can tell, was never used again by anyone else. The universal mapping problem remained in a sense open. Samuel's axiomatization is a failure if only because it does not succeed as an axiomatization: it does not clarify the basic concepts involved in the problem, it does not explain why these maps have these properties, it does not open the door to new applications and developments, it does not clearly systematize the different cases according to clear and precise principles. The axioms seem to be *ad hoc* for the most part and fail to go at the heart of the problem. To be fair to Samuel, it would take quite a while before someone understood the relations between universal arrows, representable functors and adjoint functors. The first steps were taken by Mac Lane at the same time Samuel was writing his paper and we now turn to this important step.

3.1.1 Mac Lane: Doing Duality without Elements

> In 1948, Mac Lane drew attention to categories themselves. He observed that many statements about abelian groups were equivalent to statements about the category of abelian groups. (One can prove that *all* statements about abelian groups can be so translated.)
>
> ([89], 9.)

Between 1945 and 1950, Mac Lane was trying to clarify one aspect of Eilenberg and Steenrod's work in the foundations of algebraic topology. He had also read Samuel's paper and Bourbaki's work on linear algebra where the universal mapping problem is also mentioned, since he wrote reviews of these two works. He was therefore well aware of the universal mapping problem, together with Samuel's proposal as well as Bourbaki's proposal. However, Mac Lane was using the categorical language and was working on a different problem.

As we have seen, Eilenberg and Steenrod published an axiomatic presentation of homology theory in 1945. The axioms stipulate certain properties of functors from certain topological spaces into Abelian groups. Already at the end of their preface, Eilenberg and Steenrod note that "homology theory and cohomology theory are dual to one another. We treat them in parallel." ([80], xiii.) These statements are clarified in chapter 1, §3c, where the axioms for *co*homology are introduced, immediately after the axioms for homology. It appears immediately that the axioms are obtained by *reversing* the arrows between the Abelian groups in the axioms for homology. In other words, whereas the construction of a homology group is a covariant functor, the construction of a cohomology group is a contravariant functor. Thus, duality is first presented as a property of *diagrams* in this context. But this is *not* what Eilenberg and Steenrod mean by the fact that homology and cohomology are dual to one another. For duality is a well-known phenomenon that goes back to the early 19th century in projective geometry. It is worth giving the whole quote to see what they mean by duality in this context:

> There is a duality relating homology and cohomology. It is based on the Pontrjagin theory of character groups. Precisely, let $H^q(X,A)$, f^*, δ be a cohomology theory satisfying Axioms 1c through 7c; and suppose $H^q(X,A)$ is always a discrete abelian group (R = the ring of integers) or always a compact abelian group. Let $H_q(X,A)$ be the character group of $H^q(X,A)$, and let f_*, δ be the homomorphisms dual to f^*, δ. Then it is readily shown, by the use of standard properties of character groups, that $H_q(X,A)$, f_*, δ satisfy Axioms 1 through 6 [it is a homology theory]. It follows that the dual of each theorem about $\{H_q(X,A), f_*, \delta\}$ is a true theorem about $\{H^q(X,A), f^*, \delta\}$. When passing from a theorem to its dual, arrows are reversed, subgroups are replaced by factor groups and vice versa. ([80], 15.)

Thus, the duality has a precise mathematical meaning but rests on an important assumption: the cohomology groups always have to be either discrete Abelian groups, i.e., R-modules with R the ring of integers, or compact Abelian groups. Whenever R is unrestricted, then the duality is only partial or in the words of Eilenberg and Steenrod "semiformal".

When we step back from the algebraic topological set-up and concentrate on the duality itself, we are looking at the category of Abelian groups or a related category thereof, i.e., R-modules for some ring R different from the ring of integers. One would like to clarify the background upon which this duality seems to rest. In other words, one is interested in understanding the duality of groups in general and showing how the duality between homology theory and cohomology theory follows from this more general case. This is precisely what Mac Lane was after. In his own words:

> For a topological space the duality between homology and cohomology groups with locally compact abelian coefficient groups can be formulated in terms of character groups. Another formulation is suggested by the axiomatic homology theory of Eilenberg and Steenrod. In this formulation, the axioms for a homology theory refer not to elements of the (relative) homology groups, but only to certain homomorphisms; the dual statements are exactly the axioms for a cohomology theory. For example, any continuous mapping $\varphi: X \to Y$ of one space into a second induces a mapping in the <u>same</u> direction on the homology groups, and in the <u>reverse</u> direction on the cohomology groups of these spaces. One of our chief objectives is that of providing a background in which the proofs for axiomatic homology theory become exactly dual to those for cohomology theory. ([186], 494.)

The first sentence mentions the duality via Pontrjagin's theory of character groups with the restriction. Mac Lane then continues by saying that the elements of the groups involved do not play any part in the situation, only certain homomorphisms are involved. This is certainly a key observation that seems to be restricted to the case of groups as they appear in homology and cohomology theory, although there are some indications, e.g., the so-called isomorphism theorems of group theory, that other results and notions are on the same footing. This is, in a sense, the starting point of category theory as a general methodological framework: define concepts

3.1 Universal Morphisms

and prove results about those concepts without making reference to elements. Although MacLane was looking at a specific case, his approach opened the door to this way of thinking. In the particular case he was looking at, since only homomorphisms between groups are involved, duality is expressed by reversing certain mappings. One would like to be able to dualize all *proofs* obtained for homology theories and automatically obtain *proofs* for cohomology theories. This means that once a proof is obtained about certain homology groups, then by simply reversing the arrows one would obtain the dual result for cohomology groups. This would yield a simple *method* of proof. The language of diagrams thereby acquired a new status: it allowed one to express *directly* an important relationship between theories. But since mappings between Abelian groups form a category, it seems to make perfect sense to look at the *category* of Abelian groups in general and this is what MacLane began to do. In fact, MacLane looked at the category of Abelian groups and, whenever possible, the category of groups. The final goal is therefore to obtain an axiomatic framework of those categories in which homology theory (and cohomology theory) could be developed. Notice the important shift: one is now trying to characterize certain *categories* axiomatically. Categories *themselves* become the object of study. If categories seemed to be totally dispensable in Eilenberg and MacLane's paper, the duality between homology and cohomology developed in Eilenberg and Steenrod forced MacLane to look at categories as such and try to find some of their properties.

MacLane saw the problem as one of characterizing the class of statements "about groups which do not make reference to the elements of the groups involved." ([186], 487.) At that time, MacLane did not think that this meant looking only at mappings, for he included statements about groups which make reference also to products of homomorphisms, subgroups and quotient groups. MacLane had a set of examples in mind that clearly led him to introduce subgroups and quotient groups in his search for the right axiomatic framework. These examples, together with MacLane's desire to show how his work could be applied to universal algebra, might be responsible for the fact that, in MacLane's own words, his characterization was "clumsy". ([195], 205.)

I will present slightly different examples from the ones found in MacLane's paper. But they are essentially the same. Let us restrict ourselves to Abelian groups. Recall that there is the trivial Abelian group with one element, denoted by 0. Given an Abelian group G, there is a unique homomorphism $1\colon 0 \to G$ which sends the unique element of 0 to the unit of G. Before we go on, some elementary notions about sequences of morphisms have to be recalled. Given a homomorphism of groups $f\colon G \to H$, the kernel of f, denoted by $\ker f$, is the set of all elements of G that are sent to the unit e_H of H, i.e.,

$$\ker f = \{x \in G \mid f(x) = e_H\},$$

and the image of f is the set of all elements of H that "fall" under f, i.e.,

$$\operatorname{Im} f = \{y \in H \mid \exists x \in G\, (f(x) = y)\}.$$

Consider the following sequence of homomorphisms of groups:

$$G_0 \xrightarrow{f} G_1 \xrightarrow{g} G_2.$$

Whenever $\operatorname{Im} f = \ker g$, the sequence is said to be *exact*. Exact sequences of (co-homology) groups were used explicitly for the first time in 1941 by Hurewicz in a very short note and introduced under that name by Eilenberg and Steenrod in their announcement in 1945 and in their book.

Given these notions, the simplest duality of the type examined by Mac Lane is probably the following. If the sequence

$$0 \xrightarrow{1} G_1 \xrightarrow{f} G_2$$

is exact, then this is a different way of saying that the homomorphism f is *injective*. By reversing the arrows, one obtains the sequence

$$0 \xleftarrow{!} G_1 \xleftarrow{g} G_2$$

which says that the homomorphism g is *surjective*. It is therefore possible to express, in the category of groups, the fact that a morphism is injective with the help of arrows and the notion of surjective is then its dual.

Mac Lane opened his paper with a special case of a general notion, which had not been defined in 1948, but that was about to be introduced explicitly by Cartan and Eilenberg and play an important role in the next ten years: it is the notion of *injective* module and its dual, *projective* modules, which are defined thus. A module E is *injective* if, for every module B and every submodule A of B, every $f: A \to E$ can be extended to a map $g: B \to E$. This can be expressed by a diagram in the following manner:

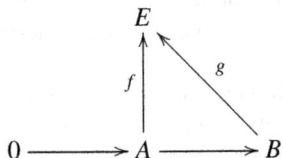

$$\begin{array}{ccc} & E & \\ {\scriptstyle f}\uparrow & \nwarrow{\scriptstyle g} & \\ 0 \longrightarrow A & \longrightarrow & B \end{array}$$

Now, by dualizing this diagram, we obtain the notion of projective module. Thus, a module P is *projective* if, for every module B and every quotient module C of B, every $f: P \to B$ can be factored through the quotient map $B \to C$. The diagram becomes:

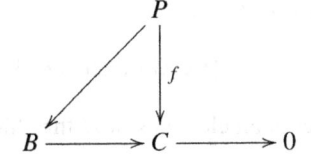

$$\begin{array}{ccc} & P & \\ \swarrow & \downarrow{\scriptstyle f} & \\ B \longrightarrow & C & \longrightarrow 0 \end{array}$$

3.1 Universal Morphisms

In his paper, Mac Lane considered free $\mathbb{Z}$-modules, which are projective. The dual notion, in this case, is said to be *infinitely divisible*. (It should be noted that nowadays, projective modules are introduced first, since there are obvious examples of these, and injective modules are then defined by dualizing the diagram, since there are no obvious examples of injective modules! However, there are plenty of injective modules; it can be shown that every (left) R-module M can be imbedded in an injective module. Furthermore, it can be shown that every injective module is divisible and that if R is a principal ideal domain, like $\mathbb{Z}$, then an R-module is divisible if and only if it is injective. Mac Lane was very close indeed to projective and injective modules, but just missed.)

Along with these examples, Mac Lane gives the following list of cases:

> Any subgroup of a free group is free, any quotient group of an infinitely divisible group is infinitely divisible. Any abelian group is isomorphic to a quotient group of a free abelian group; any abelian group is isomorphic to a subgroup of infinitely divisible group (that is, can be embedded in an infinitely divisible group). If a free abelian group F is a factor group of an abelian group, it is a direct factor; if an infinitely divisible group D is a subgroup of an abelian group, it is a direct factor. ([186], 486.)

It is from these considerations that Mac Lane arrives at a table of dualities and what will become the principle of duality. Given a statement S about groups, which does not make any reference to the elements of the groups involved, the statement dual to S is the statement obtained by interchanging domains and codomains of maps. The table looks like this:

Statement S	Statement dual to S
$f: G \to H$	$f: H \to G$
Domain $f = G$	Codomain $f = G$
f is an isomorphism into [sic]	f is an homomorphism onto
Composition $g \circ f$	Composition $f \circ g$
P is a subgroup of G	Q is a quotient of G
The injection $i: P \to G$	The projection $q: G \to Q$

In the next section, §3, Mac Lane makes a fundamental observation. He observes that the notions of free products of groups and direct products of groups can be *defined* directly with the help of diagrams satisfying a specific property. This is the first instance of concepts being defined with the help of universal mappings. As we have seen, universal mappings were known to mathematicians and at least one attempt, known to Mac Lane, had been made to provide an axiomatic characterization of the universal mapping problem, but no one before Mac Lane had thought of using the universal mapping property in order to *define* a construction. Furthermore, it turns out that free products and direct products are *dual* to one another in this setup. Here are the definitions.

Given Abelian groups A, B, a *direct product* of A and B is an Abelian group P *together with* two homomorphisms $p_A: P \to A$ and $p_B: P \to B$ such that for any Abelian group C with homomorphisms $f: C \to A$ and $g: C \to B$, there is a *unique* homomorphism $h: C \to P$ such that the following diagram commute:

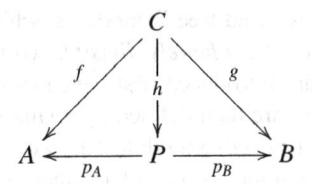

$$\begin{array}{c} C \\ {}^{f}\swarrow \;\; \downarrow h \;\; \searrow^{g} \\ A \xleftarrow{p_A} P \xrightarrow{p_B} B \end{array}$$

Reversing the arrows, we obtain the definition of a free product of two Abelian groups A, B. Thus a *free product* F of A and B is an Abelian group *together with* two homomorphisms $i_A: A \to F$ and $i_B: B \to F$ such that for any Abelian group C with homomorphisms $f: A \to C$ and $g: B \to C$, there is a unique homomorphism $h: F \to C$ such that the following diagram commutes:

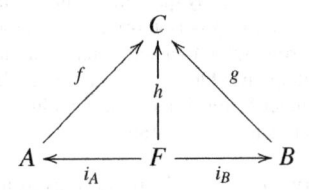

$$\begin{array}{c} C \\ {}^{f}\nearrow \;\; \uparrow h \;\; \nwarrow^{g} \\ A \xleftarrow{i_A} F \xrightarrow{i_B} B \end{array}$$

What is striking here is that we are not defining *the* direct product nor *the* free product of two groups, but *a* direct product and *a* free product. For *any* group satisfying the foregoing conditions will do. This might look like a problem, but in fact it is not. For it can be easily shown that: (1) If P with two projections p_A, p_B is a direct product of A and B, then for any Q with projections q_A, q_B satisfying the condition of products, there is a unique isomorphism between P and Q; (2) If P with p_A, p_B is a direct product of A and B and Q is isomorphic to P, then Q, with the appropriate morphisms, is a direct product of A and B. In this sense, the definition completely characterizes the concept of product. There is of course an obvious shift from the set-theoretical perspective, since we are not defining a *specific* object, like the Cartesian product of sets X and Y, which is *the* set of ordered pairs of elements of X and Y, but a *type* of object. In this case, any object satisfying the universal property given in the definition should be considered to be a *token* of the concept "direct product".

Another important feature of this approach was immediately underlined by Mac Lane: the two notions are dual to one another in the sense examined by him. It follows that if a property of one of the constructions follows from the definition and from properties of the morphisms of a category, the dual construction will have the dual property and the *proof* will be automatic, i.e., one simply has to reverse all the arrows involved. But some properties do *not* follow from properties of the morphisms involved, e.g., the *existence* of the constructions themselves. In the words of Mac Lane:

> The proof of the existence of the direct product is not dual to the proof of the existence of the free product, for both proofs involve the reference to the elements of the groups concerned. However, the proof that the direct product is unique up to an isomorphism can be phrased so as to be exactly dual to the proof of the uniqueness of the free product up to an isomorphism. Similarly, the proofs of the associative and commutative laws for direct products, formulated

in terms of diagrams like (3.1), are dual to the proofs of the corresponding laws for the free product (...) ([186], 490–491.)

Hence, at that point, Mac Lane sees clearly that *some* facts about (Abelian) groups are provable on the basis of the "algebra of mappings". Since Mac Lane's goal was to clarify the relationships between homology and cohomology theories, he did not investigate how *much* could be done with the help of the algebra of mappings alone, i.e., without reference to the elements of the entities concerned. Thus in his paper, Mac Lane does not consider the possibility of defining *for arbitrary* objects X, Y of *an arbitrary* category $\mathscr{C}$ a product for X and Y. And he does not consider the possibility of defining other concepts in this way.

In the next two sections of his paper, Mac Lane concentrates on subgroups, quotient groups and series of groups and explores various dualities between series of subgroups and quotients. I will skip the details since they are not relevant to our enterprise. Suffice it to say that at the end of §5, Mac Lane mentions that "our formulation of duality in terms of homomorphisms does not suffice to subsume all known "duality" phenomena. In particular, it does not appear to explain the duality between "verbal" and "marginal" subgroups (...), which is, however, an extension of the above duality between ascending and descending central series." ([186], 494.)

After a brief section in which Mac Lane clarifies the type of duality he is after, we finally come to the introduction of categories and what he calls "bicategories".[3] Mac Lane now has to *define* categories in such a way that the definition be self-dual; that is, by reversing the arrows in the definition, one obtains a category. This point of view leads to the following definition of a category.

Definition 3.1. [4] A *category* $\mathscr{C}$ consists of arrows $\alpha, \beta, \gamma, \ldots$ for certain pairs of which a product $\alpha\beta$ in $\mathscr{C}$ is defined and which are subject to the following three axioms:

A1. The composite $(\alpha\beta)\gamma$ is defined if and only if the composite $\alpha(\beta\gamma)$ is defined. When either is defined, they are equal (and this triple composite is written as $\alpha\beta\gamma$).
A2. The triple composite $\alpha\beta\gamma$ is defined whenever both composites $\alpha\beta$ and $\beta\gamma$ are defined.

Definition 3.2. An *identity* of $\mathscr{C}$ is an arrow u such that $\alpha u = \alpha$ whenever the composite αu is defined and $u\beta = \beta$ whenever the composite $u\beta$ is defined.

A3. For each arrow β of $\mathscr{C}$ there exist identity arrows u and u' of $\mathscr{C}$ such that $u'\beta$ and βu are defined.

It is then possible to introduce objects $X, Y, Z, \ldots$ simply by establishing a one-to-one correspondence between identity arrows and objects.

This definition satisfies a duality principle:

---

[3] Warning: Mac Lane's "bicategories" are *not* what are now called bicategories.

[4] This is *not* quite the definition given by Mac Lane in his paper. But it is equivalent to it and it is more economical.

The concept of the "dual" of a statement about homomorphisms may now be defined precisely. In a category, the only primitive statements are statements of the forms

(10.1) $$\alpha = \beta, \quad \alpha\beta = \gamma,$$

We interpret the latter to mean "the product of $\alpha\beta$ is defined and is equal to γ." All other statements can be expressed in terms of these primitive statements; in particular, we understand the statement "$\alpha\beta$ is defined" to be interpreted as "there exists a γ such that $\alpha\beta = \gamma$." A *first order statement* S in a finite number of letters (which designate mappings of the category in question) is any statement formed from a number of primitive statements of the types (10.1), combined by the standard logical connectives (including quantifiers "for all α" and "$\exists \alpha$"). The *dual* of S is the statement obtained from S by the following typographical process: replace each primitive statement $\alpha\beta = \gamma$ by the statement $\beta\alpha = \gamma$, leaving the other primitive statements, all the letters, and all the logical connectives unchanged. The dual is thus obtained by "inverting all products." ([186], 176)

Recall that Mac Lane had written his Ph.D. at Göttingen in logic. The observation stated in this passage allowed Mac Lane to establish a fundamental metatheorem of category theory, which was presented by him as a duality principle: if any statement about a category is deducible from the axioms for a category, the dual statement is likewise deducible. What Mac Lane *did not* consider explicitly is the notion of *dual category*. Thus, although categories are not merely defined by Mac Lane to assure that functors have domains and codomains, I speculate that in Mac Lane's mind at that time, categories had to be collections of structured sets with structure-preserving functions between them. For, as we will see, in a dual category, morphisms are not necessarily set-theoretical functions, they are in fact usually purely *formal* objects. They probably provide the simplest examples of categories — and of categories that play an important role in mathematics in general — in which morphisms are usually *not* set-theoretical functions.[5] In their study of duality in homological algebra, Buchsbaum and Grothendieck introduced dual categories right from the start. One could argue that the duality principle is enough and that the fact that Mac Lane did not introduce dual categories is an irrelevant detail. But I want to underline the fact that to introduce dual categories is to make a *shift* of emphasis that has an important impact on the nature and status of categories. The fact that he did not introduce them is, I believe, revealing of his attitude towards categories as mathematical objects at that time.

In order to axiomatize the various duality phenomena he was interested in, Mac Lane had to introduce more structure into categories. He therefore gave additional axioms for what he called "bicategories" and, later, Abelian categories. Bicategories are introduced so that the dualities between subgroups and quotient groups, on the one hand, and homomorphisms onto and isomorphisms into — this is Mac Lane's terminology — can be axiomatized. I will not present these axiomatizations. I will look at Grothendieck's presentation since it clearly had much more impact than Mac Lane's. But before I do so, some final remarks about Mac Lane's paper are in order.

---

[5] This point was emphasized by Colin McLarty during a talk given at Notre Dame in 2001.

1. Following his definition of bicategory, Mac Lane states without demonstration the

   Theorem 11.5. The class of objects in a bicategory is partially ordered by either of the relations

   (11.1) $S \subset B$ if and only if there is an injection $\kappa \colon S \to B$;
   (11.1') $Q \leq A$ if and only if there is a projection $\pi \colon A \to Q$.

   Mac Lane then says that if $S \subset B$, S is called a *subobject* of B, while if $Q \leq A$, Q is a *quotient-object* of A. This time, this is true in any *bicategory*. However, Mac Lane does not go as far as to define the notion of a subobject for *any category*, thus avoiding references to elements in a more general framework.
2. Mac Lane clearly still has in mind some of the goals he had with Eilenberg when they wrote their 1945 paper. But now, the emphasis is put on bicategories instead of categories. In §13 entitled "Universal algebra", which follows the section in which examples of bicategories are given — which in fact exclude the usual category of sets, but include the category of nonvoid sets where each set carries with it an equivalence relation — Mac Lane claims that:

   > These examples indicate that most types of algebraic, topological, or other mathematical systems, together with the appropriate type of transformations, yield bicategories. The bicategory language appears to be the appropriate vehicle for many of the theorems of universal algebra (...) — often giving simpler formulations, because the axiomatic formulation avoids the inevitably cumbersome explicit description of the general form of any algebraic or mathematical system. This is especially the case when universal algebra is extended to include those algebraic systems, which occur so frequently, in which several groups, homomorphisms, functions, and so forth, together constitute a single algebraic system. Using the notions of covariance and contravariance [74] one can in fact give a general definition of mathematical systems and prove, under general hypotheses, that the class of all systems of a given type is the class of objects of a bicategory. ([186], 503.)

   Thus, universal algebra would be a part of category theory, in fact even a part of bicategory theory. The connections between universal algebra and category theory would deserve a whole book in itself. Suffice it to say for the moment that bicategories in Mac Lane's sense did not play a role in the subsequent categorical approach to universal algebra.
3. Mac Lane's paper, although correct, did not have a major impact in the mathematical community. It does nevertheless constitute an important moment in the development of category theory if only because it contains, for the first time, the use of a universal property to define certain mathematical concepts and the use of categories to capture and understand a mathematical phenomenon. It is interesting to note that although Buchsbaum and Grothendieck basically had the same goal as Mac Lane, they had the chance to look at a larger class of examples than he did. Indeed, it is clear that both Buchsbaum and Grothendieck had the benefit of looking at Cartan and Eilenberg's work in homological algebra, and the duality between homology and cohomology theories as it arises from derived functors in their work. Indeed, many of their theorems are presented

twice, once for right derived functors and then again for left derived functors. (See [43], chaps. VIII–XIII.) It is these examples that guided Buchsbaum and Grothendieck in their introduction of Abelian categories. Grothendieck was also looking, in fact first and foremost, at sheaf theory.

4. The overall moral that has to be underlined again is that Mac Lane's paper makes explicit reference to the fact that various mathematical concepts and results can be defined and proved without reference to the elements of the entities involved. Mac Lane makes these remarks while he is *in* a categorical context, suggesting that category theory is the proper framework for doing mathematics in this way. Grothendieck would soon show that more homological algebra than one might reasonably expect *can* be done in this way and then Lawvere would take the bold step of trying to do *everything* in mathematics in the categorical framework.

Before we look at Grothendieck, we will leave the historical development for a very short while and look at some further examples of universal morphisms, since they occupy a key place in the categorical landscape and constitute one of the vertices of the fundamental triangle of concepts of category theory.

3.1.2 Universal Morphisms

I have already mentioned that Mac Lane's characterization of the concept of direct product of two groups holds for the concept of product in general, that is in *any* category and not only the category of groups. For the *essential* features of a product are captured by the morphisms *coming in* (which does not mean that the mappings *going out* are not important, quite the contrary),[6] *provided* one associates to the product two *universal arrows*, called in this particular case its projections, viz., $p_X: X \times Y \to X$ and $p_Y: X \times Y \to Y$, the usual projection maps in the case of the Cartesian product of sets. Indeed, given two objects, X and Y, in *any* category $\mathscr{C}$, a *product* of X and Y, if it exists, is an object P of $\mathscr{C}$ together with two morphisms $p_X: P \to X$ and $p_Y: P \to Y$ such that for any object Q of $\mathscr{C}$ with morphisms $f: Q \to X$ and $g: Q \to Y$, there is a *unique* morphism $h: Q \to P$ such that $p_X \circ h = f$ and $p_Y \circ h = g$. The associated diagram is exactly the same as the one given above for groups. Indeed, the direct product of groups is an example of the notion of products in an arbitrary category.

These data characterize a product P of X and Y in the following sense: any object Q isomorphic to P is a product of X and Y and any product of X and Y is isomorphic to P. *Thus P is characterized up to (a unique) isomorphism.* The actual elements composing P are *totally* irrelevant. It does not matter how P is constructed, what matters is that it comes equipped with universal projection arrows and that these

---

[6] Indeed, since the morphisms coming in are, so to speak its defining properties, it is the morphisms *going out* that contain non-trivial information about the construction. I want to thank Colin McLarty who emphasized to me the importance of that point.

3.1 Universal Morphisms

satisfy the universal property. Thus, the definition characterizes a *type* of object, not a particular object. This indicates that *there is* a criterion of identity at work within category theory and that this criterion is *different* from the standard set-theoretical criterion.

Indeed, in set theory, the Cartesian product of two sets X and Y is a *unique particular* set composed of certain elements: it is *the* set $\{(x,y) \mid x \in X \wedge y \in Y\}$. The axiom of extensionality, the criterion of identity for sets, forces the uniqueness of this set. In a category $\mathscr{C}$, a product P of two objects X and Y is an object belonging to a certain collection of objects of $\mathscr{C}$, which in the case of objects with universal arrows is quite special since there is a *unique* isomorphism between the objects that are products of a pair of objects. Any product of X and Y is related to X and Y by a universal arrow and any object with morphisms into X and Y factors in a unique manner through the products. This is perfectly natural if one thinks of the objects of a category as being "geometric" objects in a generalized sense. In a geometric space, one usually picks a certain figure of the space in a certain context. But in fact, any similar figure, that is any figure related to it by the underlying transformation group, would do. In a category, one picks a certain object P with certain projections as a product of X and Y, but any other similar object, that is any object appropriately related to P, would do.

Thus the concept of product is characterized in terms of a *universal property*: the latter specifies that for all objects of a certain type in the category, only certain mappings (morphisms) will be *going in* a product from these objects, in fact, only *one* in this case. This characterization holds in *any* category $\mathscr{C}$. Whether a given category $\mathscr{C}$ *has* products or not is something that has to be verified case by case. Matters of existence are sharply divorced from matters of definition. The category of sets, the category of groups, the category of topological spaces, the category of lattices: these are all categories with products. However, we should immediately note that we can easily present a *finite* category, that is, a category with a finite number of objects and a finite number of morphisms, with products or at least a product for a pair of objects, thus exhibiting a specific case of a category with products or a product. More specifically, consider the following category with four objects and five non-trivial morphisms, that is different from the obvious identities and composing in the obvious way, depicted as follows:

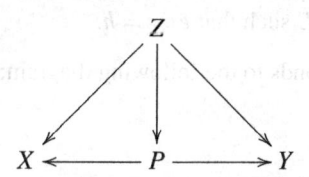

$$\begin{array}{c} Z \\ \swarrow \downarrow \searrow \\ X \leftarrow P \rightarrow Y \end{array}$$

Thus, at least for *finitary* operations, the fact that matters of existence are divorced from matters of definition is of no importance. Things are more delicate when we get to infinitary operations. The problem, of course, has very little to do with definition, but with existence, as is always the case in mathematics in general.

As we have already seen, by reversing all arrows in the foregoing definition, one obtains the notion of a coproduct. More specifically, given two objects X and Y in a category $\mathscr{C}$, a *coproduct* for X and Y in an object Q together with two morphisms $i_X\colon X \to Q$ and $i_Y\colon Y \to Q$ such that for any object P with morphisms $k\colon X \to P$ and $j\colon Y \to P$, there is a unique morphism $h\colon Q \to P$ such that $h \circ i_X = k$ and $h \circ i_Y = j$. Again, the diagram depicting the situation is the same as the one given for the free product of two Abelian groups. The category of sets, the category of topological spaces, the category of Abelian groups and the category of groups all have coproducts. Furthermore, the proof that a product is unique up to a unique isomorphism automatically yields a proof that a coproduct is unique up to a unique isomorphism. It suffices to reverse all arrows in the first proof and the second result is automatic. Many other important concepts can be defined in a similar manner.

An *initial* object in a category $\mathscr{C}$ is an object 0 of $\mathscr{C}$ such that for every object X of $\mathscr{C}$ there is a unique arrow $0 \to X$. Thus in the category of sets, the empty set is an initial object and in the category of groups, *any* one-element group is initial. The dual concept is that of a terminal object. Hence, a *terminal* object in a category $\mathscr{C}$ is an object 1 of $\mathscr{C}$ such that for every object X of $\mathscr{C}$ there is a unique arrow $!\colon X \to 1$. For instance, in the category of sets, any singleton set is terminal and in the category of groups, any one-element group is terminal.[7]

This last case brings us naturally to the concept of a *zero* object: it is an object which is both initial and terminal. The category of sets does not have a zero object whereas a one-element group is a zero object in the category of groups. Notice that the concept of zero object is self-dual: dualizing it yields the same concept. Notice also that as categories, the category of sets has to be different from the category of groups. Although we do not have a criterion of identity for categories at our disposal, it makes intuitive sense to say that **Set** is different from **Grp** if only because the latter has a zero object whereas the former does not. By doing so, we start treating categories as genuine objects with genuine properties, although, once again, we still do not know what is a categorical property is, as such.

Other concepts are definable by universal morphisms.

Definition 3.3. Given parallel arrows and $f\colon X \to Y$ and $g\colon X \to Y$ in a category $\mathscr{C}$, an *equalizer* for f, g is an object E together with an arrow $e\colon E \to X$ such that $f \circ e = g \circ e$ and for any object T with arrow $h\colon T \to X$ such that $f \circ h = g \circ h$, there is a unique arrow $v\colon T \to E$ such that $e \circ v = h$.

This definition corresponds to the following diagram:

---

[7] Mathematicians speak of "the" one-element group. There is a simple reason behind this: there is only one up to isomorphism. That is, if the criterion of identity adopted is given by the notion of isomorphism, there is indeed essentially only one one-element group. Again, this is as in geometry. Once the criterion of identity has been adopted, e.g., as in Euclidean geometry, one can certainly talk about the properties of "the" circle of radius one as if there was only one such circle. One knows that it is enough to pick one instead of working with the whole (equivalence) class of such circles.

3.1 Universal Morphisms

$$E \xrightarrow{e} X \underset{g}{\overset{f}{\rightrightarrows}} Y$$

$$v \uparrow \quad \nearrow h$$

$$T$$

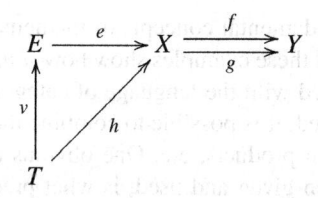

An equalizer is the arrow-theoretic formalization of the following idea. Very often in mathematics one defines certain functions from X to Y and wants to know when they are equal. For instance, consider the function of two variables $x^2 + y^2$ and the constant function which sends all pairs (x,y) to 1, both with domain the product $\mathbb{R} \times \mathbb{R}$ and codomain $\mathbb{R}$. The set $\{(x,y) \mid x^2 + y^2 = 1\}$ is a subset of $\mathbb{R}$ which contains the pairs (x,y) of reals such that the function $x^2 + y^2$ equals the constant function on 1, i.e., it is an equalizer for these two functions. Thus, more generally, in the category **Set**, given two functions $f, g \colon X \to Y$, an equalizer for f and g is given by the subset $E = \{x \in X \mid f(x) = g(x)\}$ together with the inclusion function $i \colon E \to X$ defined by $i(x) = x$. The same construction holds in many other categories of structures, for instance the category **Top**, the category **Grp**, the category **AbGrp**, just to mention the most obvious, provided the object E is endowed with the structure induced by that of X.

The dual notion is that of a coequalizer, which is just as important since it corresponds to taking a quotient of a certain object, a fundamental mathematical operation.

Definition 3.4. Given parallel arrows $f, g \colon X \to Y$ in a category $\mathscr{C}$, a *coequalizer* for f, g is an object Q together with an arrow $q \colon Y \to Q$ such that $q \circ f = q \circ g$ and for any object T with arrow $h \colon Y \to T$ such that $h \circ f = h \circ g$, there is a unique arrow $v \colon Q \to T$ such that $v \circ q = h$.

The corresponding diagram is:

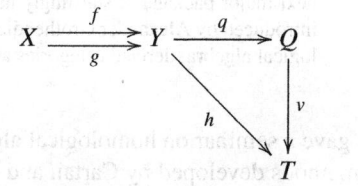

$$X \underset{g}{\overset{f}{\rightrightarrows}} Y \xrightarrow{q} Q$$

$$h \searrow \quad \downarrow v$$

$$T$$

In the category **Set**, the coequalizer of f and g is the quotient of Y by the equivalence relation generated by the pairs $(f(x), g(x))$ for every $x \in X$. In the category **Top** of topological spaces, the coequalizer is constructed as in the category of sets and provided with the quotient topology. In a category of algebraic structures, the coequalizer is somewhat more involved, but just as important. For instance, in the category **AbGrp** of Abelian groups, the coequalizer of two morphisms $f, g \colon X \to Y$ is the coequalizer of $f - g \colon X \to Y$ and the zero morphism, i.e., the morphism sending all elements of X to the identity element of Y, that is, the quotient of Y by the subgroup $(f - g)(X)$.

These concepts are fundamental concepts of mathematics defined with the help of a universal property. All these examples show how *some* fundamental concepts of mathematics *can be* defined with the language of category theory. Needless to say, once they have been defined, it is possible to combine these concepts, e.g., take the equalizer of maps between products, etc. One obvious question to consider, once these definitions have been given and used, is what precisely they cover. In other words, what are the concepts defined in this way? What is the expressive power of the categorical language? Do we actually cover all fundamental constructions of mathematics? The last question is certainly too vague, for it is not clear that we can determine what constitutes a fundamental construction of mathematic in general. But a general understanding of the concepts definable this way is certainly called for. We will come back to this question later.

Let us now return to the historical thread we have been following. We now turn to Grothendieck's introduction and application of Abelian categories, for Grothendieck's paper, and his subsequent work in algebraic geometry, represents a fundamental development, not only for category theory, but also for homological algebra, algebraic topology and algebraic geometry.

3.2 Grothendieck and Abelian Categories

> This work began with the attempt to exploit the formal analogy between the cohomology theory of a space with coefficients in a sheaf (...) and the theory of derived functors of functors of modules (...) in order to find a common framework to cover these theories and others.
>
> ([103], 119. My translation.[8])
>
> After the creation of the modern notion of a topological space and the discovery of limiting procedures basic to measure theory, the next major package of startlingly new infinitary constructions was introduced by Alexander Grothendieck with his treatment of homological algebra, derived categories and functors, topos and sites.
>
> ([210], 165.)

In 1955, Grothendieck gave a seminar on homological algebra in which he generalized the concepts and methods developed by Cartan and Eilenberg to treat with the same tools what he was mostly interested in at that time, namely the cohomology of sheaves (about which I will have to say more in chapter 7). His results were then published in 1957 in a long paper — 100 pages —, entitled *Sur Quelques Points d'algèbre homologique*, in which he gave a characterization of Abelian categories with original applications. As the foregoing quote indicates, Grothendieck's motivation *was* considerably different from Mac Lane's and even Buchsbaum's. As we

---

[8] Ce travail a son origine dans une tentative d'exploiter l'analogie formelle entre la théorie de la cohomologie d'un espace à coéfficients [*sic*] dans un faisceau (...) et la théorie des foncteurs dérivés de foncteurs de modules (...), pour trouver un cadre commun permettant d'englober ces théories et d'autres.

3.2 Grothendieck and Abelian Categories

have seen, Mac Lane's motivation was to understand a type of duality that arose in a certain treatment of algebraic topology. Mac Lane's motivations are again explicitly mentioned in the opening paragraph of the section of Abelian categories:

> Our objective in this chapter is that of providing a self-dual set of axioms for abelian groups and their homomorphisms sufficient to prove all categorical theorems which refer to a finite number of such groups — and hence adequate to explain the apparent perfect duality present for such theorems on abelian groups. ([186], 507.)[9]

Buchsbaum's main motivation was essentially similar: his goal was to develop the proper abstract framework to understand in full generality the type of duality appearing in homological algebra, basically in Cartan and Eilenberg's book. In his own words:

> Throughout this book [Cartan & Eilenberg], the authors dealt with functors defined on categories of modules over certain rings and whose values again were modules over a ring. It will be shown here that the theory may be generalized to functors defined on abstract categories that will be described below, and whose values are again in such abstract categories. The advantages of such an abstract treatment are manifold. We list a few:
>
> (1°) The dualities of the type
>
> $$\begin{aligned} \text{kernel} &\;\text{---}\; \text{cokernel} \\ \text{projective} &\;\text{---}\; \text{injective} \\ Z(A) &\;\text{---}\; Z'(A) \end{aligned}$$
>
> that were observed throughout the book may now be formulated as explicit mathematical theorems.
>
> (2°) In treating derived functors, it suffices to consider left derived functors of a covariant functor of several variables; all other types needed may then be obtained by a dualization process.
>
> (3°) Further applications of the theory of derived functors are bound to show that the consideration of modules over a ring Λ will be insufficient. Rings with additional structure such as grading, differentiation, topology, etc. will have to be considered. With the theory developed abstractly, these generalizations are readily available. (Buchsbaum, in [43], 379.)

Both points (1°) and (2°) have to do with duality. Only the third point touches on possible generalizations and they are clearly different from what Grothendieck was after. It should be noted that, however, Buchsbaum's axiomatization of Abelian categories is essentially the same as the one given by Grothendieck. Furthermore, in §8 of his paper, Buchsbaum indicates that his approach covers the duality found in Eilenberg and Steenrod's work in algebraic topology, in other words Mac Lane's target.

> The axiomatic homology and cohomology theories of Eilenberg–Steenrod (...) may be defined using an arbitrary exact category [his name for an abelian category] $\mathscr{A}$ as the range of

---

[9] I should point out a small technical difference between Mac Lane and Buchsbaum, on the one hand, and Grothendieck, on the other. Whereas, as the quotation from Mac Lane indicates, both Mac Lane and Buchsbaum defined products and coproducts for finitely many objects, Grothendieck defined these notions for arbitrary collections of such objects.

values of the theory. Thus, replacing $\mathscr{A}$ by $\mathscr{A}^*$ [the dual of $\mathscr{A}$] replaces a homology theory by a cohomology theory, and vice versa. This duality principle simplifies the exposition of the theory. Furthermore, the uniqueness proof (...) remains valid for such generalized homology and cohomology theories. (Buchsbaum, in [43], 385.)

3.2.1 Abelian Categories

Grothendieck's motivation was to develop a more general and abstract framework such that (1) a formal analogy would become mathematically precise, (2) the new framework would allow a unified treatment of both cases at hand and possibly others as well. Thus, Grothendieck's goal was not only to clarify and understand a certain phenomenon that happened to be at work, but to obtain a more precise and complete framework in which the techniques of homological algebra could be developed and applied to new cases. Thus his attitude towards categories is somewhat different from Mac Lane's and Buchsbaum's. Clearly, Grothendieck's characterization of the concept of Abelian category is a generalization going in a direction different from Mac Lane's and Buchsbaum's: Mac Lane was concentrating on Abelian groups and duality in that context; Buchsbaum was covering more, i.e., categories of Λ-modules, and he had in mind generalizations along the lines of the properties of the modules involved; Grothendieck had an informal analogy in mind between algebraic situations, i.e., those described by Cartan and Eilenberg, and geometric situations, namely the newly invented theory of sheaves. In fact, Cartan and Eilenberg were aware of the analogy with sheaves, but there was a technical hurdle that they could not overcome. Grothendieck succeeded were they failed by using the categorical language in an essential way.

Grothendieck's paper constitutes a landmark not only in homological algebra, but in category theory as well. It is striking in many different ways. First, it opens with a section on general category theory and introduces immediately dual categories. Thus, in contrast with Mac Lane who began his paper with considerations about groups, Grothendieck makes it clear that category theory is not only a useful tool, but constitutes the proper system of concepts with the appropriate structure to develop a fully general and abstract framework in which the various theories can be "melted" together. Second, Grothendieck explicitly introduces the notion of *equivalence* of categories and underlines the fact that it is different from the notion of isomorphism of categories and that the latter is in practice useless. This is no surprise since Grothendieck was working with *functor categories* and it turns out that many such categories, which are not isomorphic, are nonetheless "the same" from a categorical point of view. Thus considerations surrounding a *criterion of identity* for categories appear; categories are becoming genuine objects of mathematics. Third, and along the same lines, the very *status* of category theory changes with Grothendieck's paper. Grothendieck is treating a category as a whole in itself with specific properties. Grothendieck is not trying to identify properties of Λ-modules or Abelian groups, but common properties of categories of Λ-modules, categories of Abelian groups and categories of sheaves of Λ-modules over a topological space. Thus, *Abelian categories had to be seen in a new and different light*. For an Abelian category — and in fact other types of categories introduced by Mac Lane, Buchsbaum and Grothendieck, like additive categories — is not merely a special type of category. In the same way that a group is not merely a special type of set, an Abelian category is not merely a special type of category. An Abelian category has additional structure and that structure is of the utmost importance, just as a group has an additional structure and that structure is what a group is all about. Furthermore, in the same way that the concept of group of transformations is, in a sense an abstraction of a geometric situation, the concept of Abelian categories is in a sense an abstraction of an algebraic situation. Thus, with the introduction and use of Abelian and related categories, categories are no longer dispensable, merely present so that functors can have domains and codomains. They encode basic mathematical facts and categorical concepts have now a useful mathematical meaning. Fourth, it is somewhat surprising to see Grothendieck *and* Buchsbaum work with the algebra of mappings, the Hom-sets, and add structure on these components. They do not use the purely abstract definition of a category, as in Eilenberg and Mac Lane's paper, or Eilenberg and Steenrod's book. However, the explicit use of the algebra of mappings is very much in the spirit of the 1945 paper by Eilenberg and Mac Lane, and to that extent, it is also an generalization of Klein's program. We will come back to this point, and even see how much this is true for other aspects of Grothendieck's work, in particular toposes, later.

Grothendieck's definition of a category is similar to the one we have considered in chapter 2. His definition is immediately followed by the definition of the dual category.

Definition 3.5. Given a category $\mathscr{C}$, the *dual category* $\mathscr{C}^\circ$ is defined as follows:

1. The objects of $\mathscr{C}^\circ$ are the objects of $\mathscr{C}$;
2. An arrow $f^\circ : X \to Y$ of $\mathscr{C}^\circ$ is an arrow $f : Y \to X$ of $\mathscr{C}$;
3. Given two arrows $f^\circ : X \to Y$ and $g^\circ : X \to Y$, their composite is defined by

$$f^\circ \circ g^\circ = (g \circ f)^\circ.$$

Clearly $\mathscr{C}^{\circ\circ} = \mathscr{C}$. If a statement holds in a category $\mathscr{C}$, then the dual statement, obtained from the original by reversing all the arrows appropriately, holds in the dual $\mathscr{C}^\circ$. Since $\mathscr{C}^{\circ\circ} = \mathscr{C}$, if a statement holds in all categories, its dual holds as well in all categories. In other words: if a statement of the elementary, i.e., first-order, theory of category is a consequence of the axioms, so is the dual statement. The conceptual significance of the duality principle — for its practical advantages are obvious — is the same as any duality principle in geometry: it indicates the presence of a global structural phenomena, of the fact that what matters is not so much the choice of particular objects and principles as basic objects and principles, but rather of an overall organization that itself determines basic facts about objects and principles. Furthermore, dual categories constitute explicit examples of categories in which morphisms are *not*, in general, structure-preserving functions. Consider, for instance, the dual category of the category of Abelian groups, **AbGrp**$^\circ$. Its objects are taken to be Abelian groups, but in fact, they should be treated as symbols *indexed* by Abelian groups. Clearly, its morphisms are not group homomorphisms. A morphism $f^\circ\colon G \to H$ in **AbGrp**$^\circ$ is going in the opposite direction from $f\colon H \to G$ in **AbGrp**. Thus, a dual category cannot, in general, be interpreted as a category of structured sets with structure-preserving functions between them. But as we will see, dual categories and duality theorems in general play a crucial role in category theory, and as is revealed by the theory, in many other areas of mathematics.

Another important notion given by Grothendieck early in his paper is the notion of equivalence of categories. Eilenberg and Mac Lane had given the notion of *isomorphism* of categories, which is just a special case of isomorphism applied to categories. Thus a functor $F\colon \mathscr{C} \to \mathscr{D}$ is an *isomorphism* if there is a functor $G\colon \mathscr{D} \to \mathscr{C}$ such that $G \circ F = 1_\mathscr{C}$ and $F \circ G = 1_\mathscr{D}$. However, it turns out that in practice, the notion of isomorphism is *not* the appropriate notion. It does not provide a criterion of identity for categories. The correct notion is given by an equivalence of categories, which Grothendieck defined as follows:[10]

Definition 3.6. An *equivalence* between two categories $\mathscr{C}$ and $\mathscr{D}$ is given by a system (F, G, φ, ψ) where $F\colon \mathscr{C} \to \mathscr{D}$ and $G\colon \mathscr{D} \to \mathscr{C}$ are covariant functors and $\varphi\colon 1_\mathscr{C} \simeq G \circ F$ and $\psi\colon 1_\mathscr{D} \simeq F \circ G$ are natural isomorphisms.

Two categories are said to be *equivalent* when there is an equivalence between them.

In other words, the functors F and G are quasi-inverses to each other, the difference lying in the fact that the equality between their composites and the identity functors has been replaced by natural isomorphisms. The replacement of an identity by an isomorphism is a key aspect of category theory in general with important philosophical implications, to which we will return.

In order to define Abelian categories, Grothendieck, like Mac Lane and Buchsbaum before him, needed to define various concepts in purely categorical terms. I will depart slightly from Grothendieck's presentation, given in terms of Hom-sets, and stay in the spirit of arrows only to introduce these notions. We will come back to Hom-sets in the next section.

Definition 3.7. An arrow $f\colon X \to Y$ is a *monic* arrow or a *monomorphism* when for every pair of arrows $h\colon T \to X$ and $g\colon T \to X$ such that $f \circ h = f \circ g$, we have $h = g$. In algebraic terms, an arrow f is monic if it is left cancelable. The diagram is:

$$T \underset{g}{\overset{h}{\rightrightarrows}} X \xrightarrow{f} Y.$$

---

[10] This is not quite true. Surprisingly, what Grothendieck defines in his paper is the notion of an adjunction between two categories. It might very well be that the expression "homomorphismes de foncteurs" on page 125 ought to have been "isomorphismes de foncteurs", in which case the definition is correct, although unnecessarily complicated. What is clear, although one ought to show it in detail, is that the definition given corresponds to Grothendieck's needs in that paper. See [149].

3.2 Grothendieck and Abelian Categories

The dual concept is that of an *epi arrow* or *epimorphism*: an arrow $f: Y \to X$ is an *epi* arrow or an *epimorphism* when for every pair of arrows $h: X \to T$ and $g: X \to T$ such that $h \circ f = g \circ f$, we have $h = g$. In algebraic terms, an arrow f is epi if it is right cancelable. Simply reversing the arrows of the preceding diagram yields the appropriate graphic representation of the situation:

$$X \underset{g}{\overset{h}{\rightrightarrows}} Y \xrightarrow{f} T.$$

A morphism f is a *bimorphism* if it is both monic and epi.

In the category of sets, monic arrows are injective functions, epi arrows are surjective functions and bimorphisms are isomorphisms.[11] In the category of groups, monic arrows are injective homomorphisms and epi arrows are surjective homomorphisms (this last claim is *not* trivial). In the category of topological spaces with continuous mappings, monomorphisms are injections and epimorphisms are surjections. But this need not be the case in general. Let us consider some examples. Let $i_\mathbb{Q}: \mathbb{Q} \to \mathbb{R}$ be the standard inclusion function, $i_\mathbb{Q}(r) = r$, with $\mathbb{Q}$ and $\mathbb{R}$ considered as Hausdorff topological spaces with the usual topology. This map is clearly not surjective. However it *is* an epimorphism. For let X be any Hausdorff topological space and $f, g: \mathbb{R} \to X$ such that $f \circ i_\mathbb{Q} = g \circ i_\mathbb{Q}$. Suppose there is an $x \in \mathbb{R}$ such that $f(x) \neq g(x)$. Then there are neighborhoods U, V of in X of $f(x)$ and $g(x)$ respectively such that $U \cap V = \emptyset$. But then f and g are different on the whole neighborhood $f^{-1} \cap g^{-1}$ of x, which necessarily includes rational numbers, i.e., $f \circ i_\mathbb{Q} \neq g \circ i_\mathbb{Q}$, contradicting our hypothesis. The key property of $\mathbb{Q}$ here is that it is a dense subset of $\mathbb{R}$. The results hold for arbitrary Hausdorff spaces satisfying this property.

Let us now give an example of a monic map that is not an injection. Consider the category **CTop.** of pointed and connected topological spaces. An object of that category is a pair (X,x), where X is a connected topological space and $x \in X$ is a base point. A morphism $f: (X,x) \to (Y,y)$ is a continuous mapping such that $f(x) = y$. Consider now the projection p of the circular helix H on the circle S^1, $p: (H,h) \to (S^1,s)$, with $p(h) = s$. If $f: (X,x) \to (S^1,s)$ is a morphism which admits a lifting, i.e., if there is a morphism $g: (X,x) \to (H,h)$ such that the triangle

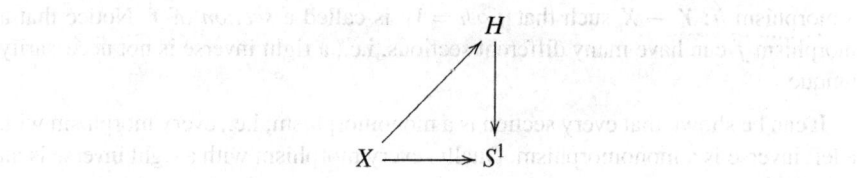

$$\begin{array}{ccc} & & H \\ & \nearrow & \downarrow \\ X & \longrightarrow & S^1 \end{array}$$

commutes, then it can be shown that this lifting is necessarily unique. But this is simply the claim that p is a monomorphism and it is clearly not an injection.

Finally, let us give an example of a bimorphism that is not an isomorphism. Let X denote the real line $\mathbb{R}$ with the discrete topology and Y denote the real line $\mathbb{R}$ with

---

[11] Notice that isomorphisms, as they were defined earlier in the categorical setting, are bimorphisms.

the usual topology. Let $i\colon X \to Y$ be the obvious inclusion $i(x) = x$. It is trivially a continuous map since the domain has the discrete topology and it is obviously surjective. It is also clearly an injection. This is enough to conclude that it is a bimorphism. But it is clearly not an isomorphism, for there is no continuous function $j\colon X \to Y$ such that $i \circ j = 1_Y$ and $j \circ i = 1_X$.

There is a general moral here: the language of arrows allows one to define concepts that are, in the category **Set** of sets and functions, coextensive with standard set-theoretical concepts. However, the categorical concepts are definable in any category and in many cases, they differ from the set-theoretical concepts, as the foregoing examples show. We have thus introduced a division or a bifurcation of concepts. It is common for bimorphisms to be different from isomorphisms; epimorphisms are rarely surjections in categories of topological structures, although they almost always are in categories of algebraic structures; monomorphisms are generally injections. Now, in categories where the objects are structured sets with structure-preserving functions, both sorts of concepts can be defined and used appropriately. But from a purely categorical perspective, things have to be looked at differently. In an arbitrary category, one cannot assume that the objects are structured sets and structure-preserving functions. The fact that in the case of the category of sets the categorical concepts are coextensive with the set-theoretical ones is taken as the indication that the arrow-theoretical concepts constitute a proper generalization of the set-theoretical concepts and as such should be considered as constituting an adequate way of capturing the corresponding informal notions. After all, the set-theoretical definitions of injective and surjective maps are nothing more nor less than one way of making precise properties of *mappings*. If category theory is taken to be primarily a theory of mappings, then it is reasonable to consider the definitions of monomorphism, epimorphism, bimorphism and isomorphism as adequate since they are defined solely in terms of properties of mappings.

A few more definitions are required. The following definition, although not required for Abelian categories, is of general interest and might as well be given at this point, since it is closely related to the previous definitions.

Definition 3.8. Given a morphism $f\colon X \to Y$ in a category $\mathscr{C}$, a morphism $g\colon Y \to X$, such that $g \circ f = 1_X$ is called a *retraction* of f and X is called a *retract* of Y and a morphism $h\colon Y \to X$ such that $f \circ h = 1_Y$ is called a *section* of f. Notice that a morphism f can have many different sections, i.e., a right inverse is not necessarily unique.

It can be shown that every section is a monomorphism, i.e., every morphism with a left inverse is a monomorphism. Dually, every morphism with a right inverse is an epimorphism.

Let us now go back to Grothendieck. Grothendieck first isolated the notion of an *additive* category.

Definition 3.9. A category $\mathscr{C}$ is an *additive* category if it is a category and it satisfies the following conditions:

AD1. for each pair of objects X, Y of $\mathscr{C}$, the collection of morphisms $\mathrm{Hom}_{\mathscr{C}}(X, Y)$ is equipped with the structure of an Abelian group, such that, for all morphisms

3.2 Grothendieck and Abelian Categories

$u\colon Z \to X$, $v\colon Z \to Y$, and all pairs of morphisms (f,g) of $\mathrm{Hom}_\mathscr{C}(X,Y)$,

$$(f+g)\circ u = f\circ u + g\circ u, \quad v\circ(f+g) = v\circ f + v\circ g.$$

In words, the last condition expresses the fact that the composition of morphisms is bilinear;

AD2. $\mathscr{C}$ has a zero object 0;

AD3. Products $X \times Y$ exist for every pair X, Y of objects of $\mathscr{C}$.

Notice immediately that the dual category of an additive category *is* also an additive category. In other words, the characterization is self-dual. It can also be shown that the axioms imply the existence of a coproduct $X+Y$ for any pair of objects X, Y of $\mathscr{C}$ and that $X \times Y$ is isomorphic to $X+Y$. Needless to say, the category of Abelian groups is an additive category, and so is the category of Λ-modules and the category of sheaves of modules over a topological space X. By duality, it can be shown that the category of compact topological Abelian groups is additive (it is essentially the same as the dual of the category of Abelian groups). The category of sets and functions is *not* an additive category, for it does not have a zero object. For the same reason, the category **Top** of topological spaces is not additive. More interestingly, the category **Grp** of groups is *not* additive. Indeed, in **Grp**, the product of two groups is not isomorphic to the coproduct of these two groups.

Let us look at the axioms more carefully. Not surprisingly, the axioms are about the properties of the morphisms within a category. They can be divided into two groups: the first axiom stands by itself and the last two can be looked at together. For the first axiom, we can go back to Eilenberg and Mac Lane's original analogy between Klein's program and category theory. If we consider a category as a "space" with its algebra of mappings, what we find is that instead of having a space X with its group of automorphisms, we have all objects X, Y, Z, ... with their (Abelian) group of morphisms (satisfying an additional condition). The last two axioms state that the category has to have certain universal morphisms, i.e., that certain constructions satisfying a universal property have to be available in the space. What has to be seen here is that in the same way that a group of transformation encodes basic facts about a geometry — it literally captures the basic geometric properties and constructions of a geometry and the structure of the group of transformation encodes the structure of the geometric space — an additive category encodes basic properties and constructions of an area of mathematics and its categorical structure encodes the structure of a type of mathematical object. This means that within an additive category, certain concepts can be defined, others can be constructed and certain results can be demonstrated, and by purely categorical means. These facts result from the additive structure of the category itself. As I have mentioned, the category of Abelian groups is additive. Thus there are certain relationships between these categories that clarify how certain mathematical theories are related to one another. Once more, the parallel with the role of groups in geometry is striking. Finally, I want to emphasize once more that the notion of additive category should be compared, from an epistemological and ontological point of view, to the notion of Lie group. In the same way that a Lie group is not just a special type of group, i.e., in the way that an Abelian group might be taken to be, an additive category — and, I might as well mention it forthwith, an Abelian category — is not just a special type of category.

To be true to the categorical spirit, functors between additive categories have to be defined: a functor $F\colon \mathscr{C} \to \mathscr{D}$ is an *additive functor* if for every pair of morphisms $u,v\colon X \to Y$ in $\mathscr{C}$, $F(u+v) = F(u) + F(v)$.

Let us now turn to Abelian categories.

Before I give the definition, some preliminary notions are required. Recall that Grothendieck's goal is to develop homological algebra for various types of mathematical entities. It amounts to doing homological algebra without reference to the elements of the groups involved, more specifically, we need to define some of the constructions that are usually defined for modules. The notions of monomorphism, epimorphism, isomorphism, products, and coproducts are necessary, as well as the notions of kernel and cokernel. We have seen above the traditional definition of the kernel of a function in terms of sets and elements. Here is the categorical definition.

Definition 3.10. Given an arrow $f\colon X \to Y$ in a category $\mathscr{C}$ with a 0 object, the *kernel* of f, denoted by $\ker f$, is, whenever it exists, the equalizer of f and the 0 arrow between X and Y, i.e.,

$$\ker f \xrightarrow{k} X \underset{0}{\overset{f}{\rightrightarrows}} X.$$

Dually, given an arrow $f\colon X \to Y$ in a category $\mathscr{C}$ with a 0 object, the *cokernel* of f, denoted by $\operatorname{coker} f$ is, whenever it exists, the coequalizer of f and the 0 arrow between X and Y, i.e.,

$$X \underset{0}{\overset{f}{\rightrightarrows}} Y \xrightarrow{c} \operatorname{coker} f.$$

Thus, the kernel and the cokernel of a morphism are specific instances of universal morphisms, namely equalizers and coequalizers. Hence, as usual with universal morphisms, the kernel and the cokernel of a morphism is characterized up to a unique isomorphism.

Definition 3.11. [12] A category $\mathscr{C}$ is *Abelian* if it is additive and satisfies the two properties:

AB1. Every arrow of $\mathscr{C}$ has a kernel and a cokernel;
AB2. Given a morphism $f\colon X \to Y$, and the following canonical decompositions:

$$\ker f \xrightarrow{k} X \xrightarrow{i} \operatorname{coker} f; \qquad \ker c \xrightarrow{j} Y \xrightarrow{c} \operatorname{coker} f,$$

there is an isomorphism $l\colon \operatorname{coker} k \simeq \ker c$ such that $f = j \circ l \circ i$.

---

[12] This is Grothendieck's definition. Equivalent definitions were discovered soon after, notably by Freyd.

3.2 Grothendieck and Abelian Categories

The morphism $j\colon \ker c \to Y$ is called the *image* of f and the morphism $i\colon X \to \mathrm{coker}\,k$ is called the *coimage* of f. Thus, in words, axiom AB2) says that the morphism from the coimage of f to the image of f is an isomorphism.

Again, notice that these axioms are readily self-dual. Thus the dual category of an Abelian category is also Abelian. Grothendieck immediately emphasizes that in any Abelian category $\mathscr{A}$, every bimorphism is an isomorphism. This is a consequence of AB2), since there are additive categories which satisfy AB1) but in which a bimorphism is not necessarily an isomorphism. Grothendieck gives the examples of the category of separated topological modules over a topological ring Λ with continuous homomorphisms and the category of *filtered* Abelian groups. These automatically yield examples of additive categories that are not Abelian. Another example is provided by the category of topological Abelian groups with continuous homomorphisms: it is additive but not Abelian. In fact, we have already seen this example above: the morphism $i\colon \mathbb{R} \to \mathbb{R}$, where in the domain $\mathbb{R}$ is a topological Abelian group with discrete topology and in the codomain, $\mathbb{R}$ has the usual topology. The morphism has zero kernel and cokernel, it is a bimorphism but not an isomorphism.

Needless to say, the category of Abelian groups and the category of Λ-modules are Abelian categories. But so is the category of sheaves of Abelian groups over a fixed topological space X. Thus, as Grothendieck himself observes, one can work in an Abelian category *as if* one were working with homomorphisms of Abelian groups, *provided* the properties used have a "finite character", i.e., infinite products or coproducts are not used. Of course, this is at best a heuristic device and a more precise statement of that correspondence, in the form of a representation theorem for Abelian categories, is needed to know precisely in what sense one can work in an arbitrary Abelian category as if it were the category of Abelian groups or Λ-modules. The precise answer was provided around 1960 in the form of various representation theorems by Freyd, Lubkin and Mitchell. (See [184], [227], and [91].) In essence, what these theorems asserts is that one can work in an arbitrary Abelian category as if one were working in a category of Λ-modules, for some ring Λ.

A few remarks about the axioms are in order. AB1) basically says that certain universal morphisms exist, since it simply states that certain equalizers and coequalizers exist in the category. The last axiom seems to be more specific, although it is very powerful. It does talk about morphisms and asserts the existence of a certain isomorphism. Freyd's axiomatization replaces AB2) with the following:

AB2') Every monic arrow is a kernel and every epi arrow is a cokernel.[13]

Again, this axiom is self-dual. Moreover, in this form it says that monic arrows *are* universal morphisms and epi arrows *are* universal morphisms too. Now all the axioms, except for the axiom on the structure of the Hom-sets, specify the existence of certain universal morphisms.

Interestingly enough, we do not need to define any other kind of functors between Abelian categories, for in practice, the most significant functors are additive. The last property is enough to carry the remaining elements of the structure required.

---

[13] In fact, it is possible to define Abelian categories with the axioms AD2), AD3), AB1) and AB2') and then *prove* that an Abelian category is additive.

Grothendieck then introduces four additional axioms, and their duals, all dealing with infinite families of objects.[14] These axioms are required since the behavior of various Abelian categories varies when infinite operations are considered. For instance, in the category of Abelian groups, infinite products are not isomorphic to infinite coproducts, in contrast with the finitary case. Furthermore, the existence of these infinitary operations is crucial for the proof of the existence of injective and projective objects in an Abelian category. In turn, the latter are required for the construction of derived functors, a basic ingredient of homology and cohomology theories, and, hence, the application of many results obtained by Cartan and Eilenberg in the category of Λ-modules. It is precisely at this point that Grothendieck saw a categorical path to the solution of the problem with respect to sheaves of Abelian groups. More specifically, if in an Abelian category $\mathscr{A}$, for an object X, there is a monomorphism of X into an injective object (respectively an epimorphism of a projective object onto X), then $\mathscr{A}$ is said to have *enough injective* (respectively enough projective). These distinctions were crucial for Grothendieck, since the category of sheaves of Abelian groups over a fixed topological space X has enough injective but not enough projective in general. Grothendieck presented axioms guaranteeing the existence of enough injective (respectively enough projective). He did not make a specific construction (no one knew how to do it, in any case), but relied instead on a more general categorical property. This was a *tour de force* and showed how categories could be used in a constructive and effective manner. As we have said, it was obvious to Cartan, Eilenberg and many others that there was more than an analogy between various constructions in homological algebra and the case of sheaves over a topological space X. However, no one before Grothendieck knew how to show that there were enough injectives in the latter case. It is by defining Abelian categories satisfying a certain condition and showing that in that case these categories had enough injectives *and* that the various required constructions were possible that Grothendieck solved the problem.

As with additive categories, Abelian categories encode fundamental aspects of a mathematical domain. Various concepts can be defined and constructed and various theorems can be proved in Abelian categories. An important portion of homological algebra consists of defining and studying functors between Abelian categories.[15]

We can now see more clearly the shift that accompanied the work done by Mac Lane, Buchsbaum, Heller and, in particular, Grothendieck. I claim that it is similar to the shift between an abstract Lie group and a concrete transformation group. As I have already said, in Eilenberg and Mac Lane, Eilenberg and Steenrod and Cartan and Eilenberg, categories are considered as heuristics for organizing mathematical data. They are merely useful devices, required for the use of functorial methods. Very little use is made of the global structure of the categories themselves. The mathematical objects are treated as a whole, that is, they form a category, *because*

---

[14] These aspects of Abelian categories were not considered by Buchsbaum and are crucial for the applications Grothendieck had in mind.

[15] In the sixties, other categories appeared on the scene and still play an important in homological algebra: derived categories, which led to triangulated categories and, for the purpose of homotopical algebra, closed model categories.

3.2 Grothendieck and Abelian Categories

some constructions are functorial, i.e., uniform for all objects of that kind. But with the axiomatic characterization of Abelian categories, categories are considered as being mathematical objects *as such*. Mac Lane wanted to understand certain types of properties of Abelian groups, those that satisfied a kind of duality, and saw that these properties could be in fact encoded by a *category* of Abelian groups, and in fact, by a category defined axiomatically. Furthermore, the properties Mac Lane was interested in *are* definable without any reference to the elements of the objects involved. They are defined by properties of the *morphisms*, in other words by certain *transformations* of the objects. Hence it appears that at least a portion of mathematics can not only be developed without reference to set-theoretical concepts, but is *clarified* by developing it without those concepts. Buchsbaum, Heller and Grothendieck provided more evidence of the usefulness and relevance of categories. In other words, the global structure of a category was taken to be epistemologically significant in itself, as providing the essential information about the objects of the category. In fact, Grothendieck would go one step further in this direction and develop a way of working with categories that was entirely new.

> One recognizes the Master's touch in the idea that the problem is not to define a motif: the problem is to define the category of motives, and to unearth the structures that it carries. ([55], 17. My translation.[16])

The basic idea is this: whenever you want to define a structure of a certain kind K, you do not define a set X together with the structure you are interested in and then move to the category of these structures. Instead you define a *category* **K** such that objects with a structure of kind K and structure-preserving morphism will constitute an example of a category **K**. Furthermore, in principle one wants to define an abstract category (e.g., Abelian categories), not by stipulating the kinds of objects and morphisms between them, but by stipulating purely categorical properties right from the start. The parallel with transformation groups is straightforward: instead of starting with, say, a metric space and then moving to the automorphism group of that space, one starts with a (Lie) group right from the start and characterizes the space and its properties from the structure of the group, as we saw in section 1.2.5. This is a radical shift from the traditional perspective, one in which sets and set-theoretical language is, at least from a practical point of view, abandoned. Of course, the foundation of category theory itself still remains an open question and we will see that, in dealing with foundations, Grothendieck fell back on set theory.

We can go back to Eilenberg and Mac Lane's original claims. Recall that they said that

> The invariant character of a mathematical discipline can be formulated in these terms. Thus, in group theory all the basic constructions can be regarded as the definitions of co- or contravariant functors, so we may formulate the dictum: The subject of group theory is essentially the study of those constructions of groups which behave in a covariant or contravariant manner under induced homomorphisms. ([74], 237.)

---

[16] On reconnaît la patte du Maître [i.e., Grothendieck] dans l'idée que le problème n'est pas de définir ce qu'est un motif : le problème est de définir la catégorie des motifs, et de dégager les structures qu'elle porte.

As we have seen in the previous chapter, what they had in mind was basically the functoriality of certain constructions. Now a different notion of what plays the key role is emerging. Although it was not entirely clear from Grothendieck's axiomatization, it would not take long before mathematicians realize that the basic constructions in the categorical contexts are given by universal morphisms and equivalent notions. As we will see, these can be presented as functors satisfying certain properties, thus as transformations of the space, again in the spirit of Klein's program.

Categories and their structure are now moving to centre stage. From 1957 on, they are not only defined so that functors and natural transformations can be defined in full generality: they are genuine mathematical objects with a definite mathematical role. Ironically perhaps, it is from this point on that the analogy with Klein's program appears in full force, albeit in a programmatic way. I claim that from 1957 onwards, it became possible to think that abstract categories are to mathematical structures what transformation groups are to geometric structures. In a sense, the notion of Abelian category captures the *invariant content* of the form a category has to have to be the context in which a homology theory will have value. Furthermore, this invariant content *is* expressed by certain transformations, more precisely functors, over certain categories. (We will see in a short while how this is so.) This was still somewhat weak since at that point Abelian categories were the only example on offer. Just as when Lie wanted to apply transformation groups to differential equations, it became clear that a new *conceptual* method was available, and that various programs could be considered. One of these programs was Grothendieck's reformulation of the foundations of algebraic geometry in which this method — i.e., don't define a single structure, define a category to characterize those structures — was to play a key role. In this vast and ambitious enterprise, one concept was to become pivotal: the notion of representable functor.

3.2.2 Representable Functors

> The main goal of this presentation is to develop a general technique allowing to recognize if such a functor F is representable, and to study the properties of the corresponding S-pre-schema X with the help of those of F.
>
> ([104], 11. My translation.[17])

One of Grothendieck's main targets was nothing less than the foundations of algebraic geometry. As we can see by looking at his lectures in the *Séminaire Bourbaki* from 1957 until 1962, the notion of representable functors became one of the main

---

[17] Le but principal de ces exposés est de développer une technique générale permettant de reconnaître si un tel foncteur F est représentable, et d'étudier les propriétés du S-pré-schéma X correspondant à l'aide de celles de F.

3.2 Grothendieck and Abelian Categories

tools he used in this enterprise.[18] It is far from clear why Grothendieck decided to use this notion instead of, say, adjoint functors, but it is reasonable to believe that it is because representable functors make a connection to sets, something which might be reassuring or comprehensible to a large audience, that is, outside categorical circles. It is also clear from the various seminars that Grothendieck thought in terms of universal "problems", that is he tried to formulate the problems he was working on in terms of a universal morphism: finding a solution to the given problem amounted to finding a universal morphism in the situation. Grothendieck saw that the latter notion was subsumed under the notion of representable functor. Still another reason might be that, in a way, representable functors, *in algebraic geometry* at least, constitute an epistemological tool. They are, in a sense to be clarified, a systematic and geometric generalization of the idea that to know an object X is to know its points.

In many cases, to know *the underlying* set of an object X — let us denote it by $U(X)$ — of a category $\mathscr{C}$, it is enough to consider the morphisms from a universal object of the category $\mathscr{C}$ to X. For instance, in the trivial case of the category of sets, to know the underlying set of a set X (yes, that is what I have written), that is to know X, it is enough to consider the collection $\mathrm{Hom}_{\mathbf{Set}}(1,X)$, the set of all functions from a terminal object 1 into X. Indeed, it is easy to see that in this case, there is a bijection

$$\varphi \colon \mathrm{Hom}_{\mathbf{Set}}(1,X) \simeq X.$$

In words: the set of functions from a terminal object 1 onto X is in one-to-one correspondence with the set X. In fact, an element $x \in X$ can be identified with a morphism $x \colon 1 \to X$. We can interpret this situation as follows: in the category of sets, there is a canonical, or universal, point, denoted by 1, and all the other points, i.e., the elements of the sets, are in fact representations or transformations of that canonical point. Thus, one knows a set when one knows all the transformations of the canonical point in it.

The same idea works in different categories, but sometimes with a different universal object. In the category of topological spaces **Top**, if 1 denotes the space consisting of one point with the obvious topology, then for any space X, we have again a bijection

$$\varphi \colon \mathrm{Hom}_{\mathbf{Top}}(1,X) \simeq U(X).$$

Thus, in this case too, one can say that there is a canonical point such that all other points are representations of transformations of it.

Things are somewhat more interesting in the category **Grp** of groups. Clearly, in this case, one *cannot* take the zero object 0, since there is a unique morphism of groups between 0 and a group G. However, the group of integers $\mathbb{Z}$ is a canonical point in that category. Indeed, there is for any group G, a bijection

---

[18] The references are: seminars 149, 182, 190, 195, 212, 232 and 236. The key seminar is 195, presented in February 1960. These lectures were then transformed into the book *Éléments de géométrie algébrique* co-written with Dieudonné. The first paragraph of the first chapter, called chapter 0, deals with representable functors. Categories, functors and natural transformations are not defined in full generality.

$$\varphi \colon \mathrm{Hom}_{\mathbf{Grp}}(\mathbb{Z}, G) \simeq U(G).$$

Thus, the number of homomorphisms from the group of integers $\mathbb{Z}$ to G is in one-to-one correspondence with the elements of G.

The preceding situation is a special case of a functor we have already encountered earlier. Indeed, for a fixed object Z of a category $\mathscr{C}$, we have a functor $\mathrm{Hom}_\mathscr{C}(Z, -) \colon \mathscr{C} \to \mathbf{Set}$. In the foregoing cases, whenever Z is chosen appropriately, for any object X of the category $\mathscr{C}$, the collection $\mathrm{Hom}_\mathscr{C}(Z, X)$ is isomorphic to the set of elements of X. Now, if $\mathrm{Hom}_\mathscr{C}(Z, X)$ is in bijection with the collection of elements of X, then it ought to behave as the collection of elements of X, that is, it should be related to the morphisms $X \to Y$ in an obvious manner: the map of sets $f \colon \mathrm{Hom}_\mathscr{C}(Z, X) \to \mathrm{Hom}_\mathscr{C}(Z, Y)$ should determine the morphism $f' \colon X \to Y$. The technical condition capturing this idea is that the functor $\mathrm{Hom}_\mathscr{C}(Z, X)$ should be *faithful*. Indeed, recall that a functor $F \colon \mathscr{C} \to \mathscr{D}$ is *faithful* if for every pair of parallel morphisms $f, g \colon X \to Y$, if $F(f) = F(g)$, then $f = g$. Thus in our particular case, this means precisely that if $f, g \colon X \to Y$ and $\mathrm{Hom}_\mathscr{C}(Z, X)(f) = \mathrm{Hom}_\mathscr{C}(Z, Y)(g)$ then $f = g$. But this condition is not always satisfied by the functor Hom. In particular, in Grothendieck's main case, that is the category of schemes, this condition is not satisfied.

Grothendieck's strategy to circumvent this difficulty is to use the categorical setup fully and to look at an object X as a geometric object. Thus, instead of trying to know the *points* of X — although it is very often what one wants to know about X, especially in the context of algebraic geometry where the points might be the solutions to a set of polynomial equations — one can know the *parts* of X by looking at all the projections into X of objects of the same kind, that is by looking at $\mathrm{Hom}_\mathscr{C}(Z, X)$ for *all Z* in $\mathscr{C}$. In other words, we are using the algebra of mappings once more. Instead of looking at the algebra of automorphisms, we are looking at the algebra of all morphisms coming into an object and its relationships to the other algebra of mappings.

This idea can now be expressed precisely as follows. For any category $\mathscr{C}$ and any object X of $\mathscr{C}$, consider the *contravariant* functor

$$h_X \colon \mathscr{C}^\circ \to \mathbf{Set}$$

defined by
$$h_X(Y) = \mathrm{Hom}_\mathscr{C}(Y, X)$$

and for any morphism u in $\mathrm{Hom}_\mathscr{C}(Y, Z)$,

$$h_X(u) \colon \mathrm{Hom}_\mathscr{C}(Z, X) \to \mathrm{Hom}_\mathscr{C}(Y, X)$$
$$h_X(u)(v) = v \circ u.$$

Furthermore, for any morphism $w \colon X \to X'$ of $\mathscr{C}$, there is a natural transformation

$$\eta_w \colon h_X \to h_{X'}$$

3.2 Grothendieck and Abelian Categories

defined by, for any object Y of $\mathscr{C}$,

$$\eta_w \colon \mathrm{Hom}_{\mathscr{C}}(Y,X) \to \mathrm{Hom}_{\mathscr{C}}(Y,X')$$
$$n_w(Y)(v) = w \circ v.$$

Moreover, these functors can be linked to general functors as follows. Given a contravariant functor $F \colon \mathscr{C}^{\circ} \to \mathbf{Set}$ and $x \in F(X)$, a natural transformation $\eta_x \colon h_X \to F$ is defined as follows: for any $v \in \mathrm{Hom}_{\mathscr{C}}(Y,X)$, we have the map $\eta_X(Y) \colon \mathrm{Hom}_{\mathscr{C}}(Y,X) \to F(Y)$ given by $\eta_X(Y)(v) = (F(v)(X))$.

It is now extremely tempting to think of a category $\mathscr{C}$ as a space and the objects of $\mathscr{C}$ as geometric objects like circles, spheres, triangles, etc. The isomorphisms of $\mathscr{C}$ correspond to the group actions in a standard space, i.e., they are invertible, and therefore play the same role. However, in a category $\mathscr{C}$, there are other transformations and *they* are used systematically to know the properties of the objects of the space. In particular, the functors of the form h_X, for all objects X of $\mathscr{C}$, provide us with a tool for investigating the objects of a category $\mathscr{C}$. These functors are linked together in a natural way, i.e., natural transformations exist between them.

In fact, we can describe the situation more systematically as follows. There is a *covariant* functor $h \colon \mathscr{C} \to \mathbf{Set}^{\mathscr{C}^{\circ}}$ which associates to each object X of $\mathscr{C}$ the contravariant functor h_X and to each morphism $w \colon X \to X'$, the natural transformation $h_w \colon h_X \to h_Y$. The fact that the functor h_X determines the object X can now be expressed by the following lemma, first proved by Yoneda in 1954:

Lemma 3.1 (Yoneda). *If $F \colon \mathscr{C} \to \mathbf{Set}$ is a functor and X an object of $\mathscr{C}$, then there is a bijection φ between the natural transformations from h_X to F and the elements of $F(X)$, i.e., if we denote the set of natural transformations from h_X to F by* $\mathrm{Nat}(h_X, F)$*, we have*

$$\varphi \colon \mathrm{Nat}(h_X, F) \simeq F(X).$$

Furthermore, it follows that if $h_X \simeq h_{X'}$, then $X \simeq X'$.

In fact, a stronger statement can be made. The functor $h \colon \mathscr{C} \to \mathbf{Set}$ is an equivalence of $\mathscr{C}$ with a full subcategory of the functor category $\mathbf{Set}^{\mathscr{C}^{\circ}}$. This means that, from a categorical point of view, the category $\mathscr{C}$ and the full subcategory of $\mathbf{Set}^{\mathscr{C}^{\circ}}$ are essentially the same. Notice that this is a representation theorem and that it extends Eilenberg and Mac Lane's representation theorem. It says that any category $\mathscr{C}$ can be represented as a category of *functors*. This might seem innocuous but it constitutes an extremely important shift that has tremendous implications, both mathematically, that is in the way one thinks about mathematics and how it should be *done*, and philosophically, that is in the way one thinks about what mathematics is *about*. For now, the objects of a category are not fundamentally structured sets, they are first and foremost *functors*. This idea will become the guideline underlying Grothendieck's program: the basic objects of algebraic geometry will be *functors*. As we will see, Lawvere will go further and work under the hypothesis that mathematical objects in general should be functors. Of course, one could say that this was true already in Eilenberg and Steenrod's work in algebraic topology and Cartan and

Eilenberg in homological algebra. But in these cases, functors were already "given" by the context. Grothendieck and Lawvere made a very bold generalization and had to be extraordinarily ingenious to extend this idea to different fields. Notice immediately how this extends Klein's program radically to the whole of mathematics: if, in Klein's case, the fundamental objects of elementary geometry are the transformations of a space, in Grothendieck and Lawvere's cases, the fundamental objects of algebraic geometry and mathematics in general are *functors*, i.e., transformations of categories.

Whenever each natural transformation η_X is an *isomorphism*, then we have a very special situation, the one we are interested in.

Definition 3.12. A *representation* of a functor $F\colon \mathscr{C} \to$ **Set** is a pair (X, η), with X an object of X and

$$\eta\colon h_X \simeq F$$

a natural isomorphism. The object X is called the *representing object*. The functor F is said to be *representable* when a representation exists.

It should be noted that it can be proved that the representing object X is defined uniquely up to a unique isomorphism. It is in this sense unique.

We can think of the notion of representable functor as follows. A functor $F\colon \mathscr{C}^\circ \to$ **Set** can be thought of as a transformation or a representation of the category $\mathscr{C}^\circ$ into the category of sets; that is, the objects of $\mathscr{C}$ are transformed into sets and the morphisms of $\mathscr{C}^\circ$ are transformed into functions. For the functor F to be representable means that the structure of the category $\mathscr{C}^\circ$ allows this transformation to be made in a uniform manner. The representing object X and the natural transformation η play that role. Thus it is possible to understand how the functor F transforms the category $\mathscr{C}^\circ$ by looking at a specific and uniform aspect of the structure of the algebra of mappings of $\mathscr{C}^\circ$ itself. These somewhat obscure remarks will be clarified by considering some examples.

Let X and Y be two objects of a category $\mathscr{C}$. Consider the functor $F\colon \mathscr{C}^\circ \to$ **Set** defined by: for any object Z of $\mathscr{C}$,

$$F(Z) = \mathrm{Hom}_\mathscr{C}(Z,X) \times \mathrm{Hom}_\mathscr{C}(Z,Y)$$

and for any morphism $f\colon Z \to Z'$, $F(f)\colon F(Z') \to F(Z)$ is given by

$$F(f)(u,v) = (u \circ f, v \circ f).$$

What does it mean for F to be representable? It means that there is an object P of $\mathscr{C}$ together with a natural isomorphism

$$\eta\colon \mathrm{Hom}_\mathscr{C}(-,P) = h_P \simeq \mathrm{Hom}_\mathscr{C}(-,X) \times \mathrm{Hom}_\mathscr{C}(-,Y).$$

This means in turn that for any object Z of $\mathscr{C}$, there is a bijection between $\mathrm{Hom}_\mathscr{C}(Z,P)$ and $\mathrm{Hom}_\mathscr{C}(Z,X) \times \mathrm{Hom}_\mathscr{C}(Z,Y)$. Thus, for each morphism $w\colon Z \to P$, there is a unique pair of morphisms $u\colon Z \to X$ and $v\colon Z \to Y$ corresponding to w and

3.2 Grothendieck and Abelian Categories

conversely, for each pair of morphisms $u\colon Z \to X$ and $v\colon Z \to Y$, there is a unique morphism $w\colon Z \to P$ corresponding to them. In particular we have, for $Z = P$

$$\eta_P\colon \mathrm{Hom}_{\mathscr{C}}(P,P) \simeq \mathrm{Hom}_{\mathscr{C}}(P,X) \times \mathrm{Hom}_{\mathscr{C}}(P,Y)$$

and

$$\eta_P(1_P) = (p_X, p_Y).$$

We can now apply Yoneda lemma to obtain the following commutative diagram:

$$\begin{array}{ccc} \mathrm{Hom}_{\mathscr{C}}(P,P) & \xrightarrow{\sim} & \mathrm{Hom}_{\mathscr{C}}(P,X) \times \mathrm{Hom}_{\mathscr{C}}(P,Y) \\ \downarrow & & \downarrow \\ \mathrm{Hom}_{\mathscr{C}}(Z,P) & \xrightarrow{\sim} & \mathrm{Hom}_{\mathscr{C}}(Z,X) \times \mathrm{Hom}_{\mathscr{C}}(Z,Y) \end{array}$$

Recall that the functor $\mathrm{Hom}_{\mathscr{C}}(-, P)$ is contravariant. So for every morphism $u\colon Z \to P$, we get a morphism $h_P(u)\colon \mathrm{Hom}_{\mathscr{C}}(P,P) \to \mathrm{Hom}_{\mathscr{C}}(Z,P)$ and then what the diagrams says is that u is sent to $u \circ p_X$ and $u \circ p_Y$ and this mapping is a bijection. Thus, whenever the functor F is representable, the object P together with the mappings (p_X, p_Y) constitute a product of X and Y in $\mathscr{C}$. Therefore, if for any pair of objects (X,Y) of $\mathscr{C}$, the associated functor is representable, then we *can* say that $\mathscr{C}$ has binary products. This shows how the notion of representable functor subsumes the notion of universal morphism described previously.[19]

The converse relationship between universal morphisms and representable functors can also be established. Suppose that there is for two objects X, Y of $\mathscr{C}$, an object P together with projections $p_X\colon P \to X$ and $p_Y\colon P \to Y$ satisfying the universal property, i.e., X and Y have a product in $\mathscr{C}$. Define the functor $F\colon \mathscr{C}^{\circ} \to \mathbf{Set}$ as follows: for any object Z of $\mathscr{C}$

$$F(Z) = \mathrm{Hom}_{\mathscr{C}}(Z,X) \times \mathrm{Hom}_{\mathscr{C}}(Z,Y)$$

and for morphisms in the obvious way. Define the natural transformation $\eta\colon h_P \to F$ by

$$\eta_P(Z)(u) = (u \circ p_X, u \circ p_Y).$$

By the universal property of P and the projections p_X and p_Y, the natural transformation η is a bijection, as required.

It can be shown that the same strategy applies to the other universal notions we have seen in section 3.1: coproducts, equalizers, coequalizers can all be described in the language of representable functors. Similarly, each and every representable functor gives rise to a universal property. The two notions subsume one another.

---

[19] Grothendieck was aware of that connection right from the start. He states in the seminar 195: "This fact underlies the notion of solution of a Univeral Problem, such a problem always consists in finding if a given functor from $\mathscr{C} \to \mathbf{Set}$ is representable." ([104], 1. My translation.)

Representable functors occupied a fundamental position in Grothendieck's projects. Many results amounted to proving that a given functor was representable. Thus one could claim that representable functors capture the basic concepts of a field. We are, once more, back to Eilenberg and Mac Lane's original claim, although in a different guise. Now the key property is not the conditions under which a concept is functorial, but rather whether a concept is a representable functor in a certain context. We have just seen how in this case the algebra of mappings (i.e., the functors of the form $\operatorname{Hom}_{\mathscr{C}}(-,-)$) occupies a key position.

Chapter 4
Discovering Fundamental Categorical Transformations: Adjoint Functors

> The multiple examples, here and elsewhere, of adjoint functors tend to show that adjoints occur almost everywhere in many branches of Mathematics. It is the thesis of this book that a systematic use of all these adjunctions illuminates and clarifies these subjects.
>
> ([195], 103.) *MacLane*

> Nowadays, every user of category theory agrees that this is the concept [i.e. adjointness] which justifies the fundamental position of the subject in mathematics.
>
> ([258], 367.) *Taylor*

> But more importantly, it [the notion of adjoint functor] also captures an important mathematical phenomenon that is invisible without the lens of category theory. Indeed, I will make the admittedly provocative claim that adjointness is a concept of fundamental logical and mathematical importance that is not captured elsewhere in mathematics.
>
> ([8], 179.) *Awodey*

As we have seen, Eilenberg and Mac Lane tried to use functors in order to classify mathematical concepts and in particular determine the "basic" concepts of a mathematical field. Their specific approach was a dead end. Soon afterwards, Mac Lane started using universal properties to define various concepts and hoped that the concept of a bicategory could have a role to play in universal algebra. After he had defined and used Abelian categories in a novel way, Grothendieck pursued his project of redefining the foundations of algebraic geometry and put representable functors at the center of the new foundational framework. At about the same time, that is around 1954–55, Daniel Kan was elaborating abstract foundations for homotopy theory with the help of category theory and in the process he discovered what is undoubtedly *the* fundamental notion of category theory. The notion involves two functors, $F: \mathscr{C} \to \mathscr{D}$ and $G: \mathscr{D} \to \mathscr{C}$ related to one another in a specific manner, namely for every X in $\mathscr{C}$ and A in $\mathscr{D}$, there is an isomorphism

$$\mathrm{Hom}(F(X), A) \xrightarrow{\phi} \mathrm{Hom}(X, G(A))$$

which is natural in both X and A, in a sense to be explained below. Following the suggestion of Eilenberg, who immediately saw the analogy between these functors and adjoint operators in functional analysis, Kan called functors related to one another in the specified way "adjoint functors".

Clearly, and as we will see, the notion was already *implicit* in Eilenberg and Mac Lane's original papers. They had many examples of adjoint functors at hand but they did not *see* them. The notion was also present in other well known theorems of that time, for instance Stone's representation theorem for Boolean algebras, Pontrjagin duality, and many others. What was required was to look at them in a certain manner, which no one did for 14 years or so. In fact, even those results were not immediately seen as being special cases of adjoint functors.

The fact that the notion of adjoint functors was not discovered before, for instance by Eilenberg and Mac Lane themselves or by any other mathematician working with categorical tools afterwards, is sometimes presented as being mysterious. Here is how Mac Lane puts it:

> This situation is a striking instance of a historical question which can be raised about the time of appearance of many mathematical concepts: Given the available formalism of adjoint operators and the numerous examples of adjoint functors, why was the general notion so late in arising?
> We [i.e. Mac Lane] have little expertise in answering such questions in the history of mathematics and any attempt at an answer can only be speculation (...). It is my own view that the climate of mathematical opinion in the decade 1946–1956 was not favorable to further conceptual development. Investigation of concepts as general as those of category theory were heartily discouraged, perhaps because it was felt that the scheme provided by Bourbaki's structures produced enough generality. It is to be noted that Kan, when developing adjoint functors, came at the time from a solitary position more or less outside active mathematical circles. ([193], 234.)

Two things are striking in Mac Lane's claim. First, Mac Lane mentions *sociological* reasons that might explain the delay in the apparition of the concept, e.g., the climate of mathematical opinion, the war and certain distrust with respect to category theory or mathematics of that level of generality. Second, Mac Lane suggests that it was left to a *younger* and *solitary* man to make the discovery, although they and others who used categories afterwards, were fairly young when Eilenberg and himself introduced categories. Again, this suggests a sociological factor: the influence of intellectual fashions in important academic circles that might have prevented the discovery. A year later, Mac Lane went a little further.

> One may also speculate as to why the discovery of adjoint functors was so delayed. Ideas about Hilbert space or universal constructions in general topology might have suggested adjoints, but they did not; perhaps the 1939–1945 war interrupted this development. During the next decade 1945–1955 there were very few studies of categories, category theory was just a language, and possible workers may have been discouraged by the widespread pragmatic distrust of "general abstract nonsense" (category theory). Bourbaki just missed. His definition of universal construction was clumsy, because it avoided categorical language (...) This formulation lacks the symmetry of the adjunction problem, (...) — and so was left to a younger man, perhaps one less beholden to tradition or to fashion. ([195], 103.)

Besides the sociological factors already mentioned, there is a new element in this quote. Now Mac Lane suggests that the concept might have come from functional analysis and general topology. A slightly different explanation is offered by Stone in his comment on Mac Lane's paper.

> If one seeks to explain why the notion of an adjoint functor appeared in category theory as much as fifteen years after its inception, it seems to me that two reasons have to be suggested. In the first place, the concept had its origins and early development in analysis, in the theory of differential equations. It was not a vital part of the mathematical experience of the algebraically-oriented pioneers in category theory. The other historical factor that has to be cited is the interruption of mathematical research imposed by World War II and the many professional readjustments that followed it. ([256], 236.)

A little later, Stone clarifies this opening remark somewhat.

> By analogy, I am equally certain that Eilenberg, Mac Lane, and the other early pioneers in category theory would have hit quite early upon the idea of adjoint functor if they had been more fully aware of contemporary work in Hilbert space theory and its background in the theories of differential equations and topological groups. As it was, they drew mainly upon algebra and algebraic topology – especially the theory of Abelian groups – for sources of the new theory. ([256], 237.)

This is certainly wrong. It is far from clear that the concept of adjoint functor has its origins in the development of analysis. It might be possible to trace it back to that field, but in actual history, it seems unlikely. Certainly, the concept of adjoint *operator* does have its origins in analysis and some mathematicians using category theory in the early years knew about adjoint operators and their properties. Eilenberg and Mac Lane themselves used Banach spaces in their 1945 paper to illustrate the concepts of functor and natural isomorphism. Grothendieck, for one, started his career in functional analysis and was considered as one of the leading functional analyst in the fifties. As we have seen, he followed Bourbaki and used representable functors instead. Thus, these mathematicians knew functional analysis very well and knew about adjoint operators. Certainly Eilenberg knew them well, for as I have already mentioned, when Kan showed him the draft of his paper on adjoint functors, it was Eilenberg who suggested the name and underlined the parallel between adjoint functors and adjoint operators. Kan himself had not seen the parallel. Of course, once the parallel is seen, it is entirely conceivable that the notion of adjoint functors could have been discovered by someone working in functional analysis.[1] Mac Lane says so explicitly in his paper:

> We now indicate how Kan's notion of an adjoint functor may have been adumbrated in the analytical study of adjoint transformations. (...) The formal analogy to the definition (1) of adjoint transformation is striking (...) The basic formal properties of adjoint functors are strikingly parallel to those of adjoint linear operators. ([193], 230–232.)

And Mac Lane gives a list of properties that are formally analogous. The analogy is so compelling that Mac Lane suggests that there might be a more general concept "which would subsume both adjoint functors and adjoint operators with the just

---

[1] In fact, it almost did. Lawvere developed his understanding of universal mappings while he was teaching a course in functional analysis (Lawvere, personal communication). Mac Lane consulted Lawvere on the issue when he wrote his tribute to Stone. It is not unlikely that Lawvere's experience and testimony influenced Mac Lane to a large extent. (Lawvere, personal communication.)

noted corresponding formal properties". ([193], 232.) The fact that the notion was not discovered in the context of functional analysis cannot be attributed to ignorance on the part of the mathematicians developing category theory from 1945 until 1955.

There is no doubt that the second World War had a major influence on mathematical research, but it is far from clear that it prevented the discovery of the concept of adjoint functor. After all, Eilenberg and Mac Lane's paper appeared in 1945 and one would expect that adjoint functors would have been discovered soon thereafter. Furthermore, Stone is somewhat loosing sight of what Eilenberg, Mac Lane and others were doing during the first ten years following the introduction of category theory. They were *not* developing category theory for its own sake. Quite the contrary, category theory was merely a useful language *for* algebraic topology and homological algebra, as Mac Lane himself readily admits. Stone returns to the impact of the Second World War later in his paper.

> Even if such be the case, one could still ask with Professor Mac Lane why, as a matter of the purely internal development of category theory, adjoint functors did not appear considerably earlier upon the cene [sic]. After all, the well-known dualities for abelian groups and for vector spaces could have provided clues without any need of venturing very far away from the initial concerns of category theory. If an answer is to be found here I think we have to look at another factor, the impact of the Second World War. ([256], 237.)

Of course, Stone is correct in saying that various examples were already available to many mathematicians right from the beginning. The role played by the Second World War is, according to Stone, the following: many mathematicians working in various areas of mathematics might have tried to apply category theory to their own research, including Stone himself, had they had not been interrupted in that research by the war. This may very well be, but it is far from clear that one of them would have *seen* adjoint functors in these applications. Be that as it may, we can see in these quotations the influence (albeit indirect) of the historian and philosopher of science Thomas Kuhn in this historical analysis.

Stone makes a different suggestion in his remarks on Mac Lane's paper and this time it seems to be, at least in its positive aspect, correct:

> Of course, it will never take very long for such a central and essentially simple concept as that of adjoint functor to emerge from research carried on strictly inside a growing branch of mathematics. While I do not know the private history of Kan's introduction of the adjoint functor, I suspect that it may have come about in just this manner.

This is not very far from what might have happened. But we have to be more precise here. I will try to clarify what I believe is required in this context.

First, it should be clear that discovering a concept like that of adjoint functors is *not* a typical problem-solving activity. It is certainly not like proving a conjecture or finding a counterexample to a conjecture, or solving a system of equations, or generalizing a known result, etc. I claim that it is more like a problem of pattern recognition, but in this case, when one is *not* necessarily looking for a pattern. Second, it seems reasonable to assume that for the process of pattern recognition to take place, the pattern has to appear implicitly in a branch or various branches of mathematics. In other words, the branch (or branches) of mathematics has to have certain properties for the pattern to be discernable. As we will see, Kan was working in *one* of the fields where the concept could be found. Third, it requires a certain type of activity or a certain frame of mind to *see* the pattern. As we will see, Kan asked the right question and tried to *understand* certain aspects of a problem. Four, once the pattern is seen, one has to have *reasons* to believe that it is a *general* and *useful* concept and *reasons* to believe that it deserves a whole and independent development. As we will see, Kan had reasons to *use* the concept in his work and found other, more general reasons, while writing his paper on the concept. It is clear that adjoint functors came to Kan during his research in homotopy theory.

I will therefore suggest an alternative reading and will give indications as to why Eilenberg, Mac Lane and others could not *see* the notion of adjoint functors and how the notion became clear to Kan.

As we will see, adjoint functors are closely related to universal morphisms and representable functors. In fact, the three notions form a circle of ideas, although, in a certain sense, adjoint functors are more faithful to the categorical spirit. Indeed, whereas universal morphisms are defined with respect to the algebra of morphisms of a unique category and representable functors are defined with respect to the category of sets, adjoint functors are functors satisfying certain conditions.

Adjoint functors bring us directly back to Klein's program. My main claim is that adjoint functors are to categories what automorphisms are to geometric spaces. To be more precise, in the same way that automorphisms encode a criterion of identity and a criterion of meaningfulness for geometries, adjoint functors encode *a criterion of identity* and *a criterion of meaningfulness* for categories. Or, to use an expression that we will come back to later on, combinations of adjoint functors encode the invariant content of various mathematical domains. Although this analogy has to be clarified and expanded, and various nuances and subtleties concerning criteria of identity and meaningfulness have to be added, it is a good first approximation. Had Eilenberg and Mac Lane discovered adjoint functors right from the start, I believe that they would have presented the global significance and the connection between category theory and Klein's program differently. For the basic concepts of a field do not have to do with whether or not they are functorial — although this is certainly a necessary condition in this context — but rather with whether or not they belong to an adjunction. We will now briefly look at the background of Kan's work (for the notion first appeared naturally in that context), and then look at some aspects of Kan's work that prepared his mind for the discovery. Once we have done this, we will step back from the historical record and concentrate on the notion of adjointness as such, returning again to Klein's program.

4.1 The Background: Homotopy Theory and Category Theory

> The philosophical emphasis here is: to solve a geometrical problem of a global nature, one first reduces it to a homotopy theory problem; this is in turn reduced to an algebraic problem and is solved as such. This path has historically been the most fruitful one in algebraic topology.
>
> ([99], vii.)
>
> Within algebraic topology, homotopy theory is generally regarded as more geometric than homology theory.
>
> ([83], xxxiii.)

Kan started his mathematical career by publishing four notes in the *Proceedings of the National Academy of Science–U.S.A.* These four notes have a very simple but revealing title: "Abstract Homotopy I, II, III, IV". ([129–132]) Thus, Kan was working in homotopy theory. He started out by noticing how homotopy groups could be defined directly on abstract cubical complexes which satisfy a certain condition, now called the Kan condition. This was the subject of his first note. Categories are used in this first note, but acquire a more specific role in the second note and a crucial role in the papers published later in 1958. We will come back to this aspect of Kan's work in due course. Before we look more closely at what Kan did, we will rehearse the basic notions of homotopy theory, not only because they played a key role in Kan's work, but also because they continue to play a fundamental role in the development of category theory even to this day.

The basic ideas of homotopy theory can be traced back to research in analysis, in fact it goes back to Lagrange's work![2] However, the precise definition of a homotopy between two continuous maps was first given by Brouwer in 1912. Brouwer's definition is in essence what is presented nowadays:

Definition 4.1. [3] Given two topological spaces X, Y, a subspace A of X, two continuous mappings $f, g \colon X \to Y$ such that $f|_A = g|_A$ are *homotopic relative to* A if there exists a continuous mapping $F \colon X \times [0,1] \to Y$ such that $F(x,0) = f(x)$, $F(x,1) = g(x)$ and $F(a,t) = f(a) = g(a)$ for all $a \in A$ and $t \in [0,1]$. The mapping F is called a *homotopy* and we write $f \cong g$ (rel A).

Informally, what the definition says is that it is possible to start from the image $f[X]$ in Y and "deform" it continuously into the image $g[X]$ in Y. When the subspace A is the empty subspace $\emptyset$, we get the original definition of a homotopy, that is a continuous deformation of f onto g parametrized by the unit interval and this case is usually called a free homotopy. As the notation indicates, for a subspace A of X, the relation "being homotopic rel A" is an equivalence relation.

Brouwer showed that a certain property, the concept of the degree of a map, is invariant under homotopy; that is, if f and g are homotopic, then they have the same

---

[2] For a detailed and interesting exposition of the early years of homotopy theory, see [262].

[3] I am defining relative homotopy, a notion introduced implicitly by Hurewicz in 1935 and not by Brouwer. As I will indicate, it is more general than Brouwer's definition.

degree. This result made it possible for him to prove his fixed-point theorem.[4] This is in fact typical and indicates the usefulness of the notion: in general, a functor F defined on a category of topological spaces (or a more abstract category) is called *homotopy invariant* if $f \cong g$ always implies $F(f) = F(g)$. This last condition was in fact included as an axiom by Eilenberg and Steenrod in their axiomatization of homology theory. Thus a homology theory has to be homotopy invariant. Informally, what this invariance means is that for sufficiently small deformations of a map, the associated algebraic morphism is not changed.

But homotopy *theory* appeared only in the 1930s with the work of Hopf and Hurewicz.[5] In the mid-thirties Hurewicz introduced higher homotopy groups (which constituted the first introduction of algebraic objects in homotopy theory), and the key notion of homotopy type, and proved various key results about higher homotopy groups that are in the background of Kan's paper on adjoint functors. Let us start with the notion of homotopy type.[6]

Definition 4.2. Two spaces X and Y have the same *homotopy type* if there are mappings $f: X \to Y$ and $g: Y \to X$ such that the composed mappings $f \circ g$ and $g \circ f$ are (free) homotopic to the identity mappings.[7] The mapping f is called a *homotopy equivalence*.

This definition makes sense since the notion of homotopy is an equivalence relation on the set $\text{Hom}(X,Y)$. In fact, given a mapping $f: X \to Y$, the equivalence class $[f]$ of mappings homotopic to f is called its *homotopy class*. The family of all such homotopy classes of mappings between X and Y is usually denoted by $[X,Y]$. This leads to the definition of a new category, the homotopy category **hTop**, with objects topological spaces $X, Y, Z, \ldots$ and a morphism from X to Y is now a homotopy class $[f]$ of mappings from X to Y. This is a natural example of a category in which the morphisms are *not* structure-preserving functions.

It is easy to see that the notion of homotopy type is in fact a *derived* notion once we are in that category, since two spaces X and Y have the same homotopy type if and only if they are isomorphic in the category **hTop**. In other words, f is a homotopy equivalence if and only if $[f] \in [X,Y]$ is an isomorphism in **hTop**. This category is in fact fundamental in algebraic topology since most of the invariants of algebraic topology depend upon the homotopy type of the space examined. One can say that the goal of homotopy theory is to *classify* geometric objects that can be continuously deformed into each other. Once again, notice that we are using *mappings* to *classify* spaces.

The classification of topological spaces by homotopy types is different from the classification based on homeomorphisms. Homeomorphic spaces have the same ho-

---

[4] For more on this, see [59].

[5] Čech had introduced higher homotopy groups in 1932 in a short note, but since they were Abelian, it did not attract much attention.

[6] For more on the history of homotopy, homotopy types and fibrations, see [214].

[7] Hurewicz introduced the notion of homotopy type and the related notion of homotopy equivalence in the third paper of a series of five. The definition came after the introduction of higher homotopy groups.

motopy type but spaces with the same homotopy type are not necessarily homeomorphic. For instance, a disk or a full sphere have the same homotopy type as a point (and they are clearly not homeomorphic). The classification by homotopy types is therefore coarser than the classification based on homeomorphisms.

Homotopy theory really got off the ground when Hurewicz introduced higher homotopy groups and connected them to known homology groups, in a series of five papers published in the mid-thirties.[8] The first homotopy group, also known as the fundamental group, was introduced earlier by Poincaré. Let I denote the usual unit interval $[0,1]$. To define the fundamental group, we need a few preliminary notions:

Definition 4.3.

1. A *path* in a space X is a continuous map $f: I \to X$. If $f(0) = a$ and $f(1) = b$, we say that f is a path *from a to b*. We say that a is the *origin* of f, written $a = \alpha(f)$, and b is the *end* of f, written $b = \omega(f)$;
2. A space X is *path connected* if, for every $a, b \in X$, there exists a path in X from a to b.
3. $a \sim b$ if there is a path in X from a to b.

It can be shown that the latter definition is an equivalence relation. The equivalence classes of X under the relation given in c) are called the *Path component* of X.

This is enough to introduce the first homotopy invariant structure: Let $\pi_0(X)$ be the *set* of path components of X. It can be shown that $\pi_0 \colon \mathbf{hTop} \to \mathbf{Set}$ is a functor. Thus, if f is homotopic to g, then $\pi_0(f) = \pi_0(g)$ and if X and Y have the same homotopy type, then they have the same number of path components. Although this is not much as a functor, since it takes its values in **Set** and the only information that can be extracted from the latter is their size (i.e., one can only *count* the number of path components), it is enough to show that the circle S^1 and the unit interval I are not homeomorphic: deleting a point in I yields two path components whereas deleting a point in S^1 yields one path component. Therefore, they are not topologically the same.

Definition 4.4.

1. Let x_0 be a point of X and $f: I \to X$ a path; f is said to be *closed at x_0* if $\alpha(f) = x_0 = \omega(f)$;
2. Let $f, g: I \to X$ be paths with $f(1) = g(0)$. Define a path $f * g: I \to X$ by

$$(f*g)(x) = \begin{cases} f(2t) & \text{if } 0 \le t \le 1/2 \\ g(2t-1) & \text{if } 1/2 \le t \le 1 \end{cases}.$$

It can be shown that $f * g$ is a path.

---

[8] For Hurewicz's papers, see [150]. These papers were highly influential. A young Polish mathematician would come to admire these papers greatly: his name was Samuel Eilenberg. "Mathematically speaking, I was profoundly influenced by Hurewicz's papers in homotopy theory. (...) Reprints of these five papers I had bound together, and I reread them from time to time. They are an important relic to me." ([70], xlvi.)

4.1 The Background: Homotopy Theory and Category Theory

Informally, the last definition says that given two paths such that the end of the first is equal to the origin of the second, it is possible to start with the first and then continue with the second in such a way that the result is a path.

Definition 4.5.

1. Let x_0 be a point of X and $f: I \to X$ a closed path at x_0. The equivalence class of homotopic mappings to f is called the *path class* of f and is denoted by $[f]$.[9]
2. If $x_0 \in X$, then the constant mapping $i_{x_0}: I \to X$ with $i_{x_0}(t) = x_0$ for all $t \in I$ is called the *constant path at x_0*.
3. If $f: I \to X$ is a path, its *inverse* path $f^{-1}: I \to X$ is defined by

$$f^{-1}(t) = f(1-t).$$

Again, the last two definitions are very natural: the constant mapping simply "stays" at the same point "all the time". The inverse path is simply going through the same points, but in the opposite direction.

With these definitions at hand, it is easy to show that, given a space X with a point x_0, called its *base point*, the following set:

$$\pi_1(X, x_0) = \{[f] \mid [f] \text{ is a path class with } \alpha[f] = x_0 = \omega[f]\}$$

is a group with the binary operation

$$[f] \circ [g] = [f * g].$$

This is the fundamental group of the space X with base point x_0.

For instance, it can be shown, although it takes more work than one might expect and it is certainly *not* a routine calculation, that the fundamental group of the circle is isomorphic to the integers, i.e.,

$$\pi_1(S^1, p) \simeq \mathbb{Z}.$$

One way to see this is to consider the punctured plane with a base point, which is essentially the same as the circle from the point of view of homotopy theory. The paths in the punctured plane can be classified in the following manner: those paths which do not go around the hole (these can be deformed continuously into the base point and are therefore homotopically equivalent to the constant path based at that point); and paths that go around the hole. These in turn are classified with respect to two properties: their direction, and the number of times they go around the hole. If a path goes around the hole once in a clockwise direction, then it is sent to 1 in $\mathbb{Z}$. If a path goes around the hole once in a counter clockwise direction, then it is sent to -1 in $\mathbb{Z}$. If a path goes around the hole n times in a clockwise direction, then it is sent to n in $\mathbb{Z}$. And so on. Of course, this is far from the actual proof, which is considerably more involved. (See, for instance, [243], pp. 50–53.)

It can also be shown that for $n > 1$,

---
[9] The path class of f is not, strictly speaking, the same thing as the homotopy class of f.

$$\pi_1(S^n, p) \simeq 0$$

where p is of course a base point of the n-sphere. Furthermore, the fundamental group of the torus is isomorphic to the group $\mathbb{Z} \times \mathbb{Z}$, the fundamental group of the projective plane P^2 is isomorphic to a group of order 2 and the fundamental group of the double torus is not Abelian. It follows from these results that the sphere S^2, the torus, the projective plane P^2 and the double torus are topologically distinct. (See [232], chap. 8.)

There is an alternative definition of the fundamental group that leads naturally to Hurewicz's generalization. The pair (X, x_0) is called a *pointed space*. It is easy and useful to define the category of pointed spaces **Top$_\bullet$**. Its objects are pointed spaces (X, x_0). A morphism $f \colon (X, x_0) \to (Y, y_0)$ is a pointed map, that is a continuous map preserving the base point, $f(x_0) = y_0$. The *pointed homotopy category* **hTop$_\bullet$** has as its objects pointed spaces (X, x_0) and morphisms $(X, x_0) \to (Y, y_0)$ are relative homotopy classes $[f]$, where $f \colon (X, x_0) \to (Y, y_0)$ is a pointed map. Given two pointed maps f, g of **Top$_\bullet$** such that their composition exists, then $[g] \circ [f] = [g * f]$. In **hTop$_\bullet$**, Hom-sets are denoted by $[(X, x_0), (Y, y_0)]$. Now, consider the pointed space (S^1, p), the unit circle S^1 with a specified base point p, say $p = (1, 0)$. A pointed map $f \colon (S^1, p) \to (X, x_0)$ is in fact the same as a closed path at x_0. Thus, the set of all pointed maps $[(S^1, p), (X, x_0)]$ can also be turned into a group and it can be shown that

$$\pi_1(X, x_0) = [(S^1, p), (X, x_0)].$$

Informally, this means that the fundamental group can be seen as (pointed) homotopy classes of (pointed) maps from S^1 into X. In categorical terms, which of course, were not Hurewicz's terms, the fundamental group $\pi_1(X, x_0)$ can be seen as a (covariant) functor

$$[(S^1, p), -] \colon \mathbf{hTop_\bullet} \to \mathbf{Grp}.$$

Again, this captures how the fundamental group encodes invariant properties of a space. For it now follows from the functorial aspect of the construction that if

$$f \colon (X, x_0) \to (Y, y_0)$$

is a homotopy equivalence, then the induced map

$$\pi_1(f) \colon \pi_1(X, x_0) \to \pi_1(Y, y_0)$$

is an isomorphism of groups.

The basic idea underlying Hurewicz's generalization is to consider spheres of higher dimensions, S^n, in the above definition (that is, to consider (pointed) maps of S^n into a space X), and their homotopy classes are then elements of the homotopy group $\pi_n(X, x_0)$. It is relevant to note that in doing this, Hurewicz started from basic facts about function spaces that he then interpreted in the context of homotopy theory, facts that were mentioned again by Kan in his paper on adjoint functors and

4.1 The Background: Homotopy Theory and Category Theory

presented as examples of adjoint functors.[10] Furthermore, these adjunctions occupy a key role in category theory in general and deserve to be presented immediately in the context of homotopy theory.

Let us briefly go back to *sets*. Given three sets X, Y and Z and a function $f\colon Z \times Y \to X$, it is well known that to f one can associate a *unique* function $f^*\colon Z \to \mathrm{Hom}(Y,Z)$, by putting $f^*(z) = f(z,-)$. Thus, there is a natural bijection:[11]

$$\mathrm{Hom}(Z \times Y, X) \to \mathrm{Hom}(Z, \mathrm{Hom}(Y,X)).$$

This is called the *exponential law* for sets (and for cardinals), since it can be written as:

$$X^{Z \times Y} \simeq (X^Y)^Z.$$

There is an important universal morphism in this context: if X and Y are sets, then the *evaluation map* $e\colon X^Y \times Y \to X$ defined by $e(f,y) = f(y)$ is universal. The universality is expressed by the following condition: for any function $f\colon Z \times Y \to X$, there is a unique arrow $f^*\colon Z \to X^Y$ that makes this triangle commute

---

[10] There *is* at this point a very important connection with functional analysis. At the end of the chapter on function spaces of his book *General Topology*, published in 1955, Kelley refers to Grothendieck's work in functional analysis (problem D), Stone's work on $C(X)$, the algebra of all continuous real-valued functions on X, X a compact topological space (problem R), Eilenberg and Steenrod's and Eilenberg and Mac Lane's paper on categories (problem S). Although Eilenberg and Steenrod's book is in the bibliography, Eilenberg and Mac Lane's paper is not. It is interesting to read how Kelley refers to the subject: "However, the pattern used above is, in part, an example of a general method. To each member of a certain collection of objects (in this case compact Hausdorff spaces X) there is associated another object (in this case the Banach algebras $C(X)$). Moreover, to each of a specified class of maps of the original objects (continuous maps in the case at hand) there is assigned an induced map satisfying certain conditions [...] This general method of investigation has been used most successfully by Eilenberg and Steenrod in their axiomatic treatment of homology theory. The method itself was first studied by Eilenberg and Mac Lane. The study of objects and maps might be called the galactic theory, continuing the analogy whereby the study of a topological space is called global." ([136], 246–247.)

Two points have to be made. As Bill Lawvere had suggested to me, a student reading Kelley's book might have been led to look at category theory for problems in functional analysis and that might have led to the discovery of adjoint functors along lines suggested by Mac Lane. Second, Kelley describes a *method*, not a theory, and talks about functors without mentioning them explicitly. The terminology is interesting: categories could have been called "galaxies", it makes perfect sense, and a collection of categories would have been called, perhaps, a "cluster" and finally the set up in which these things are situated would have been a "universe". One wonders whether this terminology would have led to different representations, in the psychological sense, of what categories are and a different reception by the various communities involved. In fact, Bill Lawvere informed me that he used the term "galactic clusters" after Kelley in his introductory letter to Eilenberg in 1959 to denote what Grothendieck called "fibered categories". (Lawvere, personal communication, December 2002.)

[11] In his textbook on set theory, Vaught goes so far as to refer to category theory when he describes this bijection: "(...) in category theory there is even an attempt to single out maps like F above from just any old one-to-one onto map by a technically defined notion: 'F is a *natural* map'." ([263], 37) In the context of the proof, where Vaught says that the first part is done "*without thought*", one wonders whether this is not a case of mathematical irony.

$$X^Y \times Y \xrightarrow{e} X$$

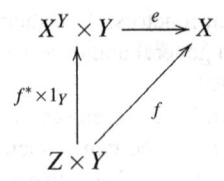

with $f^* \times 1_Y$ from $Z \times Y$ upward and $f: Z \times Y \to X$.

This is another example of a universal mapping problem and it plays a key role in mathematics in general since it deals with a basic construction: exponentials. What the bijection shows is that it is connected to products in a category in a systematic manner. In fact, many important properties of exponentials follow directly from the connection with products and the universal morphism e.

How does this fit with homotopy theory? First, the foregoing bijection between sets can be extended to topological spaces, provided that an appropriate topology is defined on function spaces, as was noted by Eilenberg and Mac Lane in their 1945 paper where they present the exponential law for topological spaces as an example of a natural isomorphism.

Let X and Y be topological spaces. Then X^Y is the set of all continuous functions from Y to X. If K is a compact subset of Y and U an open subset of X, define

$$(K;U) = \{f \in X^Y \mid f(K) \subset U\}.$$

The topology having a sub-basis consisting of all subsets $(K;U)$ is the compact-open topology on X^Y; that is, a typical open set of the space is an arbitrary union of finite intersections of sets of the form $(K;U)$.[12]

Let X and Z be topological spaces and Y be locally compact Hausdorff; let X^Y have the compact–open topology. Then the evaluation map $e: X^Y \times Y \to X$ is continuous and a map $f: Z \times Y \to X$ is continuous if and only if its associate $f^*: Z \to X^Y$ is continuous. (For the proofs, see [232], 287–288.) In other words, we have

$$\mathrm{Hom}_{\mathbf{Top}}(Z \times Y, X) \to \mathrm{Hom}_{\mathbf{Top}}(Z, X^Y)$$

or to put it in the same form as above:

$$\mathrm{Hom}_{\mathbf{Top}}(Z \times Y, X) \to \mathrm{Hom}_{\mathbf{Top}}(Z, \mathrm{Hom}_{\mathbf{Top}}(Y, X)).$$

In fact, when Z and Y are locally compact Hausdorff spaces, the exponential law holds:

$$X^{Z \times Y} \simeq (X^Y)^Z.$$

---

[12] Hurewicz does not use this definition: he assumes that X and Y are separable metric spaces, Y is compact and X is connected and locally contractible. Hurewicz mentions that the function space X^Y is usually defined as a metric space and thus relies on a metric defined on X. But, he continues, for Y compact, the topology defined is independent of the metric, which is one of the essential properties of the compact-open topology. The definition of the compact-open topology is explicitly given by [74], 243, where they present the function space construction as a functor. However, the result I am about to present seems to have appeared explicitly for the first time in the paper by Fox, published in 1945. See [86]

4.1 The Background: Homotopy Theory and Category Theory

The proof that it is a *natural homeomorphism* is given by Eilenberg and Mac Lane in their 1945 paper. (See [74], 254.)

Consider now the following special case. Let Z above be the unit interval $I = [0,1]$ and Y be locally compact Hausdorff. Assume that $f, g \colon Y \to X$ are homotopic maps, i.e., there is a homotopy $F \colon I \times Y \to X$, that is a continuous function such that $F(0,-) = f$ and $F(1,-) = g$. By the previous results, the associate map $F^{\#} \colon I \to X^Y$ is continuous. But this latter map is precisely a *path* in X^Y from f to g. Conversely, every path in X^Y determines a homotopy. Using the terminology of path components introduced above, this amounts to the observation that the homotopy classes are the path components of X^Y, that is $[Y,X] = \pi_0(X^Y)$. This is Hurewicz's starting point in his first paper.

For every $n \geq 0$, let $p_n = (1,0,\ldots,0)$ be the base point of S^n. Then, for every space (X,x_0) and every $n \geq 0$, the nth homotopy group is defined as:

$$\pi_n(X,x_0) = [(S^n, p_n), (X,x_0)]$$

Hurewicz then proved a series of important results that launched homotopy theory. For instance, he showed that:

1. $\pi_n(X,x_0)$ is independent of the base point x_0, and one can therefore write $\pi_n(X)$;
2. For $n \geq 2$, $\pi_n(X)$ is Abelian;
3. $\pi_n(S^3) = \pi_n(S^2)$ and in particular $\pi_3(S^2)$ is infinite cyclic. These last results are in fact corollaries of more general results on topological groups;
4. For $n \geq 2$, if the first $n-1$ homotopy groups of X vanish, i.e., $\pi_n(X) = 1$, then Hurewicz defined a homomorphism $h \colon \pi_n(X) \to H_n(X)$, where $H_n(X)$ is the nth homology group of X, and showed that it is an isomorphism.[13]

The last result is particularly important, it is known as the "Hurewicz isomorphism theorem", since it connects homotopy groups to homology groups. The connection is far from obvious since homotopy groups are intrinsically geometrical in their construction whereas homology groups, based on the idea of approximating a space by simpler spaces and using the latter to construct the homology groups, are more "destructive". Furthermore, and this is also crucial, Hurewicz theorem opens the possibility of computing homology groups by using homotopy groups and vice-versa.

Computing these homotopy groups turned out to be extremely difficult, and despite important developments in the understanding of relationships between homotopy groups, homology groups, and cohomology groups in the thirties and forties,[14] the next breakthrough with respect to the computation of homotopy groups had to wait for Serre's thesis in 1950, published in 1951. ([248]) This paper marked the beginning of a new development of the theory; Kan would refer to it explicitly in the

---

[13] Hurewicz used what is called integral homology, defined by Vietoris.

[14] Eilenberg, once again, played an important role in connecting cohomology and homotopy groups in his 1940 paper ([66]) and, later, by developing singular homology in a paper written with Zilber to which Kan refers explicitly in many of his papers. See [81].

bibliography of his first paper, published in 1955, and in his paper on adjoint functors. But surprisingly, perhaps, the mathematicians who inspired Kan were Eilenberg and Steenrod (via their book on the foundations of algebraic topology) and, to go by the references in his papers, Eilenberg and Mac Lane and Eilenberg and Zilber. In fact, Eilenberg and Mac Lane themselves contributed significantly to the study of the relations between homology and homotopy groups, papers on which Kan was about to build.

It is interesting to note how Eilenberg and Mac Lane used categories themselves in their collaboration that started in 1940 and ended in 1954. We have seen earlier how Eilenberg used them in his collaborations with Steenrod and Cartan and how Mac Lane used them in his paper on duality for groups. Eilenberg and Mac Lane started their collaboration on the relations between homology and homotopy groups at about the same time they were developing their ideas on categories, functors and natural transformations. Thus, in their paper entitled "Relations between homology and homotopy groups", published in 1943 (in which they introduced the spaces which were about to be named after them, the Eilenberg-Mac Lane spaces), there was no reference to categories, functors or natural transformations. In the following year, Eilenberg published a very important and influential paper in which he introduced what is now the standard version of singular homology theory. (See [68].) Using the latter theory, Eilenberg and Mac Lane wrote a second, long paper on the relations between homology and homotopy groups of spaces in which there is a very short reference to categories: "The concept of naturality is used here in a vague sense, which, however, could be made quite rigorous using the theory of functors." ([75], 502.) Thus, categories could have been used, but merely to give a precise and rigorous definition of natural isomorphism. Then Eilenberg wrote a paper entitled "Homology of spaces with operators", ([69]) quickly followed by "Homology of spaces with operators. II", ([76]) written with Mac Lane this time, presented in 1948 and published in 1949. Once more, there is a very brief reference to their work on categories when a kernel of a homomorphism is said to be a contravariant functor and a certain homomorphism is natural. ([76], 69.) Their last paper on the subject, entitled "Relations between homology and homotopy groups of spaces. II" (1950) ([77]), which is in a way the culmination of their efforts, does not mention categories, functors nor natural transformations at all. That paper is based on work done by Eilenberg and Zilber, also published in 1950, in fact in the same volume of the *Annals of Mathematics*, and both papers are referred to by Kan in 1955 and later.

However, Eilenberg and Mac Lane were certainly aware that category theory could be used in homotopy theory in a systematic fashion. Indeed, in chapter V of their 1945 paper on categories, Eilenberg and Mac Lane consider applications of categories to algebraic topology. They defined homology and cohomology groups as functors, showed that certain well-known isomorphisms are natural isomorphisms and present Čech homology theory in terms of functors. The chapter ends with a remark on homotopy theory:

> The process of setting up the various topological invariants as functors will require the construction of many categories. For instance, if we wish to discuss the so-called relative homotopy theory, we shall need the category $\mathfrak{X}_S$ whose objects are the pairs (X,A), where

4.1 The Background: Homotopy Theory and Category Theory

X is a topological space and A is a subset of X. A mapping

$$i\colon (X,A) \to (Y,B) \text{ in } \mathfrak{X}_S$$

is a continuous mapping $i\colon X \to Y$ such that $i(A) \subset B$. The category $\mathfrak{X}$ may be regarded as the subcategory of $\mathfrak{X}_S$, determined by the pairs (X,A) with $A = 0$.

Another subcategory of $\mathfrak{X}_S$ is the category $\mathfrak{X}_b$ defined by the pairs (X,A) in which the set A consists of a single point, called the base point. This category $\mathfrak{X}_b$ would be used in a functorial treatment of the fundamental group and of the homotopy groups. ([74], 292.)

The first mathematician to take up this proposal was Hu in 1947, in a paper entitled "An exposition of the relative homotopy theory" ([113]), but in an odd way. After the note by Eilenberg and Steenrod on an axiomatic approach to homology theory, it is the second published paper after the original paper by Eilenberg and Mac Lane in which categories are applied. However, as the title indicates, it is an exposition of the main results known at that time on relative homotopy theory (results mostly due to Hurewicz and Steenrod, but that had not been published systematically) and the exposition is *not* done via categories. Thus, the paper does not open with a definition of the categories involved and a presentation of various results in the framework of functors and natural transformations. In fact, categories are used in that paper in an awkward fashion. The homotopy groups are *not* defined as functors between categories.[15] Furthermore, the term "category" is used only for groups and homomorphisms. Other categories are called "general categories". Hu also introduces "local categories", which are in fact groupoids, but the latter notion, although defined, was not used at the time. Hu's paper had no effect on subsequent work.[16]

But Eilenberg and Mac Lane did finally use categories in a more systematic way, first in 1951, where various constructions in certain categories are elaborated, and then in a paper from 1952 (published in 1953) entitled "Acyclic Models". Here is how they present their application:

> There are a number of situations in algebraic topology where one establishes the existence of chain transformations and chain homotopies, dimension by dimension, using the fact that certain homology groups of a local character are zero. Most of the applications can be derived from well known theorems dealing with acyclic carriers. Other investigations (...) lead to similar proofs in situations no longer covered by theorems on acyclic carriers. The present paper formulates a general theorem, which seems to subsume all the situations of this type hitherto encountered. The theorem is formulated in the language of categories and functors. ([78], 189.)[17]

From the point of view of the history of category theory, this paper is interesting in its own right. As the foregoing quote indicates, Eilenberg and Mac Lane are us-

---

[15] In fact, even as late as 1962 Peter Hilton presents the fundamental group as a functor and has to state explicitly the advantages of doing so in an expository paper. See [112].

[16] Hu published the first textbook on homotopy theory in 1959, but there is no mention of categories, functors, etc. therein. (I do not take the notes published by Hilton in 1953 to constitute the first textbook.)

[17] Interestingly enough, in that paper Eilenberg and Mac Lane define a notion of *representable* functor. But that notion has nothing to do with the notion we considered earlier. It is restricted to functors with values in the category of Abelian groups and a functor is representable in their sense if it is related by a natural isomorphism to certain free Abelian groups in a certain way.

ing categories and functors to prove a general theorem. Thus the general set up is provided by category theory: they consider functors between an arbitrary category $\mathscr{A}$, in which a set M of objects, called the model objects, is given, and $\partial\mathbf{AbGrp}$ the category of chain complexes and chain transformations, where **AbGrp** is the category of Abelian groups. They prove three general theorems in which the category $\mathscr{A}$ is left unspecified and only properties of the functors (satisfying a certain condition) and the category $\partial\mathbf{AbGrp}$ are used. In the remaining sections of the paper, they specify a specific category for $\mathscr{A}$ and a set of models M and in this way their general theorems become known special cases. The main and original result of the paper, theorem V, establishes the equivalence between *simplicial* singular homology and *cubical* singular homology. Thus, Eilenberg and Mac Lane develop a general framework and a general method. The method is then used immediately by Eilenberg, in collaboration with Zilber, to prove an important theorem, known as the Eilenberg & Zilber Theorem, on two products of complexes. Here is how Massey comments on this theorem in his book:

> Rather than go through the details of these lengthy calculations, it seems preferable to use a more conceptual method due to Eilenberg and Mac Lane, called the *method of acyclic models*. This method makes strong use of the naturality of the chain maps ζ and η which we have defined. By making full use of this naturality, it is possible to avoid the necessity of having explicit formulas. ([215], 138.)

Once more, natural transformations are not simply presented as such, they are *used* to provide a more *conceptual* proof of a result.

All the foregoing elements constitute the background to Kan's discovery. This is not to say that this is what Kan knew about these matters nor that we know that this is what he had in mind when he was working in homotopy theory. Furthermore, there is much more to the story than what I have included, especially in homotopy theory. That being said, two elements have to be kept in mind:

1. In homotopy theory, one finds specific relations between certain concepts, even for the concept of homotopy itself. As we have seen, a homotopy can be *defined* either as a continuous map
$$F : X \times I \to Y$$
or as a continuous map
$$F^{\#} : I \to Y^X.$$
This is just one case of a more general phenomenon.
2. Category theory was starting to be used in homotopy theory in the early fifties, but not systematically. We have already seen how Eilenberg and Steenrod's book revolutionized the way to do algebraic topology. Students who learned algebraic topology from that book — and so at the same time learned category theory — would naturally think in terms of categories and functors. If, moreover, these students were in an environment where this type of thinking is not discouraged, for reasons of fashion or any other reasons, they might be more attentive to categorical aspects of various problems and situations. This is precisely the situation in which Kan found himself.

4.2 Kan's Discovery

Kan does refer to the paper on acyclic models in his first note published in 1955. However, the reference has nothing to do with acyclic models per se, but only with a certain functor having values in the category of chain complexes. There is, of course, some irony in the fact that Kan refers to many papers written by Eilenberg and Mac Lane. For Kan discovered adjoint functors by working on problems in algebraic topology, taking up work done by Eilenberg, Mac Lane, Steenrod and Zilber, and by pushing that work in a certain direction by applying systematically categorical notions to the various problems he was interested in. It is, from that perspective, rather surprising to see Mac Lane trying to explain how adjoint functors could have been discovered in functional analysis — a context in which Eilenberg and he were not intensely involved, although they certainly knew about it — when in fact adjoint functors were discovered in homotopy theory, a field in which Eilenberg and he made important contributions, contributions that were in the background of the discovery. But again, adjoint functors might have come from functional analysis, or algebraic geometry or simply pure algebra. In each case, a story can be told.

Kan really starts off from where Eilenberg and Mac Lane had left off. There is an obvious and clear progression in Kan's work from 1955 until 1958 (in fact, from 1954 until 1956, which are the years during which this work was done). The climax of this progression is undoubtedly the paper published in the *Transactions* in 1958 which *follows* the paper on adjoint functors.[18] That paper, "Functors Involving C.S.S. Complexes", is truly remarkable. Not only is category theory used in that paper in an entirely original and powerful manner (and in this respect adjoint functors occupy center stage, with some new concepts introduced for the first time in full maturity for they are still defined that way today), but it also constitutes a beautiful conceptual synthesis of all the work Kan had done. All the important results that Kan had developed in his previous notes and paper are presented in a *unified* manner. Furthermore, Hurewicz homomorphisms are *derived* from a certain functor introduced by Kan. There is no doubt that the role played by adjoint functors in that paper justified to a large extent the writing and publication of the paper on adjoint functors. But we are getting ahead of ourselves.

We will not review Kan's first four papers in detail. I have already mentioned that Kan defines various categories in his first paper but does not use categorical concepts or methods therein. The second and the third papers are of special interest to us. The second paper, "Abstract Homotopy II", in which Kan introduced what would soon be named "Kan complexes" by other mathematicians,[19] clearly shows that Kan was following Eilenberg and Steenrod in the way he used categories at

---

[18] As we will see, Kan had all these results in 1956. These papers were reviewed only in 1962! It took four years for someone, in this case Puppe, to write reviews of the two papers in the *Transactions* and they were written in German! Puppe was probably one of the very few mathematicians who could understand the categorical framework developed by Kan. This is another indication of the qualitative change these papers constitute in the field.

[19] Already in 1958, in his review of Kan's "Abstract Homotopy I, II, III, IV" published in 1957, Moore called them "Kan complexes".

that time. In fact, Kan learned category theory from Eilenberg and Steenrod's book on the foundations of algebraic topology.[20] Kan identifies some of the main categorical properties required by a category to be used in homotopy theory and gives a purely abstract definition of a category with homotopy and for two mappings to be homotopic in that context. Using a construction he had introduced in his first paper, the category of cubical complexes, Kan defines a family of functors from a category with homotopy to the category of cubical complexes which, in turn, allows him to define in a purely abstract manner what is called the cubical singular functor and to state its most important properties. In the remaining sections of the paper, Kan considers specific cases of categories with homotopy and state that in each case the abstract definitions become the standard definitions, in particular for the category ∂**AbGrp** of chain complexes. The analogy with Eilenberg and Steenrod and Eilenberg and Mac Lane's paper on acyclic models is striking. Recall that in their book, Eilenberg and Steenrod used categories as an abstract framework to define various homology and cohomology theories. They started off with what they called "admissible categories", which are simply categories of pairs of spaces and maps satisfying certain conditions, but they ended up in chapter IV with what they called "h-categories" which are abstract categories having specific properties. They then showed in the remaining chapters that specific categories are h-categories and that therefore certain general results apply immediately to various specific homology and cohomology theories. The way Kan uses categories in this paper is similar to Eilenberg and Steenrod and Eilenberg and Mac Lane's usage.

The next paper, "Abstract Homotopy III", communicated in April 1956, marks the appearance of the first specific case of an adjoint pair, although still not named as such by Kan. The date is important, since it is clear that it is during that period that Kan discovered the notion of adjoint functors. Indeed, it appears in a paper following the previous note, the paper "On C.S.S. Complexes", under the name "Adjoint Functors" and that paper was received by the *American Journal of Mathematics* in September 1956. The two other papers involving adjoint functors, both published in the *Transactions* in 1958, were also received in September 1956. It therefore seems reasonable to believe that Kan discovered adjoint functors in the winter of 1956 and spoke to Eilenberg about the notion after April 1956.[21]

In the winter of 1956, Kan was attending a seminar given by Eilenberg in homological algebra.[22] In one of these seminars, Eilenberg introduced one of the key constructions of homological algebra (and algebraic topology in general), the tensor products of two Abelian groups. In order to *define* the tensor product of A and B, Eilenberg *used* the bijection already indicated in the joint papers written with Mac Lane in 1942,[23] and to which we will return shortly:

$$\mathrm{Hom}(A \otimes B, C) \to \mathrm{Hom}(A, \mathrm{Hom}(B, C))$$

---

[20] Kan, personal communication, February 2003.

[21] This is indeed confirmed by Kan himself. (Kan, personal communication, February 2003.)

[22] Kan, personal communication, February 2003. This is more or less how Kan recalls the story.

[23] Indeed, in the paper on group extensions and homology, this approach is explicitly presented in its most general form. See [71], 788.

4.2 Kan's Discovery

Upon seeing this, Kan wondered whether one could not just as well define the right-hand side using the left-hand side instead. He asked Eilenberg, who answered that he did not think that it could be done. Kan left the seminar thinking about the possibility and worked on the problem for some time. He then discovered that it could be done. There is therefore a perfect symmetry or duality between the two concepts: it is possible to *define* one concept with the other, given the proper connection between them. Furthermore, Kan proved that they determine each other up to a unique isomorphism. Once Kan understood this relationship, he saw that he already knew many other cases in his own work that could be treated similarly and he proceeded to look at them. This led first to an observation in the third paper on abstract homotopy theory submitted in April 1956 and to the three papers written in the spring and summer of 1956 and published in 1958.

Thus, in the paper "Abstract Homotopy III", Kan considers what are called complete semi-simplicial complexes, or c.s.s. complexes, and basically transfers the results he had obtained in his first paper on cubical complexes to c.s.s. complexes. Towards the end of the paper, in §6.1, we read:

> 6.1 Let $K, L \in S$. For every integer $n > 0$, there exists (in a natural way) a one-to-one correspondence between the c.s.s. maps $Sd^n K \to L$ and the c.s.s. maps $K \to Ex^n L$.

This is the first occurrence of an adjunction in Kan's work. It appears once again in the sequel to that note, a paper published in 1957 in which the results announced in 1956 are proved explicitly. In the 1957 paper, the fact that this result is a special case of adjointness is mentioned and there is a reference to the forthcoming paper on adjoint functors.

> The main tool used in the definition of the functor Ex^∞ is what we call the *extension* $Ex K$ of a c.s.s. complex K, which is in a certain sense dual to the *subdivision* $Sd K$ of K. More precisely: let K and L be c.s.s. complexes, then there exists (in a natural way) a one-to-one correspondence between the c.s.s. maps $Sd K \to L$ and the c.s.s. maps $K \to Ex L$. In the terminology of [Abstract functors] this means that the functor Ex is a right adjoint of the functor Sd. ([133], 450.)

What Kan has in his hands now is a pair of functors, Ex and Sd, *going in opposite directions* and related to one another in a systematic manner. Kan says that they are *dual* to one another and this can be interpreted in two compatible ways: first, one is on the left of the arrow whereas the other is on the right; thus although the direction of the arrow is not reversed, they stand in opposition in a certain way; second, the operations themselves, extension and subdivision, are in a certain manner dual in the sense that one does the opposite of the other. Thus, they are *conceptual* "inverses" although they are not inverses in the usual mathematical sense of the expression. When Kan comes to the proof of the one-to-one correspondence, he claims that the result is an immediate consequence of his paper "Functors Involving C.S.S. Complexes" or that it can be verified by a straightforward computation. (See [133], 460.) It is not necessary for my purposes to look at the details of this particular construction. I will however briefly present some of the definitions and a crucial aspect of the paper on functors involving c.s.s. complexes, for it contains not only some of the ingredients that allow us to see how Kan thought of adjoint functors but also a novel way of defining mathematical concepts and proving theorems.

Kan starts his paper with a definition that is now standard but which was clearly a conceptual leap at that time (although a small step by contemporary standards).[24]

Definition 4.6. For each integer $n \geq 0$, let $[n]$ denote the ordered set $\langle 0,\ldots,n \rangle$. A function
$$\alpha \colon [m] \to [n]$$
is said to be *monotone* if $\alpha(i) \leq \alpha(j), 0 \leq i \leq j \leq m$.

It can easily be seen that the ordered sets together with monotone functions between them form a category. It is nowadays called the *simplicial category* and is usually denoted by Δ. The simplicial category now plays a very important role in mathematics in general.

The following concept, although originally defined by Eilenberg and Zilber in 1950 and mentioned earlier, is also given a categorical definition by Kan. It should be mentioned that it is this concept that occupied Kan in his early research.

Definition 4.7. [25] A *complete semi-simplicial (c.s.s.) complex* K is a *contravariant functor* $K \colon \Delta^\circ \to \mathbf{Set}$, i.e., an object of $\mathbf{Set}^{\Delta^\circ}$. A *c.s.s. map* $f \colon K \to L$ is a natural transformation from K to L, i.e., a map of the category $\mathbf{Set}^{\Delta^\circ}$.

The elements of the set are called *n-simplices* of K. Note that the category of c.s.s. complexes and c.s.s. maps is denoted by $\mathscr{S}$. What Kan called a c.s.s. complex is now simply called a *simplicial set*.

The step taken by Kan in defining c.s.s. complexes this way is important for two reasons. First, a c.s.s. complex becomes a *functor*. Although it seemed natural to describe homology and cohomology theories in terms of functors, defining c.s.s. complexes as functors is a conceptual leap, for one has to identify the simplicial category first and then *see* that a contravariant functor from the simplicial category to the category of sets *is* just a c.s.s. complex. To *see* this is not entirely trivial. For the non-categorical definition of a simplicial set is:

Definition 4.8. A *simplicial set* K consists of a collection of sets $\{K_n\}_{n \geq 0}$ together with set maps $d_i \colon K_n \to K_{n-1}$ for $i = 0,\ldots,n$ and $s_j \colon K_n \to K_{n+1}$ for $j = 0,\ldots,n$ satisfying the simplicial identities:

$$d_i d_j = d_{j-1} d_i \quad \text{for } i < j;$$
$$s_i s_j = s_{j+1} s_i \quad \text{for } i \leq j;$$
$$d_i s_j = s_{j-1} d_i \quad \text{for } i < j;$$
$$d_i s_j = 1_{K_n} \quad \text{for } i = j, j+1;$$
$$d_i s_j = s_j d_{i-1} \quad \text{for } i > j+1.$$

---

[24] All the data appear in [81], in §8. However, they do not say explicitly that the data form a category. Clearly, Eilenberg knew that they did but refrained from making it explicit.

[25] In his seminar given in December 1956, Henri Cartan presented some of Kan's results, in particular categories with a simplicial structure. The references given by Cartan are "Abstract Homotopy III", "Abstract Homotopy IV" and some "secret papers" of Kan!

4.2 Kan's Discovery

The maps d_i are called *face maps* and the maps s_j are called *degeneracy maps*.

In order to see that these definitions are the same, one has to prove certain properties of monotone functions and derive from those properties the foregoing identities. (See [195], 172–173.)

Second, there is nothing sacred about the category **Set**. Generalizing, Kan replaces the category **Set** with various other categories and defines what we now call a *simplicial object* X in each case. Thus, Kan considers ([135], 334) a special *covariant* functor

$$\Sigma \colon \Delta \to \textbf{Top}$$

defined by: for each $[n]$, $\Sigma[n]$ is a *(Euclidean) n-simplex* with ordered vertices $p_0, p_1, \ldots, p_n$. For each map $\alpha \colon [n] \to [m]$, $\Sigma\alpha \colon \Sigma[n] \to \Sigma[m]$ is a simplicial map defined by $(\Sigma\alpha)(p_i) = p_{\alpha(i)}$.

It can be seen that this amounts to the usual definition of a (Euclidean) n-simplex and simplicial map, but it is much more compact and "abstract" than the usual definition. All the work is done by the categorical structure and the functorial properties. Indeed, a (Euclidean) n-simplex is, once again, a *functor*. Kan proceeds to do the same with *standard n-simplexes* which are defined as functors in the category of simplicial sets,[26] *c.s.s. groups*, which are covariant functors $G \colon \Delta \to \textbf{Grp}$, and *c.s.s. Abelian groups* which now take values in the category **AbGrp** of Abelian groups. C.s.s. groups had been the topic of his paper "Abstract Homotopy IV" and were there defined directly.

Now the definitions all had a *unified* presentation, which immediately indicated what these objects have in common. If Eilenberg and Mac Lane had seen how categories and functors could be used to give a uniform definition of natural equivalences, and a uniform proof for various results in homology theory in their paper on acyclic models, if Eilenberg and Steenrod had found a unified treatment of various homology and cohomology theories and Cartan and Eilenberg did the same for various applications of homological methods in algebra, Kan discovered how to define in a uniform manner various constructions with the categorical language. One can only be stunned by the beauty and simplicity of the approach when it is contrasted with the earlier papers on the subject, including Kan's own prior publications.

But what is even more striking about the paper "On Functors involving c.s.s. complexes" is that the *method of proof* is also unified via the concept of adjoint functor. This aspect is certainly crucial, for it must have convinced Kan of the importance of the concept *in general*, that is, outside homotopy theory. The general strategy calls to mind Eilenberg and Steenrod again. Kan considers a general case, proves a result for that situation and then chooses specific parts involved in this situation to obtain standard results. By doing so, once again, Kan shows what various categories have in common: the category of topological spaces, the category of c.s.s. complexes, the category of chain complexes, the category of c.s.s. groups. It is at the categorical level and with the help of adjoint functors that one can exhibit precisely what these objects share. However, he does more: the general situation allows him not only to recover known results in a unified and systematic way, but also to prove

---

[26] The definition is a little more tricky in this case.

and make new connections between known concepts. Furthermore, the general case is presented in a certain format that is directly connected to the discovery of adjoint functors: Kan can see that the general case already covers important aspects of homotopy theory and allows a vast generalization of concepts and results. The key is the *form* of the presentation: there is a pattern at work.

First, a general fact presented and proved in the paper on adjoint functors. Let

$$\mathrm{Hom}(-,-)\colon \mathscr{C} \times \mathscr{C} \to \mathbf{Set}$$

be the standard Hom-functor. Let $\mathscr{D}$ be a small category, Kan defines what he calls the *lifted functor*[27]

$$\mathrm{Hom}^{\mathscr{D}}(-,-)\colon \mathscr{C}^{\mathscr{D}^\circ} \times \mathscr{C} \to \mathbf{Set}^{\mathscr{D}^\circ}$$

defined in the obvious manner, i.e., for all X, Y in $\mathscr{C}$, $f\colon X \to Y$ and $F\colon \mathscr{D}^\circ \to \mathscr{C}$, the assignments are given by the following:

$$\begin{array}{ccc} X & \longmapsto & \mathrm{Hom}^{\mathscr{D}}(F(-),X) \\ f \downarrow & \longmapsto & \downarrow \mathrm{Hom}^{\mathscr{D}}(F(-),X)(f) \\ Y & \longmapsto & \mathrm{Hom}^{\mathscr{D}}(F(-),Y) \end{array}$$

Notice that both the functor Hom and the lifted functor $\mathrm{Hom}^{\mathscr{D}}$ yield, in a sense, function spaces, i.e., sets of maps between certain objects. As we have already seen above, function spaces are naturally related to products in topology and homotopy theory. This is the key here: there is a product which will be related to the foregoing functor in the same manner. There is a tensor product functor

$$- \otimes -\colon \mathbf{Set}^{\mathscr{D}^\circ} \times \mathscr{C}^{\mathscr{D}} \to \mathscr{C}$$

whose exact definition we will leave aside, since it is rather involved and is given again in the paper on adjoint functors. It is clearly not a "product" in the set-theoretical sense of that expression! In fact, as it can be seen from its domain, it is a product of *functors* which yields an object in the category $\mathscr{C}$. Notice also that the domain of the tensor functor $\otimes$ is itself a product of functor categories. One might wonder where this object "lives". But there is not a single word about foundational questions in Kan's papers!

Using these constructions, Kan shows that if a category $\mathscr{C}$ satisfies a very weak condition (i.e., it has small limits), then for any fixed covariant functor

$$\Sigma\colon \mathscr{D} \to \mathscr{C}$$

the functors $-\otimes -$ and $\mathrm{Hom}^{\mathscr{D}}(-,-)$ yield a pair of functors

---

[27] The definition is entirely general in the paper. I present here a special case. These are now called *Kan extensions*. See [198], chap. X.

4.2 Kan's Discovery

$$-\otimes \Sigma \colon \mathbf{Set}^{\mathscr{D}^{\circ}} \to \mathscr{C} \quad \text{and} \quad \mathrm{Hom}^{\mathscr{D}}(\Sigma,-)\colon \mathscr{C} \to \mathbf{Set}$$

such that there is a natural bijection

$$\alpha\colon \mathrm{Hom}(F \times \Sigma, X) \to \mathrm{Hom}(F, \mathrm{Hom}^{\mathscr{D}}(\Sigma, X)),$$

where $F\colon \mathscr{D}^{\circ} \to \mathbf{Set}$ and X is an object of $\mathscr{C}$.

These two functors and the natural bijection between them are then applied to different contexts to yield fundamental results of homotopy theory. In the words of Kan himself: "By a suitable choice of the category $\mathscr{C}$ and the object $\Sigma \in \mathscr{C}^{\mathscr{D}}$ the above functors and natural transformations reduce to well known ones."[28] ([135], 332) In fact, in Kan's paper, the category $\mathscr{D}$ is already chosen to be Δ, the simplicial category, but the construction, as he himself notes, is entirely general. Kan then considers (§4) the case when $\mathscr{C}$ is the category **Top** of topological spaces and Σ an Euclidean n-simplex. In this case, we have that:

1. For every topological space X, $\mathrm{Hom}^{\Delta}(\Sigma, X)$ amounts to what is known as the simplicial singular complex of X;
2. For every c.s.s. complex K, $K \otimes \Sigma$ amounts to what is known as its geometric realization;
3. The natural bijection $\alpha\colon \mathrm{Hom}_{\mathbf{Top}}(K \otimes \Sigma, X) \to \mathrm{Hom}_{\mathscr{S}}(K, \mathrm{Hom}^{\Delta}(\Sigma, X))$ expresses the fact that there is a close connection between c.s.s. maps and continuous maps.

Then Kan does the same for the category of c.s.s. complexes and derives the exponential law for c.s.s. complexes, i.e., if K, L and M are c.s.s. complexes, then there is a natural isomorphism:

$$M^{K \times L} \approx (M^L)^K.$$

The result about the functors Sd and Ex mentioned above would appear at this juncture. It should be noted that even these have basically the same form. Kan then moves to Chain complexes and defines a special case of Hom^{Δ} which plays a key role in the proofs of the remaining sections. After the introduction of Kan complexes and the definition of higher homotopy groups for the latter, Kan considers the category of c.s.s. groups and of c.s.s. Abelian groups and shows that for Kan complexes, the homotopy groups are identical with the homology groups for every integer greater than or equal to zero. Using the natural bijection between the specified functors, Kan proves various properties of the functor Hom^{Δ}, and shows that the Hurewicz homomorphisms can be defined from it, as well as the Eilenberg-MacLane complexes. In other words, a vast number of crucial constructions and results follow directly from the general case and result, exhibiting the usefulness and the power of the concept of adjunction for homotopy theory.

For us, what matters more is the *form* of the relationship between the two functors. It has the same form as the one seen above between the tensor product of two

---

[28] Kan does not denote his categories by "$\mathscr{C}$", but by "$\mathscr{Z}$" and the functor category by "$\mathscr{Z}_V$".

Abelian groups and the Hom-functor *and* on product and function spaces. Thus, what Kan has at his disposal is an important number of cases that have the same fundamental form: they can all be expressed as a relationship between a product functor and a function space functor. The next step is to realize that there is nothing *specific* about the fact that it is a *product* functor and a *function space* functor and thus, that the key ingredient has to do with the fact that it is simply two functors that are related to one another in a certain manner. At this point, we are ready to look at adjoint functors.

4.3 Kan's 1958 Papers "Adjoint Functors"

> At any rate, we [Eilenberg & Mac Lane] did think that it [category theory] was good, and that it provided a handy language to be used by topologists and others, and that it also offered a conceptual view of parts of mathematics, in some way analogous to Felix Klein's "Erlanger Program". We did not then regard it as a field for further research effort, but just as a language and an orientation — a limitation which we followed for a dozen years or so, till the advent of adjoint functors.
>
> ([196], 334–335.)

What is striking in this quote is that Mac Lane does not see that the introduction of adjoint functors by Kan not only made category theory "a field for further research effort", but that it also reinforced the fact that category theory "offered a conceptual view of parts of mathematics, in some way analogous to Felix Klein's 'Erlanger Program'."

Kan's paper on adjoint functors is very different from [186], [36] and [103] and, in fact, from all the papers published from 1945 in which category theory played a role. What is striking about it is that it is a paper on pure category theory as such and, I speculate, probably the very first such paper after Eilenberg and Mac Lane's 1945 paper. One could argue that *it* marks the birth of pure category theory. Categories, functors and natural transformations are not seen as providing the proper framework to understand a problem in homology theory, to develop a framework to apply notions of homology theory to algebra, or in order to develop an analogy between aspects of homological algebra and the cohomology of sheaves of groups over a topological space. We now leave behind the idea that category theory offers a useful language, or is merely a useful tool. In Kan's paper, categories are taken as such and the purpose of the paper is to define and develop an intriguing dependence between functors.

Kan's paper can be read as a note or a fundamental improvement on Eilenberg and Mac Lane's paper. It should be noted, however, that Kan had not read Eilenberg and Mac Lane's 1945 paper before he wrote his paper on adjoint functors, but read it carefully while writing it.[29] Both papers are about natural equivalences in some

---

[29] Kan, personal communication, February 2003.

sense. At first sight, Eilenberg and MacLane's paper seems to be more important, since it develops a *general* theory of natural equivalences whereas Kan develops specific cases of natural equivalences. It is *not* at first obvious that the type of natural equivalences brought to the fore by Kan constitute the core of category theory. In order to see it as such, one has to shift considerably from the perspective developed in Eilenberg and MacLane. One aspect of the shift is that whereas one looks *primarily* at natural equivalences (isomorphisms) in Eilenberg and MacLane — they constitute what the paper is *about* — in Kan's paper, natural equivalences are secondary and *functors* constitute the object of the paper. This shift was certainly "in the air" when Kan was thinking about these notions, but it was mainly in the applications of the theory and not in the theory as such. Eilenberg and Steenrod, Cartan and Eilenberg, Grothendieck and others all had applied the categorical language to various fields and in these applications, the objects of attention are functors (although in different ways in Eilenberg and Steenrod, and Cartan and Eilenberg on the one hand and in Grothendieck on the other hand). In Eilenberg and MacLane's paper, functors constitute an organizing tool that allows a systematic and unified treatment of the notion of natural transformation. In Kan's paper, natural transformations are *used* to characterize a certain class of functors.

In the paper on adjoint functors, Eilenberg and MacLane's 1945 paper is the most-cited reference. Kan refers to various notions taken from Eilenberg and MacLane's paper, and comments and improves on various sections of the paper. Kan saw these connections while writing his paper. He started out from the various cases he knew about and intended to write a note that would serve as a preliminary step to his paper on functors involving c.s.s. complexes. In the process of writing the paper itself, Kan made additional discoveries, e.g., the relations between adjoint functors and limits and colimits, and the paper became a long exposition of the concept and of its main properties.[30]

The bibliography is surprisingly short. It contains six references. In addition to Eilenberg and MacLane's paper, one reference is made to Cartan and Eilenberg ([43]) towards the end of the paper and simply because Kan uses some of the definitions given in their book; there are three references to [80] in various examples involving the group of homomorphisms $Hom(G, H)$, the tensor product of two Abelian groups and the direct sum of groups;[31] there are two references to specific but fundamental texts in homotopy theory: a paper by James ([116]) and Serre's thesis (both are referred to in the same example); and of course there is one reference to the Kan's own aforementioned paper in the same issue of the *Transactions*.

Nothing in the paper indicates that Kan saw how fundamental his "observation" was or its *general* implications. He was certainly aware, as we have seen above, of the unifying power and the usefulness of the notion in organizing proofs, at least in homotopy theory. If the MacLane-Buchsbaum-Heller-Grothendieck line of

---

[30] Kan, personal communication, February 2003.

[31] It should be noted at this point that tensor products and Hom occupy an important place in Eilenberg and Steenrod's book. §9 of chap. V is devoted to the tensor product and §10 to groups of homomorphisms. Exercise F of that chapter asks the reader to show that there is a natural isomorphism between $Hom(A \otimes B, C)$ and $Hom(A, Hom(B, C))$.

development indicated how category theory could be applied to algebraic topology and homological algebra, Kan's work eventually led to the discovery that: (1) an unexpectedly large portion of mathematical constructions, theorems and theories can be written in terms of adjoint functors; and (2) the language of adjoint functors could be used to characterize, develop and systematize various areas of mathematics, from logic and universal algebra to algebraic topology, algebraic geometry and, nowadays, theoretical computer science and mathematical physics. But as Mac Lane observed, "the idea of adjunctions took on slowly." ([195], 103.)[32]

Kan's paper on adjoint functors contains four sections: an introduction and three chapters. In the first chapter, Kan presents the basic definitions and properties of adjoint functors together with examples. In the second chapter, Kan defines limits and colimits in full generality and considers their relationships with adjoints. Finally, the third chapter deals with constructions first expounded in Eilenberg and Mac Lane's paper — namely the "lifting principle" in the section of the 1945 paper preceding the applications — develops them in the context of adjoints (this is where Kan extensions are introduced) and looks at various applications. Kan's paper deserves to be looked at in more detail.

Kan's introduction is revealing. It opens with the example that triggered his discovery of the concept. It is taken from homology theory, and it was first presented by Eilenberg and Mac Lane and mentioned in their 1942 note as "a less obvious relation between the tensor product and the functor 'Hom' ". As we have seen, the relation is that there is a natural isomorphism (Kan speaks about a natural equivalence) of the form (this is Kan's notation):

$$\alpha \colon \mathrm{Hom}(\otimes,) \to \mathrm{Hom}(, \mathrm{Hom}(,))$$

Kan points out that the Hom-functor and the tensor functor are not independent and he adds that "there exists a kind of duality between the tensor product and the last functor Hom, while both functors Hom outside the parentheses play a secondary role." ([134], 294.) The reference to the duality shows up again. But in this case one wonders whether the choice of terminology could not refer to the situation one finds in linear algebra where adjoint operators are defined in the context of the relations between linear transformations and the *dual* space of a vector space.

The foregoing example goes back to papers published *before* the 1945 paper. However, Eilenberg and Mac Lane themselves presented it as being "less obvious" than other examples of isomorphisms between functors. The example appears in their 1942 papers and is mentioned again in the 1945 paper, but very briefly and the reader is referred to the 1942 papers for the proof of the isomorphism. It is interesting to note that this example is the *only* example given by Eilenberg and Mac Lane in their 1942 paper that has the right form. *All* the other examples of isomorphisms of functors have a different presentation. For instance, their first example is the Pontrjagin duality for locally compact Abelian groups, which Eilenberg and

---

[32] Here one might certainly invoke sociological factors. There is little doubt that the notion of adjoint functors was considered to be too abstract or too formal by most mathematicians for quite a while.

4.3 Kan's 1958 Papers "Adjoint Functors"

Mac Lane present as establishing an isomorphism of the form:

$$G \cong \text{Ch}(\text{Ch}(G))$$

where G is a locally compact Abelian group and $\text{Ch}(G)$ denotes the character group of G. There is an adjoint situation here too, but it does not seem to have the same form as the one discovered by Kan. The same remark applies to the other examples, which involve various products of groups. The situation is essentially the same in the 1945 paper. It is now worth looking carefully at the examples of natural isomorphisms given in the 1945 paper, which contains the basic examples. We have already seen some of them in chapter 2. Here is the complete list:

1. If X, Y and Z are sets, topological spaces, groups or Banach spaces, then the laws of associativity and commutativity are all seen as being natural isomorphisms between functors:

$$(X \times Y) \times Z \cong X \times (Y \times Z)$$
$$X \times Y \cong Y \times X$$

2. Consider $\mathbb{Z}$ as an additive group, and let H be any topological Abelian group, then there is an isomorphism:

$$\text{Hom}(\mathbb{Z}, H) \cong H.$$

Similarly, for $\mathbb{R}$ considered as a Banach space, it can be shown that:

$$\text{Lin}(\mathbb{R}, B) \cong B,$$

where B is any Banach space and $\text{Lin}(-,-)$ is the set of all linear transformations.

3. For X, Y and Z topological spaces, there is a functor $\text{Map}(X, Y)$ defined as the function space Y^X with the compact-open topology. Then there is a natural isomorphism:

$$\text{Map}(Z, X) \times \text{Map}(Z, Y) \cong \text{Map}(Z, X \times Y)$$

Similarly, for groups and Banach spaces, there are analogous natural isomorphisms:

$$\text{Hom}(G, H) \times \text{Hom}(G, K) \cong \text{Hom}(G, H \times K)$$
$$\text{Lin}(B, C) \times \text{Lin}(B, D) \cong \text{Lin}(B, C \times D).$$

For groups, we also have that

$$\text{Hom}(G, K) \times \text{Hom}(H, K) \cong \text{Hom}(G \times H, K).$$

And if X, Y are locally compact Hausdorff spaces and Z any space, as we already know, we have:

$$\text{Map}(X \times Y, Z) \cong \text{Map}(X, \text{Map}(Y, Z)).$$

Then comes the example that triggered the whole story:

$$\text{Hom}(G \otimes H, K) \cong \text{Hom}(G, \text{Hom}(H, K)).$$

Finally, an example which involves the same kind of relationship, we have, for D a fixed Banach space and B, C two (variable) Banach spaces, and for linear transformations with norm at most 1:

$$\text{Lin}(B, C) \cong \text{Lin}(Lin(C, D), \text{Lin}(B, D)).$$

When D is $\mathbb{R}$, then $\text{Lin}(C, \mathbb{R})$ is called the conjugate space $\text{Conj}(C)$ of C. Hence the foregoing isomorphism becomes:

$$\text{Lin}(B, C) \cong \text{Lin}(\text{Conj}(C), \text{Conj}(B)).$$

Similarly, if G and H are locally compact Abelian groups, we have:

$$\text{Hom}(G, H) \cong \text{Hom}(\text{Ch}(H), \text{Ch}(G)),$$

where $\text{Ch}(H)$ is the character group of H.

4. Finally, we have, again for locally compact Abelian groups:

$$G \cong \text{Ch}(\text{Ch}(G)),$$

And similarly, for B a reflexive Banach spaces, we have:

$$B \cong \text{Conj}(\text{Conj}(B)).$$

And the list stops there.

What is striking about this list is the *variety* of cases. Of course, they can be grouped together as I have tried to group them, so that similar cases are shown side by side. But there is no overall structure that is immediately identifiable. Eilenberg and Mac Lane noticed various natural isomorphisms, but they did not see how these isomorphisms linked various functors together ("naturally"). Again, a shift in attention was required and since Eilenberg and Mac Lane were after the notion of natural isomorphism, it is unlikely that they could have noticed a general pattern involving *functors*. As we have seen, the situation for Kan was entirely different. Whereas the relationship between the tensor product functor $\otimes$ and the Hom-functor of Abelian groups was "less obvious" and perhaps seen by Eilenberg and Mac Lane as a special case, it became the fundamental case for Kan, and as soon as the proper structure of various categories involved in homotopy theory was revealed, it was possible to see it as a paradigmatic case. The shift is striking. Not only does one have to see a sim-

4.3 Kan's 1958 Papers "Adjoint Functors"

ilarity between various cases and jump to the general pattern, it is also necessary to *ignore* a large number of cases, at least at first, for they hide the relevant aspects of the situation. The fact that Kan was working in homotopy theory was certainly helpful in this respect, since as we have seen, the proper relationship between functors shows up naturally in this context and plays an important role.

Notice that Kan's example involves the algebra of mappings again. This is essentially the role of the "external" Hom-functor: it situates the notion of adjoint functors "inside" algebras of mappings. It is within these algebras that a dependence between functors is noticed and emphasized. Of course, as Kan points out himself, the external Hom-functor has a secondary role to play in the mathematical content of the notion. However, it allows the analogy with adjoint operators in linear algebra to be structurally obvious. The dependence between adjoint functors becomes similar to the dependence between linear transformations. There is certainly a rhetorical element at work here: since adjoint operators play an important role in linear algebra, one might expect adjoint functors to play a similar role in category theory. As usual with analogies, what is known about adjoint operators can guide us and allow us to see the relevance and status of adjoint functors. However, the analogy did *not* guide Kan in his work.[33] In fact, as Mac Lane emphasized in his paper on the influence of Stone on the origins of category theory, the parallel between the two notions can be taken quite far. In the words of Mac Lane himself:

> First, when adjoint (or the adjoint functor) exists, it is unique; for functors, this of course means unique up to a natural isomorphism. Second, the composition rule $(ST)^* = T^*S^*$ for adjoint operators has an exact analogy governing the composition of adjoint functors. Third, the adjoint of a linear operator is linear, and correspondingly the adjoint of an additive functor is additive; more generally any left adjoint commutes with direct limits; Freyd's adjoint functor theorem appears to have no analog for adjoint operators.[34] In all other cited properties the parallel is so strong as to raise the evident question "Why?". ([193], 232.)

One is therefore led to look for "dualities" between functors that are written in that form. As we have seen, Kan's examples from homotopy theory are naturally written in that form because of their geometric content.

Kan then states the definition and gives two examples taken from homotopy theory: first, the fundamental relation we have seen above between the Cartesian product of a space with the unit interval I and the space of all paths; second, between the category of topological spaces and the category of c.s.s. complexes, the relationship between the geometric realization functor and the simplicial singular functor. Everyone working in homotopy theory already knew the first case, although not as a case of a more general situation, and the second was also a general phenomenon in homotopy theory. Then Kan mentions that the notion of adjoint functors can be generalized to many variables and presents the plan of the paper. It is clear from the

---

[33] Kan, personal communication, February 2003.

[34] As Peter Freyd as pointed out to me (personal communication, January 2003), this is false. The theorem for adjoint operators is that every continuous homomorphism between Hilbert spaces has an adjoint. Freyd has also suggested that Mac Lane probably had forgotten that continuity is equivalent to boundedness for linear transformations (i.e., homomorphisms) between Hilbert spaces.

introduction that Kan's motivations came from certain situations in algebraic topology. In fact, it should now be clear that the two papers published in the *Transactions* constituted a whole for Kan. It is not even inconceivable that the second paper was more significant for him than the paper on adjoint functors, since it put in clear conceptual perspective all his previous work and opened up vast horizons in abstract homotopy theory.

It is one thing to notice the "duality" between functors, their dependence, and make an analogy to linear algebra, it is another thing to jump from that case to a general case and see that it is just one instance of a very broad and important phenomenon. This is a bold and daring step. Other mathematicians might have noticed the same dependence between specific functors, as Eilenberg and Mac Lane certainly did for the basic example. But one had to see more than that to arrive at the conclusion that a genuine *concept* was underlying these instances. First, one had to see how it was related to other important notions of category theory, e.g., limits and colimits, and second and perhaps even more important, one had to see that the notion could have a real mathematical value. Having mathematical value, in this context, means essentially three things: (1) that the notion helped to clarify and prove known results that are judged significant by at least one community of mathematicians (in this case, I claim, algebraic topologists); (2) that the notion made it possible to solve problems that were still unsolved and non-trivial; and (3) that the notion was sufficiently general in some sense. Kan succeeded in doing all three. The paper on adjoint functors is a telling proof of the first and the third part. The paper that follows in the *Transactions* accomplishes the second part, as well as the first part again (Hurewicz homomorphism is certainly significant!). Thus, the elements that led to the discovery of adjoint functors can be summarized as follows.

1. Working in algebraic topology, more specifically in homotopy theory, Kan came to realize that specific constructions were best thought of as being *functorial* constructions; we have already seen how this was present in the second paper published in the *Transactions*; another important and different example is the example given by Kan himself in the paper on adjoint functors: the suspension functor and the loop space functor (see below); when one looks at these constructions as functors, then not only does their dependence becomes precise, clear and revealing (i.e., one notice that they are adjoints), but moreover, and this is a crucial fact, some of their fundamental properties *follow directly* from the adjunction (see below). In other words, the adjunction is not merely an intriguing link, it actually *does* something important; this is also shown in the paper on functors involving c.s.s. complexes.
2. The dependence, the duality between these functors, because of the nature of the constructions themselves, finds a natural expression at the level of the algebra of mappings; i.e., there is a natural isomorphism between Hom-sets in which the functors show up. In other words, when one wants to write the dependence between these functors, one is led to use the underlying Hom-sets. The dependence therefore takes a specific form (to which Kan is clearly holding throughout the paper).

4.3 Kan's 1958 Papers "Adjoint Functors"

3. Once this is seen, it is only natural to look for similar examples in different areas and Kan found them in algebra (via homology theory and homological algebra). Eilenberg, Mac Lane, Steenrod and Cartan already knew these examples. The most obvious example is the example that led to the discovery itself: it becomes a guide in the search of adjoint functors. It is important to note that this example is an example of *endofunctors*, whereas in other cases, Kan had examples of functors going in the *opposite direction*. This was certainly not obvious to Eilenberg and Mac Lane.[35]
4. The notion is sufficiently general and robust. It is not confined to a few examples, no matter how important, but belongs to the general framework of the theory; by working it out, one notices that fundamental concepts of the theory can be *defined* via that notion.

Of course, once this is done, it is not necessarily obvious that the concept circumscribed is fundamental or that it cuts through to the very heart of the theory. This takes time and work and it took a little while and quite some work before the community came to realize the importance of the concept of adjoint functors.[36]

Let us now examine the content of the chapters of Kan's paper. The first chapter assumes the notion of category, functor and natural transformation as they were defined in Eilenberg and Mac Lane 1945. Kan, however, assumes that all categories are locally small (unless specified otherwise). He then introduces the Hom-functor explicitly and defines the dual category and the dual of a functor, a notion used throughout his paper to exploit the symmetry inherent in the notion of adjoint functors. Kan defines the notion of adjoint functors in one variable thus:

Definition 4.9. Let $\mathscr{C}$ and $\mathscr{D}$ be categories, let $F: \mathscr{C} \to \mathscr{D}$ and $G: \mathscr{D} \to \mathscr{C}$ be covariant functors. Then F is called a *left adjoint* to G and G a *right adjoint* to F if there exists a natural isomorphism

$$\alpha: \mathrm{Hom}(F(\mathscr{C}), \mathscr{D}) \to \mathrm{Hom}(\mathscr{C}, G(\mathscr{D})).[37]$$

---

[35] It is interesting to read how Eilenberg and Steenrod describe homology theory in the preface of their book, for they acknowledge that in this case, the "translations" are going in one direction only: "In this respect, homology theory parallels analytic geometry. However, unlike analytic geometry, it is *not reversible*. The derived algebraic systems represents only an aspect of the given topological system, and is usually much simpler." ([80], vii. My emphasis.)

[36] Two mathematicians were ready to receive, apply and develop Kan's contribution very quickly: Peter Freyd and Bill Lawvere. Freyd had already discovered in early 1958 in his *undergraduate* thesis at Brown University the notion of reflective subcategory, a special case of adjoint functors and he proved various important facts about the existence of limits and colimits. When he saw Kan's paper, he immediately understood and saw how the notion was related to universal morphisms and representable functors. (Freyd, personal communication, January 2003.) Lawvere gave many lectures in a course on functional analysis organized by Truesdell in 1959 and saw how various concepts and results in this area could be treated in terms of universal morphisms. He found out about Kan's paper in 1960 when Eilenberg showed it to him. (Lawvere, personal communication, December 2002.)

[37] This is Kan's notation throughout his paper. The categories themselves are written as values of the functors in the definitions.

Kan then proves that an adjoint to a given functor is determined uniquely up to a unique natural isomorphism (by symmetry, they determine each other by a unique natural isomorphism), then moves to the case of functors in two variables, obtains similar results and finally shows that the case of functors in more than two variables can be reduced to the case in two variables. It is interesting to note that there is *no* example in these first four sections in which the basic definitions and theorems are proved. Examples are introduced in the following two sections of the first chapter. These two sections, §5 and §6 of the first chapter, are pivotal, for they contain Kan's main examples and reveal the extent to which the description of adjoints in terms of Hom-sets was crucial for him.

§5 describes what Kan calls the relative case, cases in which the Hom-functor is replaced by another functor. It is relevant to give the exact definition (in my symbolism but with Kan's terminology).

Definition 4.10. Let $F\colon \mathscr{C} \to \mathbf{Set}$ be a covariant functor and let $G\colon \mathscr{D}^\circ \times \mathscr{D} \to \mathscr{C}$ be a functor contravariant in the first variable and covariant in the second. The functor G is called a Hom-*functor rel.* F if there exists a natural isomorphism

$$\gamma\colon \mathrm{Hom}(\mathscr{D},\mathscr{D}) \to FG(\mathscr{D},\mathscr{D}).$$

In this definition, one can think of the functor G as being an abstract Hom-functor in a category $\mathscr{C}$. The functor G yields "Hom-objects", not Hom-sets. But when G is composed with F, the resulting functor is naturally isomorphic with the Hom-set functor $\mathrm{Hom}(-,-)\colon \mathscr{D} \times \mathscr{D} \to \mathbf{Set}$.[38]

We have already seen important examples of this notion. For instance, the functor:

$$\mathrm{Hom}_{\mathbf{Top}}(-,-)\colon \mathbf{Top}^\circ \times \mathbf{Top} \to \mathbf{Top}$$

assigns to given topological spaces X and Y, the *function space* $\mathrm{Hom}_{\mathbf{Top}}(X,Y) = Y^X$ and not the Hom-set. But by composing with the forgetful functor F, the functor $\mathrm{Hom}_{\mathbf{Top}}$ is a Hom-functor relative to F.

The foregoing definition allows the definition of adjoint functors relative to a functor F as above. This is expressed in the following manner:

Definition 4.11. Let $G\colon \mathscr{C} \to \mathscr{D}, K\colon \mathscr{D} \to \mathscr{C}$ and $F\colon \mathscr{A} \to \mathbf{Set}$ be covariant functors and let $Q\colon \mathscr{C} \times \mathscr{C} \to \mathscr{A}$ and $R\colon \mathscr{D} \times \mathscr{D} \to \mathscr{A}$ be Hom-functors rel. F. Let

$$\beta\colon R(G(\mathscr{C}),\mathscr{D}) \to Q(\mathscr{C},K(\mathscr{D}))$$

be a natural isomorphism. Then G is called *the left adjoint of K rel. F under β*.

Then Kan proves the obvious but important result, theorem 5.4, that adjointness relative to F implies adjointness and moves to the case in two variables. It is after the notion of Hom-functor relative to F is defined that Kan gives the first examples of adjoint functors.

---

[38] Nowadays, one would say that Kan is working here in what is called "enriched category theory". See, for instance [61].

4.3 Kan's 1958 Papers "Adjoint Functors"

The first example is the one given in the opening paragraph of the paper, that is, between the tensor product of Abelian groups and the Hom-functor (which in this case yields an Abelian group):

$$\mathrm{Hom}(A \otimes B, C) \to \mathrm{Hom}(A, \mathrm{Hom}(B,C)).$$

The second and the third examples go back to Hurewicz. For, X, Z topological spaces and Y locally compact Hausdorff, we have that

$$\mathrm{Hom}_{\mathbf{Top}}(Z \times Y, X) \to \mathrm{Hom}_{\mathbf{Top}}(Z, X^Y).$$

In particular we have, for Y the unit interval I, that

$$\mathrm{Hom}_{\mathbf{Top}}(Z \times I, X) \to \mathrm{Hom}_{\mathbf{Top}}(Z, X^I),$$

or that "taking the cartesian product with the unit interval is a left adjoint to taking the space of all paths" and Kan mentions that the homotopy relation for continuous maps can be defined by using either the left hand side or the right hand side of the bijection.

Next comes a very important example, namely example 5.10. It is worth considering it in detail, for if the previous examples show how the concept of adjoint functors is intimately connected to the fundamental concepts of homotopy theory, they are nonetheless trivial in some sense and simply constitute a different way of presenting known definitions and results. The next example describes an adjoint pair that had become central in homotopy theory and that was well known at the time.[39] The adjunction allows for more conceptual definitions and proofs of results contained in the very important work of Serre and James. This example is clearly in the spirit of the results presented in the paper on functors involving c.s.s. complexes.

Consider the category of pointed topological spaces $\mathbf{Top}_{\bullet}$. Given a pointed space (Z, z_0), the *(reduced) suspension* of Z, denoted by $\Sigma(Z, z_0)$, is defined as follows. Let S^1 be a 1-sphere and let $s \in S^1$ be the base point. The suspension of (Z, z_0) is a quotient space of the product space $Z \times S^1$ by shrinking to a point the subspace

$$(z_0 \times S^1) \cup (Z \times s).$$

The base point of $\Sigma(Z, z_0)$ is the image of the projection of the point (z_0, s). The suspension is in fact a *functorial* construction $\Sigma \colon \mathbf{Top}_{\bullet} \to \mathbf{Top}_{\bullet}$.

The Hom-functor relative to F is an exponential functor and is defined thus. Let Map: $\mathbf{Top}_{\bullet} \times \mathbf{Top}_{\bullet} \to \mathbf{Top}_{\bullet}$ be the functor which assigns to every two pointed spaces (X,x) and (Y,y), the function space $(Y,y)^{(X,x)}$ (with the compact-open topology) and with base point the map $p\colon (X,x) \to (Y,y)$ defined by $p(g) = y$, for every

---

[39] "Between 1955 and 1958 B. Eckmann and P. Hilton had popularized through many colloquia and lectures at seminars the idea of a duality in the category of topological spaces which allows, starting from a given notion to get another one by reversing the arrows in diagrams, by changing the functor 'product by a space' to the functor 'exponentiation by this space', and vice versa: so the reduced suspension corresponds to the loop space (...)" ([271], 623.)

$g \in X$. The functor $F \colon \mathbf{Top_\bullet} \to \mathbf{Set}$ is the forgetful functor, i.e., it assigns to a pointed space (X,x) its underlying set X. Map is a Hom-functor rel. F.

Given a pointed space (X,x_0), one can define its *loop space*, denoted by $\Omega(X,x_0)$, as $\mathrm{Map}((S^1,s),(X,x))$, i.e.:

$$\Omega(X,x_0) = \{f \colon (S^1,p) \to (X,x_0) \mid f \in \mathbf{Top_\bullet}\}$$

with the compact-open topology. It is by construction a *functor*

$$\Omega = \mathrm{Map}((S^1,s),-) \colon \mathbf{Top_\bullet} \to \mathbf{Top_\bullet}.$$

It is a standard result of algebraic topology, although not written in this form when Kan wrote his paper, that there exists a natural isomorphism

$$\beta \colon \mathrm{Map}(\Sigma(\mathbf{Top_\bullet}),\mathbf{Top_\bullet}) \to \mathrm{Map}(\mathbf{Top_\bullet},\Omega(\mathbf{Top_\bullet})).$$

Again this is Kan's notation. This means that Σ is left adjoint to Ω relative to F. Notice that this example was *not* in [74]. Let us look at the constructions more carefully. Consider two pointed spaces (X,x) and (Y,y). The construction $\mathrm{Map}(\Sigma((X,x)),(Y,y))$ takes the pointed spaces (X,x) and (Y,y) and yields a *function* space $(Y,y)^{\Sigma(X,x)}$, i.e., the space of all continuous maps which preserve the base point from the suspension of (X,x) into the space (Y,y). Similarly, the construction $\mathrm{Map}((X,x),\Omega(Y,y))$ takes the pointed spaces and yields again a *function* space $(\Omega(Y,y))^{(X,x)}$, the space of all continuous maps from (Y,y) into the loop space of (X,x). Thus, although β is an isomorphism of *spaces*, it is an isomorphism of *function* spaces. Then theorem 5.4 allows us to write the adjunction in the following manner:

$$\alpha \colon \mathrm{Hom}(\Sigma(X,x),(Y,y)) \to \mathrm{Hom}((X,x),\Omega(Y,y)),$$

that is, in terms of the algebra of mappings. Notice also that although the reduced suspension is not a product, it is nonetheless the *quotient of a product* of two spaces and the loop space is clearly a function space. Thus, even if this last example seems to be different from the previous cases, it is not that far from the general form.

The last example is attributed to P. Hilton. It is the *only* example that does not involve a function space or a Hom structure. It essentially involves two products. The first product is the so-called *wedge product*: given two pointed spaces (X,x_0) and (Y,y_0), their *wedge* $X \vee Y$ is the quotient space of the disjoint union in which the basepoints are identified.

The functors involved are:

$$\times^2 \colon \mathbf{Top_\bullet} \to \mathbf{Top_\bullet}$$

defined, for every pointed space (Y,y_0) as

$$\times^2(Y,y_0) = (Y \times Y,(y_0,y_0))$$

and the wedge

4.3 Kan's 1958 Papers "Adjoint Functors"

$$\vee^2 \colon \textbf{Top}_\bullet \to \textbf{Top}_\bullet$$

defined, for every pointed space (X,x_0) as

$$\vee^2(X,x_0) = (X \vee X, (x_0,x_0)),$$

in other words, two copies of the space X are glued together at the point x_0. It can be shown that there is a natural isomorphism

$$\beta \colon \mathrm{Map}(\vee^2(X,x_0),(Y,y_0)) \to \mathrm{Map}((X,x_0), \times^2(Y,y_0)).$$

Thus this example illustrates the variation allowed in the constructions of adjoint functors.

The pattern, the *uniformity* was obvious from the examples given by homotopy theory, including Kan's paper "On Functors involving c.s.s. complexes". Although in each case, we are dealing with spaces, when we look at the details of the isomorphisms, we are forced to look at *maps* between spaces. Hence, we are naturally put in the context of the algebra of mappings and the duality becomes more or less obvious once the connection is written in that context. As we have seen, there was no such apparent uniformity in Eilenberg and Mac Lane's original list of examples. Kan *ignored* a number of examples and *focused* on examples that had the same form. The uniformity found is a first indication that there might be an important and general underlying concept present in each particular case. But more is required to see that it *is* important.

An additional element is provided by one of the examples: the adjoint situation between the suspension and the loop space constructions contains more, and I believe that this additional fact contributed to the conviction that the notion was mathematically significant.

Indeed, in the following section of his paper, §6, Kan introduces what are now called the "unit" and the "counit" of an adjunction, two specific natural transformations that are inherent in any adjunction. (We will come back to the definitions later.) Once Kan has defined these notions, showing that they are indeed natural and that they determine completely the adjunction (as we will see, an adjunction can be defined directly in terms of the unit and the counit), he gives one example that is a direct application of these notions. Kan shows that two important homomorphisms defined and used by James and Serre in their fundamental papers can be derived directly from the adjunction between the suspension and the loop space, since they can be defined directly via the unit and the counit of that adjunction. The unit of the adjunction is the map:

$$\eta \colon (X,x) \to \Omega\Sigma(X,x)$$

whereas the counit of the adjunction is the map

$$\xi \colon \Sigma\Omega(X,x) \to (X,x).$$

The definitions of the homomorphisms found in homotopy theory constitute a direct indication that the notion of adjunction is useful in applications and that adjunctions constitute more than an intriguing duality between functors.

Chapter I ends with this application. As such, Kan already has enough indications that the notion is linked to important aspects of algebraic topology and homological algebra. But, I claim, this is not enough to conclude that it is a *general* notion worth investigating for its own sake. After all, the examples, although important and interesting, are not enough to show that the notion has wide applicability and is playing a key role in category theory *as such*. This is where the following two chapters come in. They show how general the notion of adjoint functors is and the role it plays in pure category theory and in various direct applications. Kan does this by defining in purely categorical terms the algebraic notions of limits and colimits (in his paper direct and inverse limits respectively) and shows how they are related to adjoint functors. These results apply to a wide class of categories and clearly show that the notion of adjoint functors plays a key role in the structure of categories. In other words, chapters II and III clearly justify a whole paper on the notion and show that it rightly belongs to *pure* category theory.[40]

Kan's definitions of limits and colimits in categories are now standard and we will essentially follow his presentation. Let $\mathscr{C}$ be a category and $\mathscr{J}$ a small category (which will play the role of the index category; in applications, as we will see, it is very often finite, but it need not be and, in some cases, it is infinite). Let $F\colon \mathscr{J} \to \mathscr{C}$ be a functor. Then F is called a $\mathscr{J}$ *diagram* over $\mathscr{C}$, that is, a $\mathscr{J}$ diagram is a system of objects and morphisms of $\mathscr{C}$ indexed by the objects and morphisms of $\mathscr{J}$. The *diagonal functor*[41]

$$\Delta\colon \mathscr{C} \to \mathscr{C}^{\mathscr{J}}$$

assigns to every object X of the constant functor Δ_X which maps every object of $\mathscr{J}$ into X and every morphism into 1_X, and which assigns to every morphism $f\colon X \to Y$ of $\mathscr{C}$ the natural transformation $\Delta f\colon \Delta_X \to \Delta_Y$ given by $\Delta f(j) = f$ for every j in $\mathscr{J}$.[42]

Definition 4.12. Let X be an object of $\mathscr{C}$ and let $k\colon F \to \Delta_X$ be a morphism of $\mathscr{C}^{\mathscr{J}}$. Then X is called *the direct limit of F under the morphism k* if for every object Y of $\mathscr{C}$ and every morphism $k'\colon F \to \Delta_Y$, there exists a *unique* morphism $f\colon X \to Y$ such that the following diagram

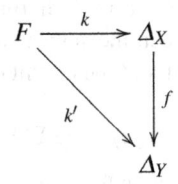

$$\begin{array}{ccc} F & \xrightarrow{k} & \Delta_X \\ & \searrow{\scriptstyle k'} & \downarrow{\scriptstyle f} \\ & & \Delta_Y \end{array}$$

commutes.

---

[40] We should however recall that Kan did not have these results before he actually sat down and wrote the paper. Thus, in a sense, he discovered post facto the generality of the notion.

[41] Not to be confused, of course, with a simplicial object.

[42] Interestingly enough, Kan denotes the category $\mathscr{C}^{\mathscr{J}}$ by $\mathscr{C}_{\mathscr{J}}$. In other words, he does not see it as being similar to a function space, although he uses the notation for function spaces in his paper.

4.3 Kan's 1958 Papers "Adjoint Functors"

Two remarks are in order. First, the notion of a limit of a system of (discrete) groups is defined in [73], [74] and [80]. There, one starts with $\mathscr{J}$ a directed *set* and the direct limit is actually explicitly *constructed* in the case of a directed system of groups. Kan generalizes the notion found in Eilenberg and Mac Lane's paper and defines the direct limit in terms of morphisms. However, and this is the second remark, there is absolutely *no* reference to a universal property in Kan's paper, although the definition is given in these terms. It is hard to tell whether Kan thought that it was clear and did not have to be made explicit or whether he (and presumably Eilenberg, who had read Kan's paper) did not see the connection.

Be that as it may, Kan then gives some examples of direct limits which show how general the notion is: a disjoint union of topological spaces (indexed by some set $\mathscr{J}$ seen as a category) can be constructed as a direct limit, so can be a direct sum of Abelian groups (again indexed by some set $\mathscr{J}$) and finally Eilenberg and Mac Lane's original example is now a special case of the general construction. Then come two theorems which link the notion of direct limit with the notion of adjoint functors: they assert that for an object in the category $\mathscr{C}^{\mathscr{J}}$ to have a direct limit under some map, it is necessary and sufficient that the diagonal functor $\Delta: \mathscr{C} \to \mathscr{C}^{\mathscr{J}}$ has a left adjoint.[43] These theorems are proved, the dual notion of inverse limits (now called "colimits") is defined and dual results are also proved. In the remaining paragraphs, Kan considers the more general case of arbitrary limits in a category and shows that a category has arbitrary limits if a certain functor has a left adjoint and a similar dual result for colimits is stated.

These remarks clearly show how general the notion of adjoint functor is: it is connected to various fundamental constructions in categories *as such*. Whereas the first chapter showed that the notion of adjoint functor arises naturally in certain mathematical situations, the second chapter demonstrates that the notion occupies a key position in category theory itself. The specific examples of the first chapter might have been important and revealing in themselves, but they are not enough to convince anyone that the concept of adjoint functor is sufficiently general. What remained to be seen is the extent to which the concept of an adjoint situation captures so many different mathematical constructs, so many different mathematical theories and theorems.

We end our examination of Kan's paper at this point but we will not leave adjoint functors immediately. The concept is so fundamental, and plays such a key role in the subsequent history and in my thesis, that it deserves to be looked at carefully before we move to the next period.

---

[43] There is a technical difficulty involved here: as Kan himself notes in the remark following the theorems: "In order to obtain the second half of Theorem 7.8 (...) a kind of axiom of choice would be needed." ([134], 311.)

Chapter 5
Adjoint Functors: What They are, What They Mean

> The theory of categories has arisen in the last twenty-five years and now constitutes an autonomous branch of mathematics. It owes its origin and early inspiration to developments in algebraic topology. When the basic concepts of *category*, *functor*, *natural transformation* and *natural equivalence* were first formulated by Eilenberg and Mac Lane they served immediately to provide the appropriate framework for describing the way in which algebraic tools were used, and could be used, in the study of topology. It was surely evident from the outset, to the inventors of these fundamental notions and to others, that their domain of application certainly extended far beyond that of algebraic topology. (...) However, it was not clear in the early stages that there was a "pure" theory latent within the domain of categories and functors which was capable of assuming substantial proportions within the body of mathematics. (...) Nevertheless, it is only in the last ten years, or less, that the source of inspiration for advances in category theory has come to any considerable extent from within the theory itself.
>
> (Hilton, 1968, v, in [37])

"The source of inspiration for advances in category theory" was provided by the development of Abelian categories and, together with these, the concept of adjoint functors. The former provided a paradigm, in the most literal sense of that expression, that is a specific model, for the development and applications of category theory, whereas the latter provided the general conceptual framework to develop the whole theory on an autonomous basis. According to Mac Lane, it was in 1963 that mathematicians realized that they could be category theorists: "Then in 1963 it suddenly became clear that general category theory (not just Abelian categories or applications of categories) was a viable field of mathematical research" ([196], 346). According to Mac Lane, some of the factors that triggered this change are: (1) the first publication of what was to become SGA4 by Grothendieck and his collaborators in which category theory and topos theory are developed for algebraic geometry; (2) Lawvere's thesis; (3) Freyd's first public presentation of his adjoint functor theorem in front of logicians at Berkeley (see [90]); (4) Ehresmann's paper on structured categories (see [63]); (5) Mac Lane's first coherence theorem (see

[188]); and (6) Mac Lane's lectures on categorical algebra as a Colloquium lecturer for the American Mathematical Society (see [189]). What *was* clear to many people is that there were many open problems in *pure* category theory. This was possible mostly because of the advent of the concept that allowed for the formulation of these problems as *autonomous* problems: the concept of adjoint functors (and its equivalent formulations).

5.1 Adjointness

It is now time to step back from historical aspects of the theory and look more closely at adjoint functors themselves. It is my belief that the claim that category theory is a generalization of Klein's program takes a new meaning when adjoint functors are introduced and when their significance is understood. We will have to go back to the history of category theory afterwards, for some more work by specific mathematicians was required before adjoint functors could seen as having this significance.

Recall the definition of adjoint functors from Chapter 4. *Adjoint functors* are functors $F\colon \mathscr{C} \to \mathscr{D}$ and $G\colon \mathscr{D} \to \mathscr{C}$ such that there is an isomorphism

$$\operatorname{Hom}(F(X),A) \xrightarrow{\phi} \operatorname{Hom}(X,G(A))$$

such that ϕ is *natural* both in X and A. Precisely, ϕ is a natural isomorphism (an invertible natural transformation) between the set-valued functors $\operatorname{Hom}(F(-),-)$ and $\operatorname{Hom}(-,G(-))$ on $\mathscr{C}^{op} \times \mathscr{D}$.

The naturality of the bijection means, informally, that ϕ does not depend on X and A. Again, this can be interpreted as a form of invariance. We represent this situation by the correspondence:

$$\frac{F(X) \to A}{X \to G(A)} \quad \Updownarrow \phi_{X,A}$$

where $\phi_{X,A}$ means that the correspondence here is between the arrows determined by X and A in the respective categories. In this case, we say that F is *left* adjoint to G or equivalently that G is *right* adjoint to F, and this is denoted by $F \dashv G$.

Technically, the bijection $\phi_{X,A}$ depends on both X and A. For ϕ to be *natural* means that ϕ is related to the family of bijections $\phi_{-,-}$ in the right way, that is, when A is transformed into B or X' is transformed into X, the correspondence between the numerator and the denominator should still hold. For instance, let $f\colon A \to B$ be an arrow of $\mathscr{D}$. Then, the following correspondences $\phi_{X,A}$ and $\phi_{X,B}$ should be related to one another as indicated by the following diagram:

5.1 Adjointness

$$\phi_{X,A} \left[\Updownarrow \dfrac{F(X) \longrightarrow A}{X \longrightarrow G(A)} \right] \xrightarrow[G(f)]{f} B \Updownarrow \;\; \Updownarrow \phi_{X,B}.$$

Hence, given the correspondence $\phi_{X,A}$, composing with f yields the correspondence $\phi_{X,B}$, indicated by the long arrows above and below. When the correspondence is natural in both variables, it means that we have indeed defined a global correspondence.

This is an interesting epistemological dimension of the way one has to work with categorical notions. We are often trying to define concepts and notions for the whole by local means. The very definition of a natural transformation is typical in this regard. The goal is to define arrows between arrows in a systematic manner, i.e., globally. One way to do it is by giving a definition for an arbitrary collection of arrows between two objects and show that the definition does not depend on the local conditions of the definition, i.e., on any specific property of the objects chosen. This is what it means for a transformation to be "natural": it is canonical in the sense that it does not depend on the presentation or the choice of the objects used in the definition of the transformation. The naturality condition says that the objects we have chosen are totally dispensable and none of their specific properties have been used in the definition of the functors involved. They are merely representatives of the objects of the category in this situation.

Let us consider products again, but this time from the point of view of adjoint functors. We will now treat a given category $\mathscr{C}$ as a whole and look at its transformations. Consider the following very simple category: it has two objects, let us call them 0 and 1, with only the identity arrow on each object and nothing else. We will denote this category by $\mathscr{P}$. It is as if we would consider two abstract copies of the one element group without the isomorphism between them.

A functor $F: \mathscr{P} \to \mathscr{C}$ from the category $\mathscr{P}$ into a category $\mathscr{C}$ simply picks out two objects of $\mathscr{C}$, namely $F(0)$ and $F(1)$, with their identity arrows. We will call such a functor F a *representation* of $\mathscr{P}$ in $\mathscr{C}$. The category of *all* such representations of $\mathscr{P}$ in $\mathscr{C}$ is of course a functor category, denoted by $\mathscr{C}^{\mathscr{P}}$, defined thus. An object of $\mathscr{C}^{\mathscr{P}}$ is a pair of objects of $\langle X, Y \rangle$ and an arrow of $\mathscr{C}^{\mathscr{P}}$ is also a pair $\langle f, g \rangle : \langle X, Y \rangle \to \langle U, V \rangle$ of arrows $f: X \to U$ and $g: Y \to V$ of $\mathscr{C}$.

There is an obvious functor $I: \mathscr{C} \to \mathscr{C}^{\mathscr{P}}$, which sends an object X of $\mathscr{C}$ to the pair $\langle X, X \rangle$ of $\mathscr{C}^{\mathscr{P}}$ and an arrow $f: X \to Y$ to the pair $\langle f, f \rangle: \langle X, X \rangle \to \langle Y, Y \rangle$ of $\mathscr{C}^{\mathscr{P}}$. The functor I can be thought of as giving a full and faithful representation of $\mathscr{C}$ in $\mathscr{C}^{\mathscr{P}}$. Looking at the image of the functor I in the category $\mathscr{C}^{\mathscr{P}}$ is like looking at $\mathscr{C}$ with one's eyes crossed: everything is simply doubled. It is reasonable to ask whether I has "inverses", since we want to think of I as being a "transformation" or an "encoding" in some sense of $\mathscr{C}$ into $\mathscr{C}^{\mathscr{P}}$. These functors would be the "dual" of I, since they would in a sense "reverse" what I does. We are therefore looking for functors $\mathscr{C}^{\mathscr{P}} \to \mathscr{C}$ satisfying certain conditions. In my previous terminology, such a functor, let us call it J, would constitute a representation of $\mathscr{C}^{\mathscr{P}}$ in $\mathscr{C}$, that is J would determine how a pair $\langle X, Y \rangle$ of objects of $\mathscr{C}$ can be represented by a *single* object in $\mathscr{C}$. For the object $J(\langle X,Y \rangle)$ to constitute a representation of a pair of objects X and Y of $\mathscr{C}^{\mathscr{P}}$ in $\mathscr{C}$, the object $J(\langle X,Y \rangle)$ has to be related to X and Y in a privileged manner. In a categorical context, this can only mean that there are arrows between $J(\langle X,Y \rangle)$ and X and Y *and* arrows between $J(\langle X,Y \rangle)$ and the objects of $\mathscr{C}$ that could be representations of pairs of objects in $\mathscr{C}$. Finally, for J to be an dual of I, the following arrows

$$X \to J(I(X)),$$
$$J(I(X)) \to X,$$
$$I(J(\langle X,Y \rangle)) \to \langle X,Y \rangle \text{ and}$$
$$\langle X,Y \rangle \to I(J(\langle X,Y \rangle))$$

should exist and should have specific properties. If J were a "real" inverse of I, that is an inverse in the usual set-theoretical sense of that expression, we would have identities between these objects, e.g., $X = J(I(X))$, etc. However, in a categorical context, identity of objects is not given by the theory. "Being identical" within a category can *only* mean "being isomorphic". Thus, in the best cases, we should have *isomorphisms* $X \simeq J(I(X))$, $I(J(\langle X,Y \rangle)) \simeq \langle X,Y \rangle$. But this is not the only possibility (and we will come back to this fundamental possibility in due time). In certain contexts, one might at best hope to obtain *universal morphisms*

$$\eta: X \to J(I(X)) \quad \text{and} \quad \xi: I(J(\langle X,Y \rangle)) \to \langle X,Y \rangle.$$

According to the preceding definitions, this means that for any morphism $f: X \to J(\langle Y,Z \rangle)$ in $\mathscr{C}$, there is a *unique* morphism $g: I(X) \to \langle Y,Z \rangle$ in $\mathscr{C}^{\mathscr{P}}$ such that the following diagram

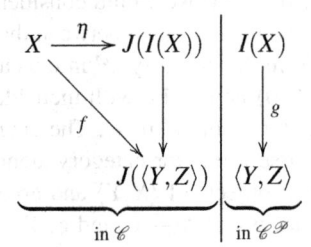

$$\begin{array}{c|c} X \xrightarrow{\eta} J(I(X)) & I(X) \\ {}_f \searrow \downarrow & \downarrow g \\ J(\langle Y,Z \rangle) & \langle Y,Z \rangle \\ \underbrace{}_{\text{in } \mathscr{C}} & \underbrace{}_{\text{in } \mathscr{C}^{\mathscr{P}}} \end{array}$$

commutes. Similarly, for any morphism $h: I(Z) \to \langle X,Y \rangle$ in $\mathscr{C}^{\mathscr{P}}$, there is a *unique* morphism $K: Z \to J(\langle X,Y \rangle)$ in $\mathscr{C}$ such that the following diagram

$$\begin{array}{c|c} J(\langle X,Y \rangle) & I(J(\langle X,Y \rangle)) \xrightarrow{\xi} \langle X,Y \rangle \\ \uparrow k & \uparrow \nearrow h \\ Z & I(Z) \\ \underbrace{}_{\text{in } \mathscr{C}} & \underbrace{}_{\text{in } \mathscr{C}^{\mathscr{P}}} \end{array}$$

5.1 Adjointness 151

commutes. In other words, this means that if J is a dual of I, then there should be a *bijection* between arrows $I(X) \to \langle Y,Z \rangle$ and $X \to J(\langle Y,Z \rangle)$.

There is yet another way to connect I and J via universal morphisms. The pair of universal morphisms obtainable might be $\eta\colon \langle X,Y \rangle \to I(J(\langle X,Y \rangle))$ and $\xi\colon J(I(\langle Y,Z \rangle)) \to X$, in which case, as can easily be verified, there is a bijection between arrows $\langle X,Y \rangle \to I(X)$ and $J(\langle Y,Z \rangle) \to X$.

These bijections can be represented with the help of the following notation:

$$\frac{\text{in } \mathscr{C}^{\mathscr{P}}}{\text{in } \mathscr{C}} \qquad \text{(I)} \frac{I(X) \to \langle Y,Z \rangle}{X \to J(\langle Y,Z \rangle)} \updownarrow \qquad \text{(II)} \frac{\langle Y,Z \rangle \longrightarrow I(X)}{J(\langle Y,Z \rangle) \longrightarrow X} \updownarrow$$

The vertical double arrows mean that to each arrow in the numerator corresponds a *unique* arrow in the denominator and vice-versa. A different illustration of the situation given on the left is as follows:

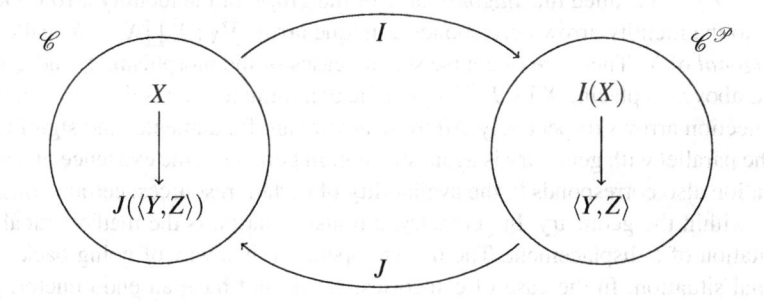

The beauty of such a correspondence is that it contains (almost) all the ingredients required to determine what J ought to be in each case. Let us first consider the case on the left, case (I). In the numerator, an arrow $I(X)$ is, by definition, a pair of arrows $f\colon X \to Y$ and $g\colon X \to Z$. Thus, the correspondence means that to such a pair corresponds a *unique* arrow $X \to J(\langle Y,Z \rangle)$. If we can show that the object $J(\langle Y,Z \rangle)$ comes with a pair of arrows $J(\langle Y,Z \rangle) \to Y$ and $J(\langle Y,Z \rangle) \to Z$, then we can conclude that $J(\langle Y,Z \rangle)$ is a *product* of Y and Z in $\mathscr{C}$. But if $J(\langle Y,Z \rangle)$ exists, then we certainly have the identity arrow $J(\langle Y,Z \rangle) \to J(\langle Y,Z \rangle)$ in the denominator, which automatically yields a unique arrow $I(J(\langle Y,Z \rangle)) \to I(J(\langle Y,Z \rangle))$ in the numerator, which by definition are a pair of arrows $J(\langle Y,Z \rangle) \to Y$ and $J(\langle Y,Z \rangle) \to Z$, as required. Thus, if J exists and satisfies the equivalence given by (I), then we can (almost) conclude that J takes an object $\langle Y,Z \rangle$ of $\mathscr{C}^{\mathscr{P}}$ to a *product* of Y and Z in $\mathscr{C}$. In other words, if J exists, then $\mathscr{C}$ has binary products. (We will come back to the missing ingredient in a short while.)

The same exercise can be performed for the equivalence on the right, namely case (II). Again, the numerator means that we have a pair of arrows $f\colon Y \to X$ and $g\colon Z \to X$ and the correspondence implies that for such a pair, there is a *unique* arrow $J(\langle Y,Z \rangle) \to X$. By an argument similar to the one sketched above, the identity arrow $J(\langle Y,Z \rangle) \to J(\langle Y,Z \rangle)$ yields a pair of arrows $Y \to J(\langle Y,Z \rangle)$ and $Z \to J(\langle Y,Z \rangle)$ and it follows that J takes an object $\langle Y,Z \rangle$ of $\mathscr{C}^{\mathscr{P}}$ to a *coproduct* of Y and Z in $\mathscr{C}$.

A few remarks are in order. Given the functor I, finding J, that is, showing that it can be defined and that it satisfies at least one of the conditions given above, is comparable to solving an algebraic equation. In this case, the problem is to determine whether the category $\mathscr{C}$ has the resources to represent within itself, that is in a uniform manner, pairs of its objects. Let us take a brief look at the composites of I and J in light of the foregoing correspondences. Replacing $\langle Y, Z \rangle$ by $I(X)$ in the numerator yields in both cases:

$$\frac{I(X) \to I(X)}{X \to J(I(X))} \updownarrow \qquad \frac{I(X) \to I(X)}{J(I(X)) \to X} \updownarrow$$

These mean that to each arrow from $I(X)$ to $I(X)$ corresponds a unique arrow from X to $X \times X$ on the left, and a unique arrow from $X \coprod X$ to X on the right. In particular, to the identity arrow $\langle 1_X, 1_X \rangle \colon \langle X, X \rangle \to \langle X, X \rangle$ corresponds a unique arrow $\Delta_X \colon X \to X \times X$, called the *diagonal* of X or the *graph* of the identity arrow. On the right, to the identity arrow corresponds a unique arrow $\nabla_X \colon X \coprod X \to X$, called the *codiagonal* of X. These are of course special cases of the morphisms η and ξ mentioned above. Replacing X by $J(\langle Y, Z \rangle)$ in the denominator yields the projection and the injection arrows respectively. *All* these arrows are fundamental and significant.

The parallel with geometry is again striking. In geometry, the existence of a transformation also corresponds to the availability of certain resources, certain constructions, within the geometry. In geometry, a transformation is the mathematical representation of a displacement. The inverse operation is a way of going back to the original situation. In the case of categories, we do not have an endofunctor, *pace* Eilenberg and Mac Lane, but a pair of functors going in opposite directions related to one another in a specific manner. The reason is that a functor is the mathematical representation of a *conceptual* transformation, of a mathematical construction performed in a certain manner; that is, a construction which is "invariant" to the type of entity on which it is applied. In a way, this is what "being functorial" means: a mathematical construction is functorial if it is invariant with respect to the *morphisms* of the category on which it is applied. A geometric transformation is a way to transform a space into itself in such a way that it is possible to go back to the original situation without losing *any* information. A functor is a way to transform a "form" of structures into (possibly different) structures and an inverse to the given functor in the sense we are interested in is a way to get "as close as possible" to the original structures from the situation we find ourselves in. In general, we might have lost information about the original structures by moving along a given functor. However, if the latter has an adjoint, then we can recover a significant part of the original structure by coming back. We will see latter how this is related to the rather subtle problem of identity of categories.

We could certainly extend the above discussion to any simple "abstract" category $\mathscr{P}$ that formally represents a property of interest. In fact, as is probably already clear to the reader, what we have just discussed is a special case of a *limit*, in the sense already introduced by Kan in full generality in 1958. Categories like the category $\mathscr{P}$ above are presented directly in an abstract fashion, in the same way that certain

5.1 Adjointness

abstract groups are given by elements and abstract relations between them. We saw in Chapter 2 some examples of very simple abstract categories, namely "the" one element category **1**, "the" two-element category **2**, "the" three-element category **3**, etc. We have also already mentioned that the functor category $\mathscr{C}^1$ is simply $\mathscr{C}$, that the functor category $\mathscr{C}^2$ is the category of arrows of $\mathscr{C}$ and that the functor category $\mathscr{C}^3$ can be seen to be the category of commutative diagrams of $\mathscr{C}$ and morphisms between them.

Functors such as I and J are of course *adjoint functors* and they are said to form an *adjunction*.

Adjoint functors $F: \mathscr{C} \to \mathscr{D}$ and $G: \mathscr{D} \longrightarrow \mathscr{C}$ are codependent functors and this, in two different ways. First, the foregoing bijection establishes a strong dependence between the way F encodes $\mathscr{C}$ in $\mathscr{D}$ and the way G encodes $\mathscr{D}$ in $\mathscr{C}$. Second, adjoints are determined up to an isomorphism: this means that if F is left adjoint to G and F' is left adjoint to G, then F and F' are related by a natural isomorphism, hence they are essentially the same. The same is true for G and G' both right adjoints to F. Notice that a functor can have both a right and a left adjoint. For instance, the category of sets has both products and coproducts, which are both adjoints to the appropriate diagonal functor.

The foregoing definition of adjoint functors is the one given by Kan in his original paper. It relies heavily on the existence of a bijection between Hom-sets, and therefore is basically set-theoretic. There is, however, an equivalent characterization that is purely algebraic. Two functors F and G as above form an *adjunction* if there exist natural transformations $\eta: 1_{\mathscr{D}} \to FG$ and $\xi: GF \to 1_{\mathscr{C}}$ such that the so-called triangular *identities* hold, that is the following diagrams commute:

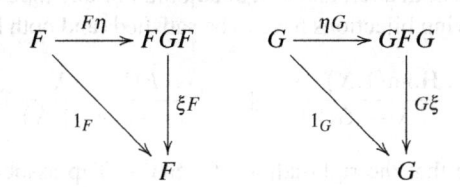

$$\begin{array}{ccc} F \xrightarrow{F\eta} FGF & \qquad & G \xrightarrow{\eta G} GFG \\ \searrow_{1_F} \downarrow{\xi F} & & \searrow_{1_G} \downarrow{G\xi} \\ F & & G \end{array}$$

Thus, we have the identities $\xi F \circ F\eta = 1_F$ and $G\xi \circ \eta G = 1_G$. Thinking about functors as giving representations of properties of the objects of categories, the foregoing characterization "says" that the representations in $\mathscr{C}$ of the functor F seen as a property given by the objects GF can be systematically transformed into the original objects of $\mathscr{C}$.

Three observations have to be made at this point. First, adjunctions pervade mathematics: they are everywhere, from logic to algebraic geometry. Second, adjoint functors capture the notions of universal property (morphism or arrow) and representable functor in all their generality. The third observation is connected to the second. As Kan already saw, very often one is given an elementary functor in a given mathematical situation, e.g., the diagonal functor above, and its adjoints turn out to be *significant* mathematical constructions. In a sense, significant mathematical constructions arise out of trivialities *if* they are related to these trivialities in an appropriate manner.

Let us first look at some standard examples of adjoint functors.

1. A whole family of fundamental cases given by free constructions of a certain type, e.g., free monoids, free groups, free vector spaces, free algebras, etc. The basic situation is this: given a category $\mathscr{C}$ of structures of a certain type, we can consider the functor $U_\mathscr{C}\colon \mathscr{C} \to$ **Set**, from $\mathscr{C}$ to the category of sets with set maps, the so-called "forgetful functor", which sends every object X of $\mathscr{C}$ to its underlying set $U_\mathscr{C}(X)$ and every morphism to the underlying map between sets. When this underlying functor has a left adjoint $F\colon$ **Set** $\to \mathscr{C}$, then $F(Y)$, for Y in **Set**, yields the free structure of the given type. Thus, we have

    a. The functor $U_{\mathbf{Grp}}\colon \mathbf{Grp} \to$ **Set**, from the category of groups and group homomorphisms to the category of sets. $F(Y)$ is the free group with generators $y \in Y$.
    b. The functor $U_{\mathbf{Vect}}\colon \mathbf{Vect} \to$ **Set**, from the category of vector spaces over a base field k and linear maps to the category of sets. $F(Y)$ is the free vector space on Y, with one basis element for each element of Y and thus has as many dimensions as Y has elements.
    c. The functor $U_{\mathbf{Top}}\colon \mathbf{Top} \to$ **Set**, from the category of topological spaces and continuous maps to the category of sets. $F(Y)$ is the discrete space on Y.

2. An important variation on the case 1c. has to be given since it generalizes to other interesting situations. Consider the functor "points of a space", as defined earlier:
$$\mathrm{Hom}(1,-)\colon \mathbf{Top} \to \mathbf{Set}.$$

This functor has both a left and a right adjoint. For any topological space X and set Y, the following bijections have to be satisfied (and both have to be natural):

$$\frac{\mathrm{Hom}(1,X) \to Y}{X \to G(Y)} \Updownarrow \qquad \frac{F(Y) \to X}{Y \to \mathrm{Hom}(1,X)} \Updownarrow.$$

It is easy to see that the right-adjoint $G\colon$ **Set** $\to$ **Top** associates to a set Y the topological space $G(Y)$ with the indiscrete topology whereas the left-adjoint $F\colon$ **Set** $\to$ **Top** associates to a set Y the space $F(Y)$ with the discrete topology. In turn, the functor F itself has a left adjoint. This means that the following bijection is satisfied (and has to be natural):

$$\frac{H(Y) \to X}{Y \to F(X)} \Updownarrow \frac{\text{in } \mathbf{Set}}{\text{in } \mathbf{Top}}.$$

It can be seen that the functor $H\colon$ **Top** $\to$ **Set** takes a space Y to the quotient space of its connected components.[1] Thus we have the following chain of adjoints:
$$H \dashv F \dashv \mathrm{Hom}(1,-) \dashv G.$$

---

[1] Recall that given a space X, we can define for all points of X the equivalence relation $x \sim y$ if there is a connected subset of X containing both x and y.

5.1 Adjointness

Similar chains of adjoints exist between other categories.

3. Instead of having a forgetful functor going into the category of sets, in some cases only a part of the structure is forgotten. Here are standard examples:

   a. There is an obvious forgetful functor $U\colon \mathbf{AbGrp} \to \mathbf{AbMon}$ from the category of Abelian groups to the category of Abelian monoids: U "forgets" about the inverse operation. The functor U has a left adjoint $F\colon \mathbf{AbMon} \to \mathbf{AbGrp}$ which given an Abelian monoid M assigns to it the "best" possible Abelian group $F(M)$ such that M can be embedded in $F(M)$ as a submonoid. For instance, if M is $\mathbb{N}$, then $F(\mathbb{N})$ "is" $\mathbb{Z}$.

   b. Similarly, there is an obvious forgetful functor $U\colon \mathbf{Haus} \to \mathbf{Top}$ from the category of Hausdorff topological spaces to the category of topological spaces, which "forgets" the Hausdorff condition. Again, there is a functor $F\colon \mathbf{Top} \to \mathbf{Haus}$ such that $F \dashv U$. Given a topological space X, $F(X)$ yields the "best" Hausdorff space constructed from X: it is the quotient of X by the closure of the diagonal $\overline{\Delta_X} \subseteq X \times X$, which is an equivalence relation. In contrast with the example 3a, where we had an embedding, this time we get a quotient of the original structure.

   c. Let us go back to modules. We have already mentioned the example given by Kan between the Hom-functor and the tensor product of Abelian groups. Here is a slightly different example. There is a forgetful functor $U\colon \mathbf{Mod}_R \to \mathbf{AbGrp}$ from a category of R-modules, where R is a commutative ring with unit, to the category of Abelian groups. The functor U forgets the action of R on a group G. The functor U has both a left and a right adjoint. The left adjoint is $R \otimes -\colon \mathbf{AbGrp} \to \mathbf{Mod}_R$ which sends an Abelian group G to the tensor product $R \otimes G$ and the right adjoint is given by the functor $\mathrm{Hom}(R, -)\colon \mathbf{AbGrp} \to \mathbf{Mod}_R$ which assigns to any group G the modules of linear mappings $\mathrm{Hom}(R, G)$.

4. Here are some elementary examples of adjoint functors that will allow us to introduce the first elements of categorical logic. As we have seen, any poset $\mathscr{C}$ can be considered as a category and $\mathbf{1}$ denotes the one-element poset. There is a unique poset homomorphism from $\mathscr{C}$ to $\mathbf{1}$, $!\colon \mathscr{C} \to \mathbf{1}$, which assigns to each X in $\mathscr{C}$ the unique element $\bullet$, i.e., $!(X) = \bullet$ and to each arrow $f\colon X \to Y$ the unique identity arrow $\bullet \to \bullet$. If this function has left and right adjoints, they have to be functors $F, G\colon \mathbf{1} \to \mathscr{C}$ such that, respectively,

$$\frac{F(\bullet) \to X}{\bullet \to !(X)} \qquad \frac{!(X) \to \bullet}{X \to G(\bullet)}.$$

Now since $!(X) = \bullet$, the arrow at the denominator of the left-hand correspondence has to be the identity arrow. Furthermore, since we are in a poset, arrows represent the order relation and we always have that $\bullet \leq \bullet$. The same is true for the numerator on the right-hand side. The bijection implies that there has to be a unique arrow $F(\bullet) \to X$, i.e., $F(\bullet) \leq X$ has to be always true on the left, and similarly $X \leq G(\bullet)$ has to be always true on the right. Therefore, if we have

$F \dashv \,!$, F picks out an initial object of $\mathscr{C}$, in this case the bottom element of the poset. Symmetrically, if $! \dashv G$, then G picks out the top element of the poset. In standard logico-algebraic notation, $\top$ denotes the top element and $\bot$ denotes the bottom element. Notice also that F can be taken to represent the property of posets "is under all elements" and G represents the property "is over all elements". A poset has the resources to encode these properties by having a bottom and a top element respectively. Notice how in this case, as in the previous cases, we start with a totally trivial functor $!: \mathscr{C} \to \mathbf{1}$ and the adjoints yield significant constructions. The same is true of the following examples, which we have in fact already seen in a more general setting.

Consider once more the diagonal functor $\Delta: \mathscr{C} \to \mathscr{C} \times \mathscr{C}$ with $\mathscr{C}$ a poset considered as a category. In this case, as we have seen, the left-adjoint to Δ is the coproduct, that is the sup, and the right-adjoint to Δ is the product, that is the inf. The correspondence takes the following special form in this special case:

$$\frac{X \vee Y \leq Z}{X \leq Z, Y \leq Z} \updownarrow \qquad \frac{Z \leq X \wedge Y}{Z \leq Y, Z \leq X} \updownarrow.$$

In order to introduce implication, we have to consider a functor with a parameter: $(- \wedge X): \mathscr{C} \to \mathscr{C}$. It can easily be verified that in the case when $\mathscr{C}$ is a poset, the function $(- \wedge X)$ is order preserving and thus a functor. A right adjoint to $(- \wedge X)$ is a functor, let us temporarily call it F, also with a parameter X, $F(X, -): \mathscr{C} \to \mathscr{C}$ such that

$$\frac{Y \wedge X \leq Z}{Y \leq F(X, Z)} \updownarrow.$$

Since we are in posets, it is easy to determine what $F(X, Z)$ ought to be. It is the largest element of $\mathscr{C}$ such that its infimum with X is smaller than Z. This element is sometimes called the relative pseudocomplement of X or, more commonly, the *implication* and it is denoted by $X \Rightarrow Z$, by $X \to Z$, or by $X \supset Z$. We can therefore rewrite the foregoing bijection thus:

$$\frac{Y \wedge X \leq Z}{Y \leq X \Rightarrow Z} \updownarrow.$$

Notice, first, that we can introduce the negation operator $\neg X$ from the last adjunction. Indeed, let Z be the bottom element $\bot$ of the lattice. Then, since $Y \wedge X \leq \bot$ is always true, we have that $Y \leq X \Rightarrow \bot$ is also always true. But also $X \Rightarrow \bot \leq X \Rightarrow \bot$ is always true, we get at the numerator that $(X \Rightarrow \bot) \wedge X = \bot$. Hence, $X \Rightarrow \bot$ is the largest element disjoint from X. We can therefore put $\neg X =_{\mathrm{df}} X \Rightarrow \bot$. Second, it is easy to prove certain facts about these operations from the adjunctions. Consider, for instance, implication. Let $Z = X$. Then we get at the numerator that $Y \wedge X \leq X$, which is always true in a poset (as it is easily verified). Hence, $Y \leq (X \Rightarrow X)$ is also true for all Y and this is only possible if $(X \Rightarrow X) = \top$.

5.1 Adjointness

We have introduced in the context of posets operations as adjoints, also called in this context *Galois connections*, which can be used to interpret logical operations. The existence of these adjoints is equivalent to certain algebraic facts. Indeed, to say that for a given poset $\mathscr{C}$, a left adjoint and a right adjoint to the terminal poset exist and a left adjoint and a right adjoint to the diagonal functor also exist, amounts to the claim that $\mathscr{C}$ is a distributive lattice. If, moreover, right adjoints to the parametrized functors $(-\wedge X)$ exist, then $\mathscr{C}$ is a Heyting algebra. Notice that we could *define* distributive lattices and Heyting algebras in this manner. Hence the existence of certain adjoints to given functors can be used to define certain structures. The interesting aspect of this strategy is that the given functors are often simple and fundamental from a conceptual point of view. In the case of lattices, there is no doubt that the diagonal functor satisfies these conditions. And so do the functors from and to the one-element lattice.

We will see in the next chapter how this analysis extends to quantifiers and how they too can be seen as adjoints. In this context, the slogan is: categorical logic *is* algebraic logic. But before we develop this slogan, let us consider other cases of adjoint functors in mathematics.

5. As I have already mentioned, all the fundamental constructions of category theory, or what Eilenberg and Mac Lane called "the basic constructions" of a category, limits and colimits, can be described as adjoints. Thus, as we have seen, products and coproducts are adjoints and so are all the other examples given in that section. Hence, whereas Eilenberg and Mac Lane's identified the basic constructions of a mathematical domain with functoriality, the development of category theory has shown that, notwithstanding the importance of functoriality, adjoints deserve to be looked at as the basic constructions of a field.[2]

If we look again at the characterization of Abelian categories, we see that the main properties in fact express the existence of adjoint functors to certain elementary functors. For instance, having a zero object can be expressed as the existence of an adjoint, being at the same time both a left and a right adjoint to a functor. The case is exactly the same as in the case of posets above: let **1** denote the category with one object and the identity morphism as the only morphism.

---

[2] I do have to insist on the importance of functoriality, for it can be taken as some sort of foundational or regulative principle. Compare Weil's and Grothendieck's approaches in the foundations of algebraic geometry. Weil looked at the foundations of algebraic geometry in the same way that geometers of the 19th century looked at geometry: there was a fundamental "universe" of geometry provided by complex projective geometry. Everything was more or less going on within this framework. As Weyl pointed out in his book on the classical groups, Klein's approach helped geometers to liberate themselves from this point of view. Grothendieck, by contrast, thought that the important idea was functoriality and what is called "a change of base". In this framework, there is no privileged universe, there is rather a "conservation principle" which is expressed by the rule that the constructions should be functorial. Perhaps categorical principles should be compared in general with conservation principles in physics. (See, for instance [20] where this comparison is taken up.) Interestingly enough, the later correspond to invariance under a transformation group.

Then a category $\mathscr{C}$ has a zero object if there is a functor $0\colon \mathbf{1} \to \mathscr{C}$ which is both a left *and* a right adjoint to the functor $!\colon \mathscr{C} \to \mathbf{1}$. As we have seen, the existence of products amounts to the existence of adjoints. We can now see more precisely how the notion of Abelian category captures the invariant content of, to put it briefly and rather sketchily, homology theory. On the one hand, it captures what is common to the categories, e.g., categories of Λ-modules for certain rings Λ, that appear in homological algebra. On the other hand, the invariant content is *expressed* by the existence of certain adjoint transformations between various categories involved. In the same way that the invariant content of Euclidean geometry is captured by the group of isometries, the invariant content of (a part of) homological algebra is captured by the notion of Abelian category. This *is* a direct extension of Klein's program.

6. Notice that an *equivalence* of categories is a special case of adjointness. Indeed, if in the above triangular identities the arrows $\eta\colon 1_{\mathscr{D}} \to FG$ and $\xi\colon GF \to 1_{\mathscr{C}}$ are natural *isomorphisms*, then the functors F and G constitute an equivalence of categories. We can now come back to duality from a different perspective. As we have seen in Chapter 2, Eilenberg, Mac Lane and their contemporaries knew about certain important dualities and at least one of them played a key role in the development of category theory, namely Pontrjagin duality between certain groups and homomorphisms of groups. This duality can be described as an equivalence of categories. Let $\mathscr{C}$ denote the following category:

> the objets fo $\mathscr{C}$ are the locally compact Abelian groups;
> the morphisms of $\mathscr{C}$ are the continuous group homomorphisms.

The Pontrjagin duality theorem amounts to the claim that the category $\mathscr{C}$ is equivalent to the category $\mathscr{C}^\circ$, that is the opposite category. Of course, the precise statement requires that we describe the functors $F\colon \mathscr{C} \to \mathscr{C}^\circ$ and $G\colon \mathscr{C}^\circ \to \mathscr{C}$ and prove that they constitute an equivalence of categories.

Another well-known and important duality was discovered by Stone in the thirties and now bears his name. (see [255].)[3] As is well known, Stone was a functional analyst studying linear operators in Hilbert space, and in particular algebras of commuting projections. It was known to Stone and others that the latter could be given the structure of Boolean algebras, although of a peculiar kind since these algebras could not be represented as algebras of subsets. Stone knew that a representation theorem for these Boolean algebras would be a tremendous tool. He achieved this result by importing topological concepts into algebraic structures, something that was entirely new. In this way, he was able to construct from an arbitrary Boolean algebra a topological space and, conversely, from a (compact Hausdorff and totally disconnected) topological space, to construct a Boolean algebra. Moreover, this correspondence is functorial: any Boolean homomorphism is sent to a continuous map of topological spaces and conversely, any continuous map between the spaces is sent to a Boolean homomorphism. In other words, there is an

---

[3] For a thorough and systematic presentation of dualities in general, see [120]. For some of their applications in logic, see [202] and [259] and [260].

equivalence of categories between the category of Boolean algebras and the dual of the category of Boolean spaces (also called Stone spaces). Stone's theorem is fundamental in more than one way. For: (1) Stone's motivations were clear right from the start: his result allowed him to prove various results by going back and forth between the algebraic setting and concepts and the topological setting and concepts. For instance, he proved the maximal compactification of a completely regular space by looking at the algebraic side of the problem and did the same for a generalization of Weierstrass' approximation theorem;[4] (2) Although Stone did not present his result in the language of category theory—he obtained his results in the mid thirties and published the full version in 1937, almost ten years before Eilenberg and Mac Lane's paper—the connection between a category of algebraic structures and the opposite of a category of topological structures established by Stone's theorem constitutes but one example of a general phenomenon that did attract and still attracts a lot of attention from category theorists. Categorical study of duality theorems is still a very active and significant field and is largely inspired by Stone's result.

In general, duality theorems like Stone's theorem or Gelfand's theorem between categories are neither more nor less than the categorical expressions of deep mathematical connections between mathematical domains that might seem to be unrelated.

We now have to make some important remarks about adjunctions.

1. As we have seen, all fundamental concepts of category theory can be described as adjoints and many important mathematical results are also described by the existence of adjoints to certain given functors. Adjointness is clearly the fundamental or the core concept of the theory. Its discovery and its role contributed directly and importantly to the emergence of category theory as a distinct and autonomous branch of mathematics. As we have seen, this came as a surprise. It was a *post-hoc* discovery. Thus, the importance of adjoints is justified in two complementary ways. Firstly, a priori, the notion is the natural expression of what it is to be a conceptual inverse to a given functor. Secondly, a posteriori, the fact that so many important mathematical concepts, including the notion of universal arrow, representable functor and logical concepts, can be presented as adjoints is a remarkable and important fact. This is some sort of "empirical" evidence of the importance and the "objective" nature of the concept.

2. In the same way that being functorial is heuristically valuable in many circumstances, looking for an adjoint to a given functor can be fruitful and can lead to surprises. It often yields the right operation with the right properties. Indeed, as we have already seen, an adjoint to a totally trivial functor very often turns out to be a fundamental mathematical construction! The best examples of this are given by products and coproducts, and, more generally, by limits and colimits.

---

[4] I should point out that Stone's paper contain a proof of universality of the construction and constitutes an early example of the phenomenon.

However, one has to reflect more seriously about adjoints. Functoriality makes a lot of intuitive sense: after all, preserving commutative diagrams amounts to preserving "identities", i.e., isomorphisms, and various properties expressed by commutative diagrams. What about adjointness? Of course, adjoints do have important preservation properties and equivalences preserve the key facts about categories. A left adjoint, *as* a left adjoint, preserves every existing colimit of its domain category and a right adjoint, *as* a right adjoint, preserves every existing limit of its domain category. Even though these preservation properties are extremely important, another and probably simpler aspect of the *conceptual* significance of adjoints can be given by going back to the conceptual significance of functors in general.

If we think of a functor as translating a certain property, or "problem", of one category into another (and that, as we have seen, allows us to conclude in certain cases that our problem has *no* solution), the existence of an adjoint tells us that our problem *has* a solution and that, in fact, it is the *best* possible solution of a certain type. It is a *generic* (or, I am tempted to say, *paradigmatic*) solution in the sense that any other solution will factor through it. An adjoint *is*, in some sense, a solution to a representation problem. If we think about a given functor as encapsulating a problem, an adjoint is one of the generic answers, the other one given by the other adjoint, if it exists. To use a metaphor: if the given functor is seen as encoding a given problem arising in a certain theory, thus in a certain language, into a different theory, thus a different language, then its adjoints can be thought of as taking the answer in the second language and decoding it to send it back in the original language in the best possible way. In other words, given a category of objects of a certain type, what are the constructions one can perform or that exist within that family? Again, these are fundamental questions from an epistemological point of view. When one is trying to know a family of objects better, one of the things one tries to do is to find a canonical decomposition of these objects into "prime" objects, or a subclass of these objects that are easier to study or to know. But a decomposition means that we can join together these objects in one way or another, thus that there are legitimate constructions available within the family of objects and these correspond to the existence of certain adjoints.

3. But here lies what may seem to be a mystery: why do almost all fundamental constructions performed in different fields of mathematics turn out to be adjoints? Is this an *epistemological* phenomenon, a fact of our way of thinking? Is it an *ontological* phenomenon, a fact about the way mathematical objects are? Can it be a mixture of the two or something else altogether? Is this a legitimate question? What is the framework one has to put oneself in to provide an answer? It might very well be that there is no "deep" answer, if this is what one is looking for. It is as if we had discovered that there is something called "conjunction" which covers so many different linguistic cases. We might marvel at it and try to find how this can be so. But it is simply a fact. What *is* wonderful is that the language of category theory allowed for the definition of adjoints, that within the categorical framework, one could find the appropriate level of abstraction.

Once adjoints have been defined, once their usefulness has been recognized, once their ubiquitous character has been acknowledged, then it becomes natural to use them whenever appropriate, to look for them whenever they might show up and to exploit them as much as possible. In other words, the way one *does* and *thinks* about mathematics has changed in an irreversible manner.

5.2 Equivalence of Categories Again

We have seen in the previous section that an equivalence between two categories $\mathscr{C}$ and $\mathscr{D}$ is given by an adjunction such that the natural transformations η and ξ are natural *isomorphisms*. Thus an equivalence provides a way to transform a category into itself without loosing in any essential manner the underlying criterion of identity. Indeed, given an arbitrary adjunction, one or both functors F and G can send objects that are *not* isomorphic to isomorphic objects, even the same object. For instance, as we have already seen, the forgetful functor $U \colon \mathbf{Grp} \to \mathbf{Set}$ sends groups to their underlying sets and group homomorphisms to their underlying set functions. Clearly, the two non-isomorphic four-elements groups will be sent to isomorphic sets, even to the same set in some cases. Thus representing groups as sets forces us to identify things that we considered distinct in the original context. As we have already mentioned, the forgetful functor U has a left adjoint F which associates to each set X the free group $F(X)$ with generators the elements of X. Thus, coming back along F into the category of groups, our original non-isomorphic groups will be sent to isomorphic free groups that will nonetheless be related to the original groups in the way specified by the adjunction. This is what an equivalence does *not* do: it does *not* send non-isomorphic objects to isomorphic objects, or put differently, it *reflects* isomorphisms in the sense that if $F(X)$ and $F(Y)$ are isomorphic, then X and Y were isomorphic in the first place. In the case of equivalences, the functors preserve *and* reflect isomorphisms.

Isomorphic categories are simply isomorphic objects in the category of categories, even if we do not know exactly what that is. The fact that isomorphism between categories is *not* the correct criterion of identity for categories is another indicator of their intrinsically geometrical nature. Indeed, the notion of equivalence between categories has the flavor of a homotopy equivalence between topological spaces. As we have seen, two spaces X and Y are homotopy equivalent if there is a homotopy equivalence between them, that is if there are continuous maps $f \colon X \to Y$ and $g \colon Y \to X$ such that $f \circ g$ is homotopic to 1_Y and $g \circ f$ is homotopic to 1_X. In turn, $g \circ f$ is homotopic to 1_X means that there is a continuous deformation of the image of $g \circ f$ and the identity map 1_X of the space X into itself. The spaces X and Y are not mapped one-to-one and onto by a homotopy equivalence. Recall that a disk in the real plane, for instance, is homotopy equivalent to a point. Analogously, two categories $\mathscr{C}$ and $\mathscr{D}$ are equivalent if there is an equivalence between them, that is if there are functors $F \colon \mathscr{C} \to \mathscr{D}$ and $G \colon \mathscr{D} \to \mathscr{C}$ such that $FG \to 1_{\mathscr{D}}$ and $GF \to 1_{\mathscr{C}}$; that is, the image of FG and GF in the respective categories can be naturally transformed into the identity functor of the category.[5] The "size" of the categories can be very different. Furthermore, in the same way that the notion of homotopy equivalence can be considered to be the fundamental criterion of identity for spaces, the notion of equivalence of categories is the correct notion for categories, at least at a certain level. However (and we will come back to this point in the conclusion), since it is possible to define a homotopy between homotopies, there is in fact a hierarchy of criteria of identity inherent to homotopy theory, and not surprisingly the same holds true for the notion of equivalence for categories.

Here are some fundamental examples of equivalent categories that will be useful later.

Let X be a set. Consider the functor category $\mathbf{Set}^X$ of all functors from X into the category of sets, that is, the category whose objects are functors $F: X \to \mathbf{Set}$ and morphisms natural transformations $\eta: F \to G$. It can be seen that every object of this category can be presented as a disjoint X-indexed family $\{Y_i\}_{i \in X}$ of sets and morphisms are functions between these families of sets. Consider now the so-called slice category $\mathbf{Set}/X$ of sets over X: an object of this category is a function $f: Y \to X$ and a morphism between two objects $f: Y \to X$ and $g: Z \to X$ is a function $h: Y \to Z$ such that $g \circ h = f$. Thus a morphism can be represented thus:

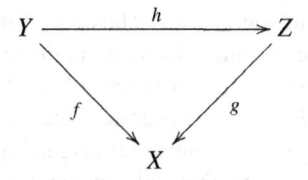

$$\begin{array}{ccc} Y & \xrightarrow{h} & Z \\ & \searrow_f \swarrow_g & \\ & X & \end{array}$$

Notice that an object $f: Y \to X$ of $\mathbf{Set}/X$ can also be thought of as being a disjoint family $\{Y_i\}_{i \in X}$ of sets: indeed for each $i \in X$, the fiber $f^{-1}(i)$ over i is a set; for $i \neq j$, $f^{-1}(i) \neq f^{-1}(j)$ since f is a function and clearly $Y = \bigcup_{i \in X} f^{-1}(i)$. There is therefore an obvious functor $\mathbf{Set}/X \to \mathbf{Set}^X$, sending an object $f: Y \to X$ to the corresponding family $\{Y_i\}_{i \in X}$ of sets and similarly for morphisms. This is an obvious equivalence between the given categories.

The previous example can be extended in a somewhat less obvious direction. We start with the category of topological spaces over a space X, denoted by $\mathbf{Top}/X$. As in the previous case, an object is a continuous map $f: Y \to X$ and an arrow $h: f \to g$ is a continuous map $h: Y \to Z$ such that $g \circ h = f$. We restrict this category further, to what are called *étale bundles* over X. In order to define étale bundles, we need to recall the notion of a local homeomorphism. A continuous map $f: Y \to X$ is a *local homeomorphism* if for each y in Y, there is an open V with $y \in V \subset Y$ such that $f[V]$ is open in X and $f|_V : V \to f[V]$ is a homeomorphism. An *étale bundle over X* is simply such a local homeomorphism. The category $\mathbf{Etale}(X)$ of étale bundles over X is defined to be the category whose objects are étale bundles over X and arrows continous functions as above. It can be shown that a continuous function between

---

[5] There is, by the way, a homotopy theory of groupoids (see, for instance, [34]), and higher-dimensional categories, to which we will turn later, can be thought of as a generalization of homotopy theory to categories in general.

5.2 Equivalence of Categories Again

two étale bundles is necessarily a local homeomorphism. For instance, when X is "the" one point space, it can be verified that an étale bundle over X is necessarily a space Y with the discrete topology. This is one part of the equivalence.

To look at the equivalent category, we have to take an algebraic point of view on topologies. Indeed, from the algebraic point of view, the topology of a space X, that is the family $\mathcal{O}(X)$ of opens of X, is a lattice, which can thus be considered to be a category, namely with objects the open sets $U, V, W, \ldots$ of the topology and arrows the inclusion functions between open sets. We can therefore consider the functor category $\mathbf{Set}^{\mathcal{O}(X)^\circ}$, from the dual of $\mathcal{O}(X)$ into the category of sets. The objects of this functor category are called *presheaves* on X. We consider a (full) subcategory of the category of presheaves on X, the category of *sheaves* on X, denoted by $\mathbf{Sh}(X)$. A *sheaf* (of sets) on X is a functor $F\colon \mathcal{O}(X)^\circ \to \mathbf{Set}$ satisfying the following condition: for each open covering $U = \bigcup_i U_i$, $i \in I$, of an open set U of X, the diagram

$$F(U) \xrightarrow{e} \prod_i F(U_i) \underset{q}{\overset{p}{\rightrightarrows}} \prod_{i,j} F(U_i \cap U_j)$$

is an equalizer, where for $t \in F(U)$, $e(t) = \{t|_{U_i} \mid i \in I\}$ and for a family, $p(t_i) = \{t_i|_{(U_i \cap U_j)}\}$, $q\{t_i\} = \{t_j|_{(U_i \cap U_j)}\}$. This rather abstract condition expresses a local property of sheaves: it "says" roughly that there is a unique way to "glue" together functions that are defined locally. In other words, for sheaves, one can systematically move from the local to the global, or the global can be "constructed" from the local in a systematic manner. The category $\mathbf{Sh}(X)$ is therefore the category of such sheaves and morphisms are natural transformations between them. The category $\mathbf{Sh}(X)$ is in fact equivalent to the category $\mathbf{Etale}(X)$. (For a proof of this basic and important fact, see [200], Chapter II, 6, 88–92.) We will come back to sheaves in Chapter 7.

Another variation on the same theme, which brings us back to transformation groups. We can define the category $\mathbf{B}G$ of continuous G-sets, for G a topological group. The objects of this category are the (right) G-actions on sets X with the discrete topology, namely sets X with a right G-action $X \to X$, and the arrows are functions which preserve this action. It can be shown that this category is also equivalent to a category of sheaves, denoted by $\mathbf{Sh}(G)$, but this time the sheaves are defined on what is called a site, which is a generalization of a topology to the categorical case. We will also come back to this case in Chapter 7. (See, [200], 150–154.)

One last example involving duality in an essential way and which we have already mentioned in my examples of adjunctions: the so-called duality theorems. Dualities are concrete examples of links made between fields which seem to be at first conceptually incommensurable. For instance, Stone duality states that there is an equivalence between the *dual* of the category of Boolean algebras and the category of totally disconnected compact Hausdorff spaces. This means that from an abstract point of view, they are the "same". Notice that it is the dual category of the category of Boolean algebras, or equivalently, the dual of the category of totally disconnected compact Hausdorff spaces, one or the other, which is involved. In practice, what this means is that it is possible to start with one of the given categories, say the category of Boolean algebras, define some properties for these algebras and then, by dualizing, obtain a property of totally disconnected compact Hausdorff spaces.

Let us note a few significant facts about these examples. First, in the first three examples, we have an equivalence between a category of "standard" mathematical objects on the one hand, and mathematical objects presented as functors on the other hand. From a categorical point of view, this functorial presentation is fundamental; as we have seen, this was emphasized by Grothendieck explicitly very early on and it is found implicitly in Kan's work. Second, two homeomorphic topological spaces X, Y will lead to equivalent categories of presheaves $\mathbf{Set}^{\mathscr{O}(X)^\circ}$ and $\mathbf{Set}^{\mathscr{O}(Y)^\circ}$ and, a fortiori, two equivalent categories of sheaves. In fact, in these cases, what matters is the structure of the lattice of open sets of the spaces: isomorphic lattices will yield equivalent categories. More generally, two equivalent categories $\mathscr{C}$ and $\mathscr{D}$ will yield two equivalent functor categories $\mathbf{Set}^{\mathscr{C}^\circ}$ and $\mathbf{Set}^{\mathscr{C}^\circ}$ (and also $\mathbf{Set}^{\mathscr{C}}$ and $\mathbf{Set}^{\mathscr{D}}$). The converse is far from being true: in general, if $\mathbf{Set}^{\mathscr{C}^\circ}$ is equivalent to $\mathbf{Set}^{\mathscr{D}^\circ}$, it does *not* follow that $\mathscr{C}$ is equivalent to $\mathscr{D}$.

Equivalence of categories is not only useful in practice, but it is also conceptually significant and constitutes the first step towards identifying he underlying criteria of identity and meaningfulness in the context of category theory. Since these occupied the front stage in my presentation of Klein's program, we have to see how they appear in a categorical context. It is now time to go back to Klein's program explicitly.

5.3 Back to Klein

As we have seen, Eilenberg and Mac Lane established a correspondence between a space and its group of transformations and a category and its algebra of mappings, e.g., the Hom-sets with the operation of composition. This is indeed a generalization, but I now want to argue that it leads to another generalization of Klein's program. I want to interpret it more literally to start with: a category *can* be thought of as a space and then one can look at its "algebra" of transformations, i.e., the "algebra" of functors. Needless to say, the last expression is meaningless, strictly speaking, for there is no such thing as "the" algebra of functors on a category. Recall that what Eilenberg and Mac Lane did in their paper, is to restrict the analysis to endofunctors, e.g., of the category of topological groups and, with this parameter fixed, the algebra of functors amounted to the relationships between subfunctors of given functors. The introduction of adjoint functors changes the scene dramatically. Because adjoint functors allow for an autonomous development of the basic concepts of category theory, not only does it become an autonomous branch of mathematics, but its general conceptual significance, already present in its original form as a general language and framework, is also greatly modified. Going back to Eilenberg and Mac Lane's original example, the category of (topological) groups, the systematic investigation of the existence of adjoints to certain "trivial" functors, e.g., the existence of limits and colimits, now constitutes an analysis of the basic construc-

5.3 Back to Klein

tions of that field. The search for adjoints to other naturally arising functors, e.g., forgetful functors, is also part of that analysis. Furthermore, from this point of view, mathematics studies functors and, among them, adjoint functors. This is simply reformulating and generalizing Eilenberg and Mac Lane's claim about group theory. Recall that in the original paper, they claimed that

> The subject of group theory is essentially the study of those constructions of groups which behave in a covariant or contravariant manner under induced homomorphisms. More precisely, group theory studies functors defined on well specified categories of groups, with values in another such category ([74], 237).

But the autonomy allowed by adjoint functors leads to an ontological leap that we have already encountered. By describing a mathematical field with the adjoint functors and their relationships that are characteristic of that field, one can forget entirely about the precise nature of the objects and morphisms involved. The objects do not have to be given in advance, they do not have to be sets, nor do the morphisms need to be given in advance, and they do not have to be functions. What *are* the objects of an Abelian category? Are they sets? They do not have to be. What are the morphisms of an Abelian category? Are they structure-preserving functions? They do not have to be. The various representation theorems guarantee that the objects of an Abelian category are at least functors and morphisms are natural transformations! But one does not have to know what the objects and morphisms of an Abelian category are to do (parts of) homological algebra. At some stage, one will have to know various representation theorems to prove some specific results. However, this does not change the fact that the characterization of that part of mathematics is now accomplished in a way that is entirely set-free (provided that a category itself is not a set or a class, something that is certainly not clear at this stage). Be that as it may—and we will come back to this important issue in due course—the nature and role of categories have now changed. I can now exploit the parallel with transformation groups. For, in the same way that a transformation group is an abstract encoding of a geometry, a category presented with the help of certain adjoints and properties of these adjoints is an abstract encoding of a piece of mathematics: it gives precise means of *definition* and *proofs*. Using this analogy, we can transfer properties of transformation groups to categories, for the status and role of transformation groups are to geometry what categories are to mathematical domains. In fact, since a transformation group *is* a category, we are simply generalizing the claims made for transformation groups to categories in general. I will make a few remarks and try to establish more properties in the following chapters.

It should be clear that the various categories described in the examples of equivalences are in some sense different. The descriptions of the objects are different and so are the morphisms. If categories were sets or classes, then we could certainly say immediately that these categories are different *as* sets or classes. Their cardinality could even be different: they could be of different sizes. These are striking cases of mathematical objects that are different in their presentation, in the way we are to think about their constituents. The equivalences between these categories indicate, however, that they are nonetheless the same in a different way. One would like to be able to find what is underlying this identity, to state precisely what they have in common. The appropriate level of analysis is provided by adjoint functors. The parallel with geometrical spaces and transformation groups that we have seen in Section 1.2 could not be more striking. In both cases we are dealing with well-defined mathematical structures and properties that seem to be different in important respects. However, from a global point of view, they are essentially the same. In the case of geometrical spaces, the different spaces can be seen as different representations of a common abstract transformation group. Not surprisingly, the same is true for categories.

Once categories defined with the help of specified adjoints have been introduced, one can consider that the subject matter of category theory is not, say, the category of groups or the category of topological spaces, but rather what deserve to be called "abstract categories". In some of its *applications*, it is possible to go back to specific categories of given structures and morphisms. In this context, the given category will be seen as an occurrence, an instance of one or more "abstract" categories and the properties of the specific category will be derived from these instantiations. Again, we are witnessing a shift analogous to the shift initiated by Klein and Lie in geometry: categories of specific entities are pushed to the background and "abstract" categories, that is categories given by a list of adjoint functors and properties of the latter, constitute the fundamental object of study, what a given mathematical field is all about. The role played by categories of specific entities becomes similar to the role played by representations of transformation groups.

I have claimed that transformation groups have a key role to play in elementary geometry, not only because they are useful to define concepts and prove various results, but also because they constitute criteria of identity and meaningfulness for geometric spaces. We ought to be able to transfer this claim to categories in some way. In other words, categories should constitute in some sense criteria of identity and meaningfulness. Here, the fact that a group is a special kind of category gives us a hint. This point is so important that it deserves a section of its own.

5.4 From Groups to Groupoids

> The idea that one can "introduce" a kind of objects simply by laying down an identity criterion for them really inverts the proper order of explanation. As Locke clearly understood, one must first have a clear conception of what kind of objects one is dealing with in order to *ex*tract a criterion of identity for them from that conception. (...) So, rather than "abstract" a kind of object from a criterion of identity, one must in general "*ex*tract" a criterion of identity from a metaphysically defensible conception of a given kind of objects.
>
> ([183], 517)

5.4 From Groups to Groupoids

> Any geometric object in a simple manifold determines a particular structure and therefore a space (...) The geometric objects and the spaces which they determine are classified by means of the pseudo-group of regular point transformations, two objects belonging to the same class if, and only if, they are equivalent. It is this pseudo-group, rather than the group G_u, which is relevant, because a geometric object is not necessarily defined over the whole of a simple manifold. This classification of geometric objects is in the spirit of the Erlanger Programm, equivalence under the pseudo-group being an inevitable generalization of equivalence in the narrower sense.
>
> ([264], 49)[6]

> This [Categories in differential geometry] is the title of the last survey article by C. Ehresmann, who in categories recognized a decisive motive force in the foundations of differential geometry. Among the roots of this recognition are Lie's theory of continuous transformation groups, and the Lie-Klein Erlangen Program. Lie etc. considered in this context structures more general than those we now call Lie groups—namely "pseudogroups" (possibly infinite-dimensional) and with multiplication only locally defined.
>
> To understand differential geometry in the spirit of the Erlangen program means thus more than understanding the formal notions of group and group actions. The necessary comprehensive notions for this were brought to the light of day by É. Cartan and C. Ehresmann, in particular with the realization that pseudogroups are categories (in fact groupoids).
>
> ([142], 551)

Eilenberg and Mac Lane's methodological motivation was clearly fruitful and more than useful. However by considering a group of automorphisms merely as a tool for investigating the properties of the underlying structure, they had put aside a crucial component of Klein's approach. Indeed, as we have seen, a group of transformations of a space is not merely a group of automorphisms. From the geometric point of view, it also provides a criterion of identity for the objects of the space. In elementary geometry, that is, in homogenous spaces, groups of transformations do more than capture global properties of a space. Given any structure, its associated group of automorphisms, if it is not trivial, captures important features of the structure, but it does not necessarily provide a criterion of identity for meaningful objects definable within that structure. If my claim that categories are to spaces of mathematical structures what transformation groups are to geometry is to hold any water, I ought to be able to show that categories provide a criterion of identity for mathematical structures. And indeed, they do.

As a first step, we have to go back to geometry and see how the group concept as a criterion of identity can be extended to cover spaces in which it cannot be applied directly. As we have seen, for the so-called Klein geometries, the spaces are homogeneous and therefore the transformations can be defined for the whole space and the group action is transitive. I have already mentioned the fact

---

[6] It has been said that this small book influenced Ehresmann in his work in the foundations of differential geometry. Ehresmann does refer to it in a note in his paper "Sur la théorie des espaces fibrés". (See [62], 133.)

that there are many spaces for which no non-trivial global transformation can be defined. But this does not mean that it is impossible to define local transformations, in particular local automorphisms, of the space and collect them into an algebraic structure. It is important to see that the resulting algebraic structure does constitute a criterion of identity for geometric objects in the space just as a transformation group yields a criterion of identity for homogeneous spaces. Indeed, it is the structure of the space itself that forces us to look at things locally. The collection of local automorphisms reduces to a transformation group whenever the space has the appropriate structure. This yields the definition of what is called a pseudogroup. Here is the formal definition: given a topological space (manifold) X, consider the collection of all local homeomorphisms (diffeomorphisms) of X, that is the set $\Gamma = \{f\colon U \to V : \forall U, V \subseteq X \ (f \text{ is an homeomorphism})\}$ such that:

1. If $f\colon U \to V$ is in Γ and $W \subseteq U$, W open, then $f|_W : W \to f[W]$ is in Γ, i.e., the restriction of an element $f \in \Gamma$ to any open set in its domain is also in Γ;
2. Let $U = \bigcup_i U_i$, U_i open sets of X, if $f|_{U_i} : U_i \to f[U_i]$ is in Γ for all i, then $f\colon U \to \bigcup_i f[U_i]$ is in Γ; this means that the property of being in Γ is local;
3. $1_U : U \to U$ is in Γ for all open sets U;
4. If $f \in \Gamma$, then f^{-1} is in Γ;
5. If $f\colon U \to V$ is in Γ and $f'\colon U' \to V'$ is in Γ and if $V \cap U'$ is non-empty, then $f' \circ f$ is a homeomorphism and is in Γ. In particular, when $V = U'$, then this conditions states that the composition $f' \circ f$ of two elements of Γ, when defined, is in Γ.

Such a collection Γ is called a *pseudogroup* of X.[7] The conditions simply ensure that all local homeomorphisms are compatible with one another and compose in the correct manner. Pseudogroups play a fundamental role in the foundations of differential geometry. (See for instance [141] from which the above definition is taken.) However, pseudogroups are defined on topological spaces. They depend on topological spaces in the same way that transformation groups depend on geometric structures. We need a more abstract definition.[8] This leads us to the concept of *groupoid*.

Although the definition of a groupoid can be given independently of category theory, it is best seen as a special type of category: a *groupoid* $\mathscr{G}$ is a category in which every morphism is an isomorphism.[9] It can be shown that a groupoid can be constructed from every pseudogroup and conversely certain subsets of a groupoid admit a multiplication that makes them the abstract counterpart of a pseudogroup. For, given a pseudogroup Γ on a space X, the associated groupoid is constructed by

---

[7] Notice that this is not how Veblen and Whitehead defined the notion in their book [264] but it is indeed the concept they are referring to in the quote above.

[8] Notice how I am using the expression "abstract" in this context. We want a definition that keeps the algebraic ingredients of the situation but ignores the inessential, topological, elements.

[9] Oddly enough, although the definition was known when Eilenberg and Mac Lane wrote their first papers on category theory, they did not include groupoids as an example of a category. Ehresmann, on the other hand, came to categories via groupoids. For a brief history of the concept of groupoid, see [34].

5.4 From Groups to Groupoids

considering the "germs" of elements of Γ. In general, germs of continuous functions are defined thus: let x be a point of a space X and F a family of functions defined on a neighborhood of x (each function in its own neighborhood). Two functions f, g are said to be equivalent (at x) if they coincide in some neighborhood of x. An equivalence class generated by this relation is called a *germ of functions of class F at* x. Hence, given a pseudogroup Γ, it is easy to see how a germ of functions is defined from Γ and since the functions are homeomorphisms, one obtains a groupoid.

A one-element groupoid, that is a category with one object in which every morphism is an isomorphism, *is* a group and in fact can be seen as a group of transformations, for the elements of the group are presented as transformations, namely morphisms. This last example can be generalized. Let $\{G_i\}_{i \in I}$ be a family of groups indexed by a set I. Define the category $\mathscr{G}$ with $\mathrm{Ob}(\mathscr{G}) = I$ and morphisms

$$\mathrm{Hom}(i,j) = \begin{cases} \emptyset & \text{if } i \neq j \\ G_i & \text{if } i = j \end{cases}$$

Composition on each $\mathrm{Hom}(i,i)$ is given by the multiplication of the group G_i. This is clearly a groupoid and whenever I is a singleton set, we get a group. Thus, the concept of groupoid is a generalization of the concept of group. Two specific examples of this construction are worth mentioning. First, consider the category $\mathbf{GL}(k)$ indexed by the natural numbers, whose morphisms $\mathrm{Hom}(i,i)$ are the groups $\mathbf{GL}_n(k)$ of invertible $n \times n$ matrices with entries in the field k and $\mathbf{GL}_0(k) = \{1\}$, the trivial group. This is clearly a groupoid. Second, let S_n be the symmetric group of all permutations of the finite set $\{1,\ldots,n\}$. Let $S_0 = \{1\}$. Then the category $\mathscr{S}$ of all symmetric groups defined as above is a groupoid.

Hence the notion of groupoid is a natural generalization of the notion of group and the notion of pseudogroup. Moreover, and this is a key fact, it is a generalization both in the logical *and* the geometric directions. Indeed a groupoid can still be thought of as providing a criterion of identity. There are at least four ways to see how a groupoid is a formal expression of a criterion of identity. Let us consider them in turn.

First, let us take a somewhat different look at the previous discussion. Instead of considering a topological space and the collection of its local homeomorphisms, consider an indexed family of structures, for instance geometrical spaces, $E = \{E_x\}_{x \in B}$ over a space B, that is a projection map $p\colon E \to B$. The *fiber* over $x \in B$ is the set $p^{-1}(\{x\})$, which is simply E_x. A fiber can be, for instance, the tangent space E_x at that point of the space B. For each $x \in B$, we have a group $G(E_x)$ of automorphisms of the fiber and these groups are connected to one another via the isomorphims $G(E_x) \to G(E_y)$ induced by the isomorphisms between E_x and E_y (another case of transport of structure). This yields the groupoid $G(E)$ associated with the given family. Notice, however, that in this generality, the resulting groupoid *is* the generalization of an arbitrary group of automorphisms, that is a "symmetry groupoid" and as such it does not necessarily reflect or yield a criterion of identity for objects definable in the family. Thus, to keep the criterion of identity, E, B and p have to have the appropriate geometric structure. Again, the basic idea is that a group will act only locally, that is on each fiber, and can be moved around in the space B systematically, according to the latter structure. In this context, we basically have a local criterion of identity that can be transported in the space according to a certain law. More specifically, E and B must be manifolds, p a smooth projection and there must be a (Lie) group H acting simply transitively (and smoothly) on each fiber E_x. In this case, a fiber can be thought of as a space of all moving frames over a point of the space B. The isomorphisms $E_x \to E_y$ are sometimes called in this context "admissible maps", that is homeomorphisms commuting with the action of H. This is the proper context in which differential geometry (developed by Élie Cartan, Herman Weyl and Charles Ehresmann) is elaborated; differential geometry constitutes a generalization of Klein's program.

Second, groupoids sit right at the core of topological spaces, via the fundamental groupoid. The latter is customarily defined as the fundamental group, but it is really a groupoid and should be treated as such. Recall that if X is a space, then there is a partial binary operation defined on the set of all path classes in X by $[f][g] = [f*g]$. It can easily be verified that it is a groupoid. When a point x_0 of X is fixed, then, as we have seen, we obtain the fundamental group of X. There are, however, numerous advantages to considering the groupoid instead. See, for instance, [35].

Third, it can easily be shown that any equivalence relation R on a set X can be presented as a groupoid on X. Thus, a groupoid is a generalized equivalence relation. The set X is first thought of as a category by taking its elements as the objects of the category. An arrow $f \colon x \to y$ exists if and only if $(x,y) \in R$. It is easily seen that the conditions satisfied by an equivalence relation yield a category, and in particular that symmetry of the relation turns the category X into a groupoid as required. The diagonal equivalence relation on X, that is the relation containing only the pairs (x,x) for all x in X, is sometimes called the *fine groupoid* on X. Seeing an equivalence relation as a groupoid can be thought of as providing a criterion of identity of elements of X as "identical as R-forms". Moreover, a groupoid not only encodes which objects are equivalent, but also the different ways ("isomorphisms") in which two objects are equivalent.

Finally, a group G acting on a set X can also be seen as a groupoid: representing the action of the elements g of G on the elements x of X as arrows $(x,g) \colon x \to x^g$, a product of two arrows $(x,g) \colon x \to x^g$ and $(x^g, h) \colon x^g \to x^{gh}$ is defined by:

$$(x,g)(x^g, h) = (x, gh).$$

Thus, as a groupoid, its objects are the elements of X and the arrows are defined as above. The laws satisfied by a group action immediately yield a groupoid. Both these constructions, which appear trivial at first, play an important role in practice. (See, for instance, [34], where these notions are presented and applications discussed.)

Let us come back to categories. As one can easily see, any category $\mathscr{C}$ has an underlying groupoid: it is simply given by the isomorphisms of $\mathscr{C}$. The foregoing discussion suggests that we should think of a category $\mathscr{C}$ as a space, in an abstract sense, whose "points" are in fact "geometric objects" and with a criterion of identity provided by the underlying groupoid. Notice that the groupoid also keeps track of

5.4 From Groups to Groupoids

the different ways two objects are the same. Thus a category is an "abstract geometry". In more familiar terms: given a family of structures of a certain kind, i.e., a category $\mathscr{C}$, the underlying groupoid of the category provides the criterion of identity for the objects seen as structures of that kind. This simply means that in a category, isomorphic objects are indistinguishable and this is precisely because they can be transformed into one another. In the same way that equal figures in a geometry are identical as geometric objects and may be distinct as sets, in a category $\mathscr{C}$, two isomorphic objects are identical as $\mathscr{C}$-objects and may be distinct as sets. When in a category $\mathscr{C}$, it is not necessary nor even sensible to work with the set-theoretical criterion of identity, e.g., the axiom of extensionality. The underlying criterion of identity of a category $\mathscr{C}$ does not depend upon an underlying set-theoretical criterion in the same way that the criterion of identity of geometric objects does not depend upon an underlying set-theoretical criterion. We will therefore say that the underlying groupoid of a category $\mathscr{C}$ is an *internal* criterion of identity, that is a criterion of identity for the entities *within* the category. This is simply a different way of saying that isomorphic objects of a category $\mathscr{C}$ are considered to be the same qua $\mathscr{C}$-objects. Notice that this implies that the identity relation is *contextual*: two things are treated as the same in a given context, whereas in a different context, they might be different. This illustrates what was obvious but implicit all along: from a categorical point of view, entities are always entities in a given context. In a categorical perspective, there is no such thing as a context-independent entity. The question "What is x?" for x a mathematical entity, even as simple an object as π, does not make sense if a context is not specified. (This is what [162] have named the "context principle".)

I have already given illustrations of this phenomenon. Applying a functor F from $\mathscr{C}$ to a different category $\mathscr{D}$ is analogous to mapping a certain geometric space into another space, and thus potentially modifying the criterion of identity, which is the associated transformation group. Thus, applying a functor amounts to a possible change of criterion of identity in the same way that modifying a group of transformations amounts to changing the criterion of identity for geometric figures. For instance, as we have seen, applying a forgetful functor $U: \mathscr{C} \to \mathbf{Set}$ amounts to a changing the criterion of identity in as much as two different $\mathscr{C}$-objects may turn out to be indistinguishable as sets. There are, for instance, various partial orders definable on a given set X of cardinality n, not necessarily finite. As partial orders, they are all different objects in the category of partially ordered sets. Applying the forgetful functor U to the category of partially ordered sets will send all these different objects to the "same" object, that is to the component, let us call it $[n]$, of the groupoid of equinumerous sets.

Just as a transformation group encodes a criterion of identity for geometrical figures *and*, at the same time, a criterion of identity for a geometry, the underlying groupoid of a category $\mathscr{C}$ encodes an internal criterion of identity for $\mathscr{C}$-objects *and*, at the same time, an external criterion of identity for the category $\mathscr{C}$ itself. But here there is a subtle point that has to be emphasized. A *specific* group action encodes an internal criterion of identity for geometric figures, *but* a group *type* (that is, not a specific group but a group up to an isomorphism) encodes an external criterion of identity for a geometry; likewise, a *specific* category encodes an internal criterion of identity, but an *equivalence type* encodes an external criterion of identity for mathematical fields.

It should be obvious that there is a connection between groupoids and equivalence of categories. Indeed, since functors in an equivalence between categories preserve and reflect isomorphisms, it follows that such functors preserve the underlying groupoids of the categories involved. It seems extraordinarily tempting at this point to claim that not only do we have a criterion of identity—which is true but incomplete when looked at carefully (we will come back this point later)—but also a criterion of meaningfulness: a meaningful categorical property is a property P preserved and reflected by equivalence functors. Here lies our first surprise: it is simply false.

This was first shown by Freyd in 1976. We should pause for a second and think about the time it took before someone actually asked the question and provided an answer! Categories had been defined thirty years earlier, adjoint functors and equivalence of categories had been around for twenty years and no one before Freyd could say precisely what a genuine categorical property was, that is, provide a systematic and general answer. In practice, everybody knew what they were and supposedly everyone thought that these properties were preserved and reflected by equivalence functors.

Freyd's counterexample is very simple: an equivalence $F\colon \mathscr{C} \to \mathscr{D}$ between categories preserves equalizers but does not reflect them. Indeed, a morphism $F(f)$ may be an equalizer of $F(g)$ and $F(h)$ in $\mathscr{D}$, without f being an equalizer for g and h in $\mathscr{C}$, and this for the most simple and, as Freyd says himself, perverse reason. For it is entirely possible, for instance, that $g\colon X \to Y$ and $h\colon U \to V$ in $\mathscr{C}$, thus prohibiting the construction of the equalizer in $\mathscr{C}$ (because the morphisms g and h don't even have the same domain and codomain), while $F(X) = F(U)$ and $F(Y) = F(V)$ in $\mathscr{D}$, making the construction of the equalizer possible in $\mathscr{D}$. But "having equalizers" is intuitively a genuine categorical property, since it is a universal property.

What is required is a finer analysis of the notion of "P is a categorical property", in the light of the criterion of identity of categories, namely equivalence of categories. "Being preserved and reflected by equivalences" will not do. Informally, what is needed is a characterization of the properties that are invariant within equivalence types. More formally:

(&) If $P(\mathscr{C})$ and $\mathscr{C}$ is equivalent to $\mathscr{D}$, then $P(\mathscr{D})$ if and only if P is a genuine categorical property.

The problem is to characterize what is meant by "P is a genuine categorical property". Freyd answers the problem with the help of a *diagrammatic* language in which it is possible to characterize genuine categorical properties in the sense that the characterization yields a proof of the statement (&). What is interesting here is that, if one were to start with the standard first-order logic and try to characterize categorical properties in this fashion, one would fail. As Freyd himself observes, free formulas are not preserved by equivalence functors, and neither are negatomic predicates. However, once the diagrammatic language has been introduced and used

5.4 From Groups to Groupoids

in the proof of the theorem (&), then it is possible to translate the diagrammatic language into a two-sorted first order language with bounded quantifiers. Here we will restrict ourselves to the first-order counterpart. My presentation will follow Freyd's.

Freyd introduces a two-sorted language: a sort O for objects $X, Y, Z, \ldots$ and a sort M for morphisms $f, g, h, \ldots$. One atomic relation, $=$, and two unary operations written $\Box f$ and $f\Box$ are also needed. Informally, one reads the formula "$X = \Box f$" as "X is the domain or source of f" and "$X = f\Box$" as "X is the codomain or the target of f". Quantifiers are "restricted" in the following sense:

$$\forall_{X \xrightarrow{f} Y}[\ldots] \text{ is the abbreviation of } \forall_f[(X = \Box f) \wedge (Y = f\Box) \Rightarrow \ldots]$$

and

$$\exists_{X \xrightarrow{f} Y}[\ldots] \text{ is the abbreviation of } \exists_f[(X = \Box f) \wedge (Y = f\Box) \wedge \ldots].$$

A sentence is called *Freyd-diagrammatic*[10] if all quantified maps are restricted in the aforementioned sense and

1. No map is quantified without its domain and codomain having been previously quantified;
2. The atomic predicates $X = \Box f$, $Y = f\Box$ do not appear other than implicitly in the restricted quantifiers;
3. The atomic predicate $(X = Y)$ does not appear;
4. If $(f = g)$ appears as an atomic predicate then the restricted quantifiers for f and g imply that $\Box f = \Box g$ and $f\Box = g\Box$;
5. If $g \circ f = h$ appears as an atomic predicate then the restricted quantifiers for f, g, and h imply that $\Box f = \Box h$, $f\Box = \Box g$, and $g\Box = h\Box$;
6. If $f = 1_X$ appears as an atomic predicate then the restricted quantifier implies $X = \Box f$ and $X = f\Box$.

The other formation rules are standard. What Freyd shows is that theorem (&) can be proven for the diagrammatic language first, and then he shows that any elementary sentence S in the diagrammatic language is invariant within equivalence types if and only if there is a Freyd-diagrammatic sentence S' such that the axioms of category theory imply $S \Leftrightarrow S'$.

Notice that there is no equality between objects X, Y of a category and that they are only needed to restrict quantifiers. This could be easily simplified by adopting bounded quantification over a multisorted language right from the start. Equality exists *only* between morphisms. This is not surprising since the identity conditions for objects are *extracted* from the categorical context in the form of isomorphisms between them. Although I haven't presented the diagrammatic language, I will simply claim that the usual universal properties are diagrammatic properties and that

---

[10] Freyd calls them "Frege-diagrammatic", but since the notation has nothing to do with Frege's notation (as Freyd is well aware), I prefer to call them "Freyd-diagrammatic". "Peano-diagrammatic" would be another obvious alternative.

therefore they are invariant within equivalence types. The interested reader can consult [93]. (For a more systematic and more general presentation of a formal system adequate for equivalence types of categories, see [206].)

I am thus in the position to make the following claims:

1. Category theory is a generalization of Klein's program in many different senses. By replacing a group with a groupoid and then with a category (which is appropriate and natural in all cases), one can use categories in the way Klein thought groups could be used in geometry.[11] But in the same way that Klein's analysis is suited to elementary geometry and that a more sophisticated analysis is required for more general geometrical settings, groupoids and categories constitute the entry point of higher-dimensional groupoids and categories, a setting in which the question of criteria of identity is clearly delicate. (We will briefly come back to this point in the conclusion.)
2. The use of adjoint functors to define various types of categories reveals a profound *unity* of mathematical concepts and methods; in the same way that the introduction of the group concept in geometry led to a systematic understanding of the links between various geometrical systems and geometrical methods, the introduction of the category concept opened the way to a better understanding of various mathematical concepts and methods. Interestingly enough, Klein's view on the use of groups in geometry was first and foremost seen as a pedagogical tool. It was also how Eilenberg and Steenrod conceived the role of categories in algebraic topology: they thought it was pedagogically valuable. Of course, both were right, but history showed that there was much more to it than a pedagogical instrument.
3. The introduction of categories and (adjoint) functors paves the way for a systematic classification of mathematical domains; this classification is well founded since there are intrinsic criteria of identity and of meaningfulness at work; furthermore, these criteria show that categories are in some sense independent from set-theoretical foundations.

Two important points still have to be clarified: (a) the very status of category theory itself; (b) the extent to which category theory can be applied to foundational issues in mathematics. We will now make some remarks about the first issue from an historical perspective and then move to the second in the following chapter, since major developments were made in the sixties, mostly by a few mathematicians, which set the stage for further developments.

---

[11] This is probably literally true of Ehresmann's work in differential geometry. There is of course a direct filiation from Klein to Cartan to Ehresmann. But Ehresmann's work still has to be analyzed carefully. It was more or less ignored by the categorical community for a long time, for reasons that are beyond the scope of this study.

5.5 The Foundations of Category Theory… Again

> Categorical algebra has developed in recent years as an effective method of organizing parts of mathematics. Typically, this sort of organization uses notions such as that of the category **G** of all groups. (…) This raises the problem of finding some axiomatization of set theory—or of some foundational discipline like set theory—which will be adequate and appropriate to realizing this intent. This problem may turn out to have revolutionary implications vis-à-vis the accepted views of the role of set theory.
>
> ([194], 231)
>
> It is a remarkable empirical fact that mathematics can be based on set theory. More precisely, all mathematical objects can be coded as sets (in the cumulative hierarchy built by transfinitely iterating the power set operation, starting with the empty set). And all their crucial properties can be proved from the axioms of set theory. (…) At first sight, category theory seems to be an exception to this general phenomenon. It deals with objects, like the categories of sets, of groups etc. that are as big as the whole universe of sets and that therefore do not admit any evident coding as sets. Furthermore, category theory involves constructions, like the functor category, that lead from these large categories to even larger ones. Thus, category theory is not just another field whose set-theoretic foundation can be left as an exercise. An interaction between category theory and set theory arises because there is a real question: What is the appropriate set-theoretic foundation for category theory?
>
> ([28], 6)

The first textbooks in category theory appeared on the scene in the sixties. They are, in order of appearance: [89], [228], [64], [37], and [235]. This is an indication that category theory, which in the sixties almost always took the form of the theory of functors, was considered by some as being a mathematical theory worthy of exposition as such. The emphasis in all these books but one, namely Ehresmann's book, was on Abelian categories and homological algebra. Freyd's book, the very first one on category theory, is even entitled *Abelian Categories*. They all introduced adjoint functors and made the connection between adjoint functors, representable functors and universal morphisms.[12] At that point, the constructions used in category theory were more intricate: functor categories were part and parcel of the machinery, both in the context of Abelian categories and in relation to adjoint functors. It was no longer possible to claim, as Eilenberg and Mac Lane had done in 1945, that one could dispense with the notion of category altogether and work with pairs of objects. The status of categories was becoming a legitimate question and its precise relationship to set theory was also an issue. It is interesting to note that there was in fact very little published during this period on the topic. It may very well be that category theory as an independent discipline was still in its infancy and that there

---
[12] Freyd introduces the material related to adjoint functors in the exercises of his book. Here is how he presents the situation: "One important area of functor theory which is not touched in the text is the theory of adjoint functors. It is too important to leave out entirely, and hence we have included a range of exercises on the subject" ([89], 10).

were more challenging categorical problems to tackle than the problem of how it relates to set theory. The books published in that period are revealing in that respect: the way categories are *defined* and the manner in which the *foundations* of category theory are treated give us a picture of category theorists' attitudes. The status of categories just did not seem to make any real difference in practice. One defined categories in a certain way, then ignored the underlying aspects of the definition and concentrated on the relevant and useful properties, and these properties are, not surprisingly, purely categorical in the sense previously explained. *Some* care had to be taken, but everyone knew how to circumvent the most obvious problems, which all seemed to have something to do with questions of size. But as I have already suggested, and as we will see later, this is probably a bit careless. There is one important exception, as far as books are concerned: Mac Lane's book published in 1971.

But the situation is not as simple as the textbooks might lead one to believe. It is certainly the case that almost all category theorists did not care much about the issue and similarly, given the recent proof by Cohen of the independence of the continuum hypothesis and the forcing method introduced therein, almost all set theorists did not care much about the issue either, since they were busy exploring further applications and modifications of Cohen's method. Here is how Cohen himself viewed the situation in 1967:

> Yet, unless I am not sufficiently aware of current trends, the set-theoretical difficulties in handling categories have not inspired many set theorists and it has had little impact in logic as a whole. Thus, although we thoroughly accepted highly impredicative set theory because we understand its internal cogency, we, as logicians, are less likely to accept category theory whose roots lie in algebraic topology and algebraic geometry. It could be retorted that the existing axioms of infinity are ample to cover formalizations of category theory, yet an obstinate categorist could say that categories themselves should be accepted as primitive objects ([49], 13–14).

This statement probably reflects the attitude of most logicians of that period. Notice that the argument is pretty weak: because the roots of category theory lie in algebraic topology and algebraic geometry, logicians are less likely to accept it. This is a clear and simple case of a genetic fallacy. The fact that set theory arose from real analysis is not viewed as weakness by logicians. Other, more interesting arguments, were given by other logicians at about the same time and we will get back to them shortly. Two mathematicians seem to have taken the problem seriously. Mac Lane had presented a paper on the topic in the late fifties in which he described the various difficulties that crop up when working in NBG for instance. (See [187].) At about the same time Grothendieck had developed his solution to the problems, at least those generated by the use of functor categories.[13] But these did not generate much discussion.

The situation drastically changed after the publication of Lawvere's widely read Ph.D. thesis and two short notes that were published soon afterwards in which he

---

[13] It seems that Grothendieck started thinking about the problem in the late fifties when the question of the foundations of category theory were brought to the fore at the meetings of the Bourbaki group. The first appearance of Grothendieck universes are in [96]. I owe this information to Ralf Krömer.

5.5 The Foundations of Category Theory... Again

proposes to found mathematics on category theory and, in particular, to reformulate set theory in categorical terms. ([165], [167]) Here is how Mac Lane reports reacting to Lawvere's thesis: "I was stunned when I first saw it [i.e., Lawvere's thesis]; in the spring of 1963, Sammy [Eilenberg] and I happened to get on the same airplane from Washington to New York. He handed me the just completed thesis, told me that I was the *reader*, and went to sleep. I didn't"[14] ([196], 346). Although he did not endorse Lawvere's program entirely—for Mac Lane never argued in favor of a purely categorical foundational stance—Lawvere's dissertation and papers contained various suggestions as to how to handle foundational issues. It is clear, as we will see, that Mac Lane took these suggestions up and investigated them. Thus, towards the end of the sixties, Mac Lane presented a series of papers on the foundations of category theory and the foundations of set theory.[15] The second paper published in 1969 was accompanied by a paper written by Feferman on the same topic. (See [84].) These were, for quite a while the *only* papers that aimed at clarifying the set-theoretical foundations of category theory. One fact that played a key part in this development is the appearance of elementary toposes in the late sixties and early seventies, due to Lawvere and Tierney. But we are getting ahead of ourselves. Let us first go back to textbooks to see how the various authors defined categories.

In all the books published between 1963 and 1970, a category is either defined as a *class M* of morphisms with the obvious axioms, as in [89], or as a *class A* of objects and a *class M* of morphisms satisfying the obvious axioms, as in [228]. It is then specified that for objects X, Y, $\mathrm{Hom}(X,Y)$ is a *set*. Thus all authors work in a set-theoretical language, usually NBG. During that period, a few distinctions were made to organize the field: (1) a *large category* is one in which the collection of morphisms is a class whereas a *small category* is a category in which the collection of morphisms is a set; (2) a distinction is introduced between *concrete* and *abstract* categories. An *abstract category* is any category satisfying the foregoing definition of a category as a class. A *concrete category* is a pair $\langle \mathscr{C}, U \rangle$ where $\mathscr{C}$ is a category and $U : \mathscr{C} \to \mathbf{Set}$ is a *faithful* functor. What this definition says is that a concrete category is a category in which objects can be treated as sets and morphisms can be treated as functions. The underlying motivation behind this distinction is that there are categories in which morphisms are *not* functions, as I have already mentioned and many mathematicians saw in this fact the fundamental *raison d'être* of category theory. For instance the category **hTop**, is *not* concrete in this sense, while the category **Top** is concrete. The proof of the non-concreteness of a category is more involved than one might think. (See [92] for details and other examples of non-concrete categories.) The first distinction guarantees that a functor

---

[14] It should be pointed out that it was also Mac Lane who communicated Lawvere's notes published in the Proceedings of the National Academy of Sciences, published in 1963 and 1964.

[15] See [190], [191], [192], [194]. Here the dates of publication are misleading. For in fact the last paper, published in 1971, was in fact the first paper presented on the issue. It is based on a paper presented at a symposium on pure mathematics in 1967. The second one was probably finished before 1968 and it is more or less a general presentation of category theory with a sketch of Lawvere's proposal. The first paper published in 1969 was in fact finished in 1968. The second was received by the editor in the spring of 1969. These dates are important since Mac Lane seems to change his mind considerably between 1968 and 1969. We will come back to this point.

category $\mathscr{C}^{\mathscr{D}}$ can be constructed provided that the category $\mathscr{D}$ is small. Although this approach seems to work for most purposes, the introduction of restrictions on the size of categories is, from the point of view of category theory, rather artificial. Furthermore, in some contexts, large categories and functors between large categories became inescapable, as in Grothendieck's work with fibred categories first presented in the early sixties. (See [105].) But there is no trace of these difficulties in the aforementioned books.

Let us now turn to alternatives available in the literature at that time. As I have already indicated, a different approach, based on ZFC with additional axioms, was developed by Grothendieck in the late fifties. It is called the *method of universes* or *Grothendieck universes*. The definition, as presented in SGA4 by Verdier, goes as follows. (See [5].)

Definition 5.1. A *universe* is a non-empty set U satisfying the following properties:

(U1) If $x \in U$ and if $y \in x$, then $y \in U$.
(U2) If $x, y \in U$, then $\{x, y\} \in U$.
(U3) If $x \in U$, then $\wp(x) \in U$.
(U4) If $\{x_i \mid i \in I \in U\}$ is a family of elements of U, then $\bigcup_{i \in I} x_i \in U$.

Clearly the empty set is a universe, called U_0. Furthermore, the collection of finite sets is also a universe, let us call it U_1. Thus these axioms are, as such, not very strong. For this reason, Grothendieck added the following axiom:

(UA) For any set x, there is a universe U such that $x \in U$.

This is a very strong axiom. Just consider the case when x is the set $\mathbb{N}$ of natural numbers. Then axiom (U3) yields all the powers of $\mathbb{N}$ and axiom (U4) then makes it possible to take the union of these powers. We will call this universe U_2. It can be shown that the axiom (UA) is equivalent to the existence of strongly inaccessible cardinals, an axiom independent from ZFC.

The advantage of this framework is that the "usual" categories, that is the categories of structured sets, are definable: given a universe U, it is possible to define the category of sets in U, or U-sets, the category of topological spaces in U, the category of Abelian groups in U, the category of categories in U, etc. For instance, in U_1, there is the category of U_1-sets, i.e., of finite sets, the category of finite topological spaces, the category of finite Abelian groups, the category of finite categories, etc. Of course, one normally considers a universe U in which there is an infinite cardinal. The "usual" categories are then elements of U.

As for the "unusual" categories (e.g., in Grothendieck's usage, presheaves and sheaves, and more generally functor categories), the method of universes gives the means to handle them. Some terminology is required. Given a universe U, we say that a set is U-*small* if it is isomorphic to an element of U. We say that a category is a U-*category* if for every pair (X, Y) of objects of $\mathscr{C}$, $\mathrm{Hom}_{\mathscr{C}}(X, Y)$ is U-small. It is easy to verify that if the categories $\mathscr{C}$ and $\mathscr{D}$ are members of U, then the functor category $\mathscr{D}^{\mathscr{C}}$ is an element of U and that if $\mathscr{C}$ and $\mathscr{D}$ are U-small, then $\mathscr{D}^{\mathscr{C}}$ is U-small. Furthermore, if $\mathscr{C}$ is U-small and $\mathscr{D}$ is a U-category, then $\mathscr{D}^{\mathscr{C}}$ is a U-category. Now, let $\mathscr{C}$ be a U-category and consider the category of all functors from

5.5 The Foundations of Category Theory... Again

$\mathscr{C}$ to U-sets, i.e., $Usets^{\mathscr{C}}$. This is clearly not, in general, a U-category. Just consider the case when U is U_1 and $\mathscr{C}$ is the category $\mathbb{N}$ with objects the natural numbers and morphisms the usual ordering relation. It is clearly a U_1-category, but the category $U_1 sets^{\mathbb{N}}$ is not. However, it is a U_2-category. Thus, there is always a chain of universes $U_1 \subset U_2 \subset U_3 \cdots$ that can be extended indefinitely as needed.

Although this framework is technically flawless, various criticisms were leveled at it. The most obvious criticism, raised for instance by Mac Lane, is that it is not possible to consider the category of *all* groups. One is always restricted to the category of U-groups, for some universe U, and this is not faithful to category theory.[16] Second, from a set-theoretical point of view, large categories require existence principles that go beyond ZFC, that is, beyond what is apparently needed for the rest of mathematics. Third, the method of universes introduces a lot of technicalities that seem totally irrelevant to the actual *practice* of category theory, e.g., keeping track of the universes and moving from one universe to the next when constructing functor categories. Presumably, one would want a foundational framework that is faithful to the practice and not one that solves a technical problem in a merely ad hoc manner. Finally (this criticism was raised much later by, e.g., [23]), although questions of size are relevant to category theory and have to be treated with care, there are other important conceptual issues that are not even touched upon by this framework.

In his book, Mac Lane starts by presenting a slight modification of the method of universes: he presents a framework in which there is only one universe U. This approach was initially presented in [192], probably following a lead offered by [167], where we read: "The last clause thus embodies the idea that only *one* inaccessible is needed for most mathematics; our world thus stops far short of the second Grothendieck universe if we assume the above axioms" ([167], 18). But in this paper Lawvere is developing a purely categorical framework and not a set-theoretical framework. He therefore leaves open the question of a set-theoretical translation of his idea, and this is what Mac Lane develops. More formally, he assumes ZFC and adds the following axiom:

(AU) There is a set U, called a *universe*, satisfying the following properties:

    (i) $x \in u \in U$ implies $x \in U$;
    (ii) $u \in U$ and $v \in U$ imply $\{u,v\}$, $\langle u,v \rangle$ and $u \times v \in U$;
    (iii) $x \in U$ implies $\wp(x) \in U$ and $\bigcup x \in U$;
    (iv) $\omega \in U$;
    (v) if $f: a \to b$ is a surjective function with $a \in U$ and $b \subset U$, then $b \in U$.

Thus, the basic idea is to add an infinite set and some closure properties on the universe right from the start. With such a universe U fixed, a set x is said to be *small* if it is a member of U. Notice that this characterization of smallness is different from Grothendieck's definition. It can easily be seen that the small sets thus defined are

---

[16] This is certainly not *entirely* convincing. For why would it not be adequate to consider *a* category of groups, deemed adequate for the purposes at hand? This *was* Eilenberg and Mac Lane's attitude in the forties after all.... This clearly shows that Mac Lane's views on the status of categories had changed in the meantime.

themselves a model of ZFC. In this way, Mac Lane believes that he is simplifying the framework greatly and that it is for most purposes sufficient for category theory.

> Our intention is that the small sets can serve as the objects of Mathematics, while the other sets, not necessarily small, may be used to describe the various categories and functor categories of these Mathematical objects ([192], 195).

It is possible to construct the category of all small sets, the category of all small groups, the category of all small topological spaces, etc. Of course, the category of all small sets is not itself small. Thus the category of *all* groups cannot be formed. What Mac Lane takes to be the key property for the development of category theory is that of being locally small: a category $\mathscr{C}$ is said to be *locally small* when the set $\mathrm{Hom}(X,Y)$ is small for each pair of objects X, Y of $\mathscr{C}$. The reason is that local smallness is the key property for the proofs of crucial results in category theory, e.g., Yoneda's lemma, Kan extensions and Freyd's adjoint functor theorem. Therefore it seems that any set-theoretical framework adequate for category theory has to make room for local smallness. All the categories mentioned above (the category of small sets, etc.), are locally small. The rub is that, as we already know, not all categories are locally small: given locally small categories $\mathscr{C}$ and $\mathscr{D}$, the functor category $\mathscr{D}^{\mathscr{C}}$ is not necessarily locally small.

Mac Lane circumvents that problem by embedding any given set S into a larger set $\bar{S}$. The set $\bar{S}$ is no longer a universe in the foregoing sense, but it is such that problematic constructions which were leading to categories that were not locally small could now be made to yield locally small categories. I won't give the detail here for it is clearly an ad hoc solution to the problem once again, and to this extent it is not satisfactory.

The foregoing approach, although weaker than Grothendieck's method of universes, is still stronger than ZFC, since it postulates a specific universe (and possible extensions which are not universes but larger). Mac Lane presented this approach in 1969. It is not his first stab at the problem. Indeed, Mac Lane had earlier considered a set theory *weaker* than ZFC, which he called a theory of *schools* and was presented under that name only once. (See [192].)[17] However, the main point of this approach is to allow for "variable" models of set theory: it portrays category theory as moving from one model of set theory to a different, presumably larger, model of set theory. The starting point, however, is a weak set theory, essentially Zermelo set theory, that can be extended in different directions, depending on one's needs. Since schools have classes as parts, the *theory of schools* has two sorts: one for items and one for classes. The axioms are extensionality for classes, ordered pair, empty class, unit classes, Cartesian product and bounded comprehension over classes. A distinction between small and large classes is introduced via the concept of a subschool. But the key concept is that of a normal subschool: a subschool N of a school S is a *normal* subschool if it is a school such that the N-classes are full S-classes and all

---

[17] Mac Lane sketches the theory once again in 1971, but not under that name. It is there presented as a framework where there are multiple systems for foundations and was certainly written *before* he had the developed the formal version of the theory, since the paper was presented in 1967. Once again, it is hard not to see Lawvere's influence at work.

5.5 The Foundations of Category Theory... Again

S-subclasses of an N-class are also N-classes. Grothendieck universes are normal subschools and so is the class of all sets in models of NBG. There is no need for us to look at the details of this approach, since it did not have any echo when it appeared and Mac Lane resuscitated the idea in a different form later, after toposes were introduced. But what we do have to keep in mind is that category theory in fact does not need a theory as strong as ZFC to be developed. Mac Lane even claimed at some point that most of mathematics, in fact all of it except set theory itself, did not need ZFC but only a form of Zermelo set theory. It *is* interesting to note that this idea appeared at the same time that he was proposing a stronger framework and *before* the advent of topos theory.

Be that as it may, Mac Lane's attitude is revealing. We see him presenting first a system *weaker* than ZFC and then, soon after, a system *stronger* than ZFC. And that is not all. As I have already mentioned, he already knew about Lawvere's program of *starting* with categories and derive sets from them instead. This amounts to the claim that categories are *neither* sets *nor* classes in the standard formal sense, although the axioms of category theory can be *interpreted* to some extent within a set theory like ZFC or a variant thereof. Mac Lane is clearly sympathetic to Lawvere's program. Indeed, Mac Lane sketches Lawvere's proposal and comments on it by saying that "In this description of a category, one can regard 'object', 'morphism', 'domain', 'codomain', and 'composite' as undefined terms or predicates; the definition as given is then independent of set theory." ([190], 287.) In the same spirit, Mac Lane opens up his textbook on category theory by defining what he calls a *metacategory*, again borrowing the terminology and the presentation from [167]. The idea is to give an axiomatic definition of category theory for which it is *not* assumed that categories are sets or classes. Here is how the definition goes:

Definition 5.2. A *metagraph* consists of objects $a, b, c, \ldots$, arrows $f, g, h, \ldots$, and two operations, as follows:

Domain, which assigns to each arrow f an object $a = \mathrm{dom}\, f$;
Codomain, which assigns to each arrow f an object $b = \mathrm{cod}\, f$.

A *metacategory* is a metagraph with two additional operations:

Identity, which assigns to each object a an arrow $id_a = 1_a\colon a \to a$;
Composition, which assigns to each pair $\langle g, f \rangle$ of arrows with $\mathrm{dom}\, g = \mathrm{cod}\, f$ an arrow $g \circ f$, with $g \circ f \colon \mathrm{dom}\, f \to \mathrm{cod}\, g$.

And these operations are subject to the usual axioms, associativity and the unit law.

Although Mac Lane does not take this step, the foregoing data can be presented by the following diagram:

$$C_2 = C_1 \times_{C_0} C_1 \underset{\pi_2}{\overset{\pi_1}{\rightrightarrows}} \,{\cdot}{\circ}{\cdot}\, C_1 \underset{\tau}{\overset{\sigma}{\underset{\leftarrow 1_{(-)}}{\rightrightarrows}}} C_0$$

where we do not specify what C_1, C_0, σ, τ and $1_{(-)}$ are, but where C_2 is the *pullback* of $C_1 \underset{\tau}{\overset{\sigma}{\rightrightarrows}} C_0$. Then, the axioms can be stated as additional diagrams that have to be commutative. Needless to say, this description is motivated by the original definition but we now use the language of diagrams to present the data required.

Mac Lane's motivation is straightforward: "A metacategory is to be any interpretation which satisfies all these axioms" ([195], 8). He comes back to this motivation in the section on foundations:

> For this reason, there has been considerable discussion of a foundation for category theory (and all of Mathematics) not based on set theory. This is why we gave the definition of a category in a set-free form, simply by regarding the axioms as first-order axioms on undefined terms "object of $\mathscr{C}$", "arrow of $\mathscr{C}$", "composite", "identity", "domain", and "codomain" ([195], 24).

Thus, after reading Mac Lane in the late sixties and early seventies, the reader comes to the conclusion that there are three logical options: category theory can be developed within a set theory weaker than ZFC, within a set theory stronger than ZFC or on its own, that is, independently of ZFC or any formal set theory. The latter approach does not mean that sets are forgotten once and for all; they are simply reintroduced within a categorical framework. In other words, all options are on the table and, at that time, none seems to be really satisfactory.[18] It seems that everything is possible but that nothing is clearly adequate! How are we to decide? At that point in time, there does not seem to have been any *reasonable* way to make a choice. In fact, judging from what Mac Lane wrote in the years following and even to this day, it seems that he stayed in this superposition of mental states and never really made up his mind about the issue.[19]

There was yet another possibility, presented in 1969 by Feferman, following a suggestion made by Kreisel. From a purely set-theoretical point of view, it can be said that what Mac Lane presents as being a problem generated by category theory is simply a special case of a general problem that set theorists had considered even before the advent of categories; it is simply one version of the usual problem of set formation, that is the problem of what totalities can be considered to be sets. In fact, the problem goes back to Cantor and Frege. When Cantor considered the totality of all ordinals, he simply mentioned that it did not constitute a set. Cantor's reflections on the absolute point in the same direction. Frege's infamous axiom V leads to a similar problem: an unrestricted principle of comprehension, intuitively convincing as it might be, yields contradictions. Set theorists are well aware of this problem and have proposed various well-known solutions. Already in the thirties, when Zermelo introduced the cumulative hierarchy of sets, he envisaged a solution for sets that reminds one of Grothendieck and Mac Lane: "Any specifically described model of set theory can in some way be described as a set, that is, as an element of a higher model of set theory" ([270], 47). This idea has been translated, in contemporary set theory, into what is called a *reflection principle*: whatever is true of the whole

---

[18] We will see in the next chapter why Lawvere's program was seen as unsatisfactory at that point.
[19] Well, it seems that he gave up the one-universe approach and argued more in favor of a weak set theory, namely Zermelo with bounded quantifiers and choice. Mathias, a set theorist who is not defending set theory as the only acceptable foundational framework, looked carefully at Mac Lane's revised weak framework as presented in [196], and showed some of its limitations. See [216] and [217].

5.5 The Foundations of Category Theory... Again

universe of sets is in fact true of a part of that universe. The underlying idea is that the whole universe of sets, whatever it is, is in fact impossible to grasp by rational means and therefore, whatever is presented or described by rational means is merely a part of that universe. Feferman's approach is based on such a reflection principle.[20]

Formally, Feferman starts with ZFC written in the usual first-order language. A constant symbol, $\underline{s}$ for 'small', is then added to the language. Since formulas will be relativized to $\underline{s}$, we introduce the following conventions: write $\exists x \in \underline{s}\,\phi(x)$ for $\exists x[x \in \underline{s} \wedge \phi(x)]$ and $\forall x \in \underline{s}\,\phi(x)$ for $\forall x[x \in \underline{s} \supset \phi(x)]$. For any formula ϕ, $\phi^{(\underline{s})}$ denotes the result of relativizing all quantifiers to $\underline{s}$. The system ZFC/$\underline{s}$ is then defined by the following axioms:

1. the axioms of ZFC;
2. $\exists x(x \in \underline{s})$;
3. $\forall x \forall y[(y \in x \wedge x \in \underline{s}) \supset y \in \underline{s}]$;
4. $\forall x \forall y[x \in \underline{s} \wedge \forall z((z \in y \supset z \in x) \supset y \in \underline{s})]$;
5. $\forall x_1 \in \underline{s} \ldots \forall x_n \in \underline{s}[\phi^{(\underline{s})}(x_1,\ldots,x_n) \Leftrightarrow \phi(x_1,\ldots,x_n)]$ for each formula ϕ of the language with free variables $x_1,\ldots,x_n$.

The second axiom says that the universe of small sets is not empty, the third amounts to the claim that the union of small sets is small, the fourth axiom guarantees that subsets of small sets are themselves small. The fifth axiom is the reflection principle: it says literally that any formula ϕ with small sets as parameters describing a property of the universe in our given language is true if and only if the description is true of small sets. Put differently, it does not matter whether the variables are interpreted as denoting small sets or arbitrary sets. From a model-theoretic point of view, it means that the universe of small sets is an elementary substructure of the universe of all sets. Notice, however, that the reflection principle 5 is *false* if we take finite sets for the universe of small sets.

Feferman shows how categories, functors and natural transformations can be defined in this framework and how the distinctions between small categories and locally small categories can be made. He then conjectures that the theorems of category theory, as of 1969, are provable in a system weaker than ZFC/$\underline{s}$, where in fact the reflection principle is not even used.[21] However, the real interest of the system is that it has the two following properties:

1. Given any "metacategory" $\mathscr{C}$, the reflection principle guarantees that there is a category $\mathscr{C}_{\underline{s}}$ which is a *set* and such that they have the same *set-theoretical* properties, i.e., they have the same properties expressible in the language of set theory under consideration. In the words of Blass, "if we prove a theorem about small sets by using large categories, then the same theorem holds for arbitrary sets" ([28], 8). Any reference to inaccessible cardinals is simply removed. This is an exact formulation of the conviction that questions of size are only used to justify certain general constructions and they do not bear on the real mathematical content of the construction and its consequences.

---

[20] Lawvere also mentions a reflection principle, referring to Bernays and Levy, as a possible framework in his paper on the category of categories in 1966.

[21] Feferman works in a somewhat weaker system at first. But this is irrelevant to the discussion.

2. Feferman proves that ZFC/*s* is a conservative extension of ZFC. This means that whatever we can prove about sets in ZFC/*s* is already provable, although perhaps in a much more convoluted manner, in ZFC. Thus, as far as results about *sets* are concerned, category theory *does not* go beyond ZFC.

Feferman's results are important for they can be interpreted as showing that *as far as set theory* is concerned, category theory does *not* raise new foundational problems. Thus, from the purely set-theoretical point of view, Mac Lane's worries seem to be simply unfounded.

As we have seen, logicians in general never showed a deep interest in the foundations of category theory. It can be said that they never saw the situation as being radically different from other foundational problems, e.g., wheter the collection of all well-orderings is itself a well-ordering. The most striking and important illustration of this attitude is provided by Kreisel, who wrote two papers in the sixties in which he voices his opinion on the subject ([145] and [146].) He also wrote a critical review of Mac Lane's 1971 paper. Kreisel's reaction gives us the opportunity to present what seems to be the first expression of a profound disagreement on the nature of a foundational enterprise, a disagreement that seems to last even to this day. Indeed, I believe that Kreisel's opinion was and still is very influential among logicians and philosophers. Although Kreisel's arguments *were* legitimate and had an undeniable force when he formulated them in the late sixties, I believe that they are no longer tenable given the developments that followed in category theory and categorical logic.

The basis of Kreisel's criticisms is a distinction between *foundations* and *organization* of mathematics. Kreisel claims that the distinction is useful "for analyzing the nature of the problems presented by (existing) category theory, and, more generally, for analyzing the role of foundations for working mathematicians" ([146], 189). The distinction between foundations and organization amounts to the fact that foundations have to do with the *validity* of mathematical principles, as opposed to their efficiency. In other words, foundations should yield a *justification* of axioms or *reasons* for axioms. How is this supposed to be done and what type of justification are we to expect?

Here Kreisel's answer appears to be somewhat mysterious and difficult to understand. Let us start with a somewhat long quote.

> For foundations it is important to know what we are talking about; we make the subject as *specific* as possible. In this way we have a chance to make *strong* assertions. For practice, to make a proof intelligible, we want to eliminate all properties which are not relevant to the result proved, in other words, we make the subject matter less specific.
>
> Foundations provide an *analysis* of practice. To deserve this name, foundations must be expected to introduce notions which do *not* occur in practice. Thus in foundations of set theory, *types* of sets are treated explicitly while in practice they are generally absent; and in foundations of constructive mathematics, the analysis of the logical operations involves (intuitive) *proofs* while in practice there is no explicit mention of the latter ([146], 192).

Foundations, according to Kreisel, make the subject as *specific* as possible. What exactly does that mean? It suggests that in a foundational analysis, we come to *fix* the referent of our discourse. For instance, in a foundational analysis, the real numbers

5.5 The Foundations of Category Theory… Again

are *identified* with certain equivalence classes of Cauchy sequences or, in a different analysis, with Dedekind cuts. Kreisel seems to suggest that the foundational analysis exhibits what we are talking *about* when we talk about the reals. These notions do *not* appear in practice. Someone doing analysis does not prove results about real functions by referring to real numbers *as* equivalence classes of Cauchy sequences, for instance. At most, the analysis is provided in a preliminary chapter and is then forgotten, as in [236]. In order to do analysis, one uses the properties of the reals that are usually stated in the axiomatic presentation: the real numbers constitute a complete ordered field. Thus, in this sense, we make the subject less specific since we simply state the properties required to prove the various theorems of analysis.

This is fair enough. But, as we will see, foundations in this sense can *also* be carried out in a categorical framework. To mention but one interesting aspect revealed by a categorical analysis of the reals, whereas Cantor's and Dedekind's constructions are logically equivalent in ZFC, they are not necessarily so in some categories, more specifically in some toposes. This already indicates that some foundational analysis performed in a set-theoretical framework can be transferred into a topos-theoretical framework. This in itself might not satisfy someone like Kreisel. For although the technical result is surprising to some extent (but not too surprising since we were already prepared for some such result by constructive analysis of real numbers), it is fundamentally the "same". What would be interesting is a genuinely "new" analysis in the proposed framework and, indeed, such innovations are possible, as was demonstrated in synthetic differential geometry.

Furthermore, it is not entirely clear that seeing real numbers as, say, Dedekind cuts, provides *reasons for* or *justifies* the axioms. Clearly, one can take the axioms for a complete ordered field and verify that indeed, Dedekind cuts satisfy these axioms, i.e., make them true. In this sense, they constitute a justification. But the axioms will be just as true in different foundational schemes, e.g., in various categories or, more specifically, toposes. Furthermore, one can reasonably question whether a set-theoretical analysis is, at bottom, more *specific* in the foregoing sense than a topos-theoretical analysis, unless one assumes to begin with that there is a *unique* universe of sets (a claim which requires its own justification, given the state of knowledge in contemporary set theory). (See, for instance, [250].) To be fair to Kreisel, he wrote these notes before the advent of topos theory and the topos-theoretical implementations of these analyses. (More on elementary toposes in Chapter 7.) Be that as it may, we seem to be forced to shift to a different issue, namely the status of category theory itself.

What Kreisel seems to be indirectly objecting to here is the fact that category theory necessarily makes "the subject less specific", thus cannot provide a proper foundational analysis of mathematics. This is underlined further in the following passage:

> Foundations and organization are similar in that both provide some sort of more systematic exposition. But a step in this direction may be crucial for organization, yet foundationally trivial, for instance a new *choice of language* when (i) old theorems are simpler to state but (ii) the primitive notions of the new language are defined in terms of the old, that is if they are logically dependent on the latter. Quite often, (i) will be achieved by using new notions with more 'structure', that is less analyzed notions, which is a step in the *opposite* direction

to a foundational analysis. In short, foundational and organizational aims are liable to be actually contradictory ([146], 192).

It is hard not to see that Kreisel's target is category theory, in particular the role it played in the foundations of algebraic topology and homological algebra. Indeed, everyone, Eilenberg and Mac Lane to begin with, presented category theory first and foremost as a language that made it possible to express certain concepts, results and problems precisely, and to prove theorems in algebraic topology. Furthermore, as everyone saw and as I have indicated, category theory also provided the language with which to systematically *organize* algebraic topology and, afterwards, other areas of mathematics and even mathematics as a whole. Finally, categories, functors and natural transformations do indeed seem to have more "structure" and to be logically dependent on "older" and more "simple" notions, i.e., set-theoretical notions. Kreisel raises the same point in his appendix to Feferman's paper:

> Organization and foundations are incomparable. Organization involves a proper choice of language; we have already seen that this is not necessarily provided by set-theoretical foundations. On the other hand, we may have a very successful organization which leaves open the verification of adequacy conditions, at least for a given foundational scheme.
>
> Generally speaking, the aims of foundations and organization will be in conflict. Being an analysis of practice foundations must be expected to involve concepts that do not occur in practice (just as fundamental theories in physics deal with objects that do not occur in ordinary life). Organization is directly concerned with practice; (...) (Kreisel, in [84], 244–245.)

Kreisel's claims recall Russell's claims in the opening page of his *Introduction to Mathematical Philosophy*:

> Mathematics is a study which, when we start from its most familiar portions, may be pursued in either of two opposite directions. The more familiar direction is constructive, towards gradually increasing complexity: from integers to fractions, real numbers, complex numbers; from addition and multiplication to differentiation and integration, and on to higher mathematics. The other direction, which is less familiar, proceeds, by analyzing, to greater and greater abstractness and logical simplicity ([245], 1).

This is certainly in the spirit of Kreisel's proposal. Category theory and the organization it provides is certainly in the realm of "higher mathematics", whereas set-theoretical analysis, although this is not what Russell had in mind, belongs to the foundational project. Clearly, Eilenberg, Mac Lane, Steenrod, Cartan, Grothendieck and Kan were primarily concerned with mathematical practice. Category theory was useful as a language to organize, clarify and deepen some mathematical concepts, problems and constructions. However, Russell goes on:

> (...) instead of asking what can be defined and deduced from what is assumed to begin with, we ask instead what more general ideas and principles can be found, in terms of which what was our starting-point can be defined or deduced. It is the fact of pursuing this opposite direction that characterizes mathematical philosophy as opposed to ordinary mathematics ([245], 1).

The introduction of the concept of adjoint functors changes the landscape and the status of category theory. Adjoint functors provide the "general idea" and the "principle" from which various starting points can be defined and deduced. In this sense, category theory becomes part of mathematical philosophy.

5.5 The Foundations of Category Theory... Again

Kreisel would probably still not agree. He would claim that he has in fact anticipated this possibility already in the early phase of category theory. Thus, we read:

> Before going further into the relation between mathematical practice and foundations, it is worth noting the obvious distinction between (i) foundational analysis (which is specifically concerned with validity) and (ii) general conceptual analysis (which, in the traditional sense of the word, is certainly a philosophical activity). As mentioned above, the working mathematician is rarely concerned with (i), but he does engage in (ii), for instance when establishing *definitions* of such concepts as length or area or, for that matter, natural transformation. For this activity to be called an *analysis* the principal issue must be whether the definitions are *correct*, not merely, for instance, whether they are useful technically for deriving results not involving the concepts (when their correctness is irrelevant). In short, it's not (only) what you do it's the way that you do it ([146], 192).

What does it mean for a definition to be *correct*? We are told what it does *not* mean: it does not mean that it is useful in applications. But we are not told what it *is*. Is it given by some a priori criterion? Or is it given by some general empirical criterion? We are left in the dark. It is no surprise to find Halpern, in his review of Kreisel's paper, saying "In the next section the distinction between organization and foundation is elaborated. In this section a further distinction is made—conceptual analysis versus foundational analysis (which the reviewer found more confusing than clarifying)"([110], 1).

Two elements are mentioned by Kreisel in all his papers on the topic: (1) precise adequacy conditions, given by a set-theoretical analysis; and (2) logical simplicity, which goes back at least to Russell. With respect to the first ingredient, we have to turn to the second appendix of the textbook Kreisel wrote with Krivine. (See [148].)

In this appendix, which is on the foundations of mathematics, Kreisel starts off by stating what we have already seen, namely that foundational studies are concerned with describing and analyzing informal mathematics. One of the main goals of foundational work is that

> (...) in foundations we try to find (a theoretical framework permitting the formulation of) good reasons *for* the basic principles accepted in mathematical practice, while the latter is only concerned with derivations *from* these principles. The methods used in a deeper analysis of mathematical practice often lead to an extension of our theoretical understanding. A particularly important example is the search for new axioms, which is nothing more than a continuation of the process which led to the discovery of the currently accepted principles. (Kreisel in [148], 161.)

Then Kreisel attacks what he calls the formalist doctrine, which was also his main target in his paper published in 1971 (presented in 1967 at Berkeley).[22] ([146]) Once this is done, Kreisel turns to what he takes to be two valuable foundational programs: the set-theoretic semantic foundations and the combinatorial foundations. According to Kreisel, both views, the first realistic and the second idealistic, contain worthy elements but both also have limitations and defects. This judgment is based on the idea that "a conceptual framework is defective if it does not allow (theoretical) explanations of facts for which an alternative theory has an explanation, one purpose of theory being the extension of the range of theoretical understanding." (Kreisel, in

---

[22] More specifically, the targets were Paul Cohen and Abraham Robinson.

[148], 228.) He then claims that there are results that allow us to see the defects of both points of view.

Kreisel claims that the set-theoretical foundations provide a realistic analysis of mathematical practice; that is, it presents mathematics as being about certain abstract objects. They do so by reducing each mathematical structure $\mathfrak{U}$ to a set, which is then called a *realization* of that structure (usually called a *model*). An *adequate* axiomatization of the reduction of a structure $\mathfrak{U}$ to set theory is a set of axioms $\mathscr{A}_\mathfrak{U}$ satisfying the following conditions (p. 172):

1. $\mathscr{A}_\mathfrak{U}$ is purely logical (in the language of predicate calculus);
2. $\mathfrak{U}$ satisfies $\mathscr{A}$ and hence, there exists a structure that satisfies $\mathscr{A}_\mathfrak{U}$;
3. All structures that satisfy $\mathscr{A}_\mathfrak{U}$ are isomorphic (and, hence, isomorphic to $\mathfrak{U}$);
4. All intuitive properties of $\mathfrak{U}$ can be expressed or defined in terms of those explicitly mentioned in $\mathscr{A}_\mathfrak{U}$;
5. All assertions about $\mathfrak{U}$ that can be proved intuitively follow logically from $\mathscr{A}_\mathfrak{U}$.

Notice that the language does not have to be first-order and the formulas do not have to be finite in all cases. (And indeed, these cases are investigated in the book.) In fact, it is clear that first-order logic is *inadequate* in the foregoing sense for various structures, in particular it does not allow for a distinction between finite and infinite. (For more on limitations of first order logic, see [16].) The claim is that all 19th century informal mathematics can be reduced adequately to set theory. Kreisel underlines the fact that the set-theoretic foundations themselves cannot be adequately reduced in this way. For that would mean that they can be realized by a *set* by 2, and by 3, this set would be in one-to-one correspondence with the collection of all sets, which is impossible.

For structures such as the collection of all sets or the intuitive notion of ordinal, a generalized notion of realization is required. In this case, predicate symbols are added but for which variables now range over *all* sets, and not only over elements of *a* set. The adequacy conditions are then adapted to this variation: "axioms are set-theoretically justified if one has a (precise) concept which satisfies the axioms in the wider sense of realization" ([148], 176). Category theory clearly falls under this generalized notion: there is a first order characterization of the concept and although it permits self-applicability, this is no different from the notion of well-ordering, according to Kreisel. Thus, from his point of view, the foundations of category theory *do not* raise any genuinely new logical or set-theoretical issues. But, I do have to point out immediately, although it certainly was not clear to most people at that time, that condition 3 above is clearly inadequate for categories. For, as we have seen, the notion of isomorphism does not provide the proper criterion of identity for categories. This is the first serious indication that there is something different going on. As we have seen, the notion of equivalence for categories is analogous to the notion of homotopy equivalence for spaces.

Kreisel does not claim, however, that category theory might not bring about some important revisions in the foundations of mathematics. For one thing, there are already, according to him, foundational notions that are not reducible to sets, e.g., the notion of rule as treated in intuitionistic mathematics, where a rule is regarded

5.5 The Foundations of Category Theory... Again

as a process, or the notion of abstract structure or abstract property. Kreisel even suggests that non-set-theoretical foundations might begin with the (primitive) notions of rule and abstract structure in terms of which the notions of category theory can be defined, and not from the technical notion of category itself. (See Kreisel in [84], 243.) In the same paper, Kreisel rightly calls attention to the fact that category theorists have not given a clear and unambiguous presentation of category theory itself.

> What was missing before one could even *begin* to consider a problem of foundations for category theory, was a formulation of the latter. In other words, the specialists in the subject should decide what they are talking about, and state properties of these concepts (it is the business of specialists because 'category' in contrast to 'number' was introduced as a technical notion). Then we can see whether or not we get an adequate set-theoretical foundation. (Kreisel, in [84], 240.)

As we have seen, the very presentation of the concept of a category had changed from Eilenberg and Mac Lane's original definition to various presentations in the textbooks. In this respect, Kreisel is entirely correct.

Thus, with respect to the question of the relationship between set theory and category theory, we can sum up Kreisel's attitude as follows:

1. Category theory (at least as it was presented in the sixties) does not raise any new set-theoretical foundational problems;
2. Instead of starting from category theory and trying to see what it can bring to the foundations of mathematics, which is what Kreisel takes Mac Lane to be doing (see Kreisel's review of Mac Lane, [147]), one should instead look more carefully at category theory from the point of view of foundations;
3. Set-theoretical foundations are not, according to Kreisel, the only possible or valuable foundations for mathematics; there are alternatives and category theory might play a role in the formulation of new foundational schemes;
4. Category theory itself (at least as category theorists formulated it in the sixties) does not provide a foundational scheme as such; it is merely an organizational tool for mathematical practice.

The last point brings us to the issue of logical simplicity. Kreisel assumes, as many people do, that a foundational analysis ought to be based on logically simple notions and that the notions of membership, collection and rules are such notions. One might then argue that the notion of category presupposes the notions of collection and rule and that, therefore, from a foundational point of view, it cannot be taken as a starting point. Here, extreme care has to be taken.

From a purely philosophical point of view, this presupposes a certain conception of the nature and role of a foundational framework. From a mathematical point of view, what can be said at this point is that the work of Eilenberg, Mac Lane, Grothendieck, Kan and others had contributed to the introduction of a precise, although not entirely formal, notion, somewhat similar to what Cantor and Dedekind had done for the notion of set. In the same way that every mathematician knows what a set is, but is probably incapable of writing down the axioms of ZFC, most mathematicians nowadays know what categories, functors and natural transformations are *as tools* in their work. What remains to be done is to give a precise and formal analysis of categories in general, and this might turn out to be somewhat different from the informal picture used by most people. The first person who thought of this was Bill Lawvere; and as I mentioned, he proposed that the category of categories be considered as an adequate foundation for mathematics. This was certainly a radical suggestion. It was bold, almost inconceivable for mathematicians used to thinking in terms of elements, collections, properties, etc. It required a qualitative shift in the way one was to think about mathematical knowledge, mathematical entities and foundations, a shift similar to the one proposed by Klein and Lie in geometry a century earlier.

We are now ready to look at this alternative.

Chapter 6
Invariants in Foundations: Algebraic Logic

> When they introduced the theory of categories in 1945, Eilenberg and Mac Lane suggested the possibility of "functorizing" the study of general algebraic systems. The author has carried out the first steps of this program, making extensive use of the theory of adjoint functors, as introduced by Kan and refined by Freyd.
>
> ([164], 869.)

We now turn to logic and the foundations of mathematics proper. Without a doubt, the whole program of thinking the foundations of mathematics *in general* in a categorical framework is due to one person: F. William Lawvere. He launched the program in his very first work, namely his Ph.D. thesis, defended at Columbia under Eilenberg's supervision in 1963. Remarkably, the thesis already contains the key ideas that have guided Lawvere throughout his career and that have influenced the categorical community so greatly. Let us start with a summary of these key ideas. The category of categories is a framework for mathematics; it should be the foundation of mathematics, but in an "open" manner, i.e., there is no ultimate and once-and-for-all foundational framework. (Whether there is *one* category of categories is still a debatable issue. But in the early years, Lawvere used those terms.) Thus, every aspect of mathematics should be representable in one way or another in that framework. In other words, categories constitute the background to mathematical thinking in the sense that, in this framework, essential features of that thinking are revealed. More specifically mathematical objects and mathematical constructions should be thought of as *functors* in that framework. Even sets, considered now as *abstract sets*, can be construed as functors. There are no such things as sets by themselves; in fact there is no such thing as a mathematical concept by itself. Sets, more specifically abstract sets, form categories and the latter categories play a key role in the category of categories, i.e., in mathematics. Adjoint functors occupy a key position in mathematics and in the development of mathematics; one of the guiding principles of the development of mathematics should be "look for adjoints to given functors". In that way, foundational studies are directly linked to mathematical practice and the distinction between foundational studies and mathematical studies is a matter of degree and direction; it is not a qualitative distinction. Invariance occupies a central position in this program. As the foregoing quote clearly indicates, Lawvere is going back to the claim made by Eilenberg and Mac Lane that the "invariant character of a mathematical discipline can be formulated in these terms" (i.e., as we saw in chapter 2, in terms of functoriality). ([74], 237.) But now, this invariance is revealed by adjoint functors and other categorical structure and properties. The invariant content of a mathematical theory is the "objective" content of that theory; this is expressed in various places throughout his published work. To wit:

> As posets often need to be deepened to categories to *accurately reflect the content of thought*, so should inverses, in the sense of group theory, often be replaced by adjoints. Adjoints retain the virtue of being uniquely determined reversal attempts, and very often exist when inverses do not. ([176], 47. My emphasis.)

Not only should sets be treated in a categorical framework, but also logical aspects of the foundations of mathematics should be treated categorically, in as much as they have an objective content. In particular, the logical and the foundational are directly revealed by adjoint functors.

As I have said, these ideas, as well as others, are more or less implicit in Lawvere's thesis. Lawvere's approach should be contrasted not only to Eilenberg and Mac Lane's, but also and probably more importantly to Grothendieck's and Kan's. Whereas the latter were using and developing category theory for specific purposes, e.g., in the foundations of algebraic geometry in order to prove the Weil conjectures or the development of combinatorial homotopy theory, Lawvere's goal is general in the sense that it aims at incorporating the whole of mathematics. Furthermore, Lawvere's *usage* of categories is such that their *status* changes progressively in the sixties and seventies. As we will see, not only does Lawvere recognizes explicitly that categories defined axiomatically constitute autonomous *kinds* or *types* and are, as such, independent of any underlying set-theoretical structures and structure-preserving functions, but they become polymorphic: in addition to their usual role, they become the algebraic descriptions of formal systems and as such can be thought of as formal systems; they provide the underlying framework for semantics and as such can be thought of as the universe of interpretations.

A summary of the main results of the thesis were communicated to the *Proceedings of the National Academy of Sciences* by Mac Lane and published in 1963. Essentially the same summary was presented at Berkeley in 1963 at a symposium on model theory and later published in 1965 in the proceedings of the meeting. ([166]) Mac Lane also communicated Lawvere's axiomatization of the elementary theory of the category of sets in the same *Proceedings* in 1964 ([165]). That axiomatization was not in the thesis as such. The following year, Lawvere presented an explicit axiomatization of the category of categories, published again in the *Proceedings* in 1966 ([167]). Another paper, published in 1968, gives an account of the main elements of the thesis together with some new extensions. ([170]) Two short notes, abstracts of talks given at meetings of the Association of Symbolic Logic, are published in the *Journal of Symbolic Logic*. These notes contain the first attempts at presenting a systematic and comprehensive categorical treatment of first-order logic, via the notion of an *elementary theory*. ([169])

Between 1965 and 1968, a certain methodological shift in attention can be detected in Lawvere's work. Going back to his original program of clarifying the conceptual content of semantics, Lawvere realizes that certain types of categories can be *defined* purely by stipulating that certain adjoint functors to given elementary functors exist. The definitions of types of categories require nothing else than the existence of left or right adjoints to specific elementary functors. This is not how adjoint functors were used. In a loose sense, defining a category via the existence of adjoints amounts to the claim that certain basic conceptual operations can be represented in that category. This in itself would probably not be of foundational relevance, were it not for the fact that the categories so defined correspond in a precise technical sense to *logical* concepts and theories. Thus the existence of certain adjoints to specific elementary functors amounts to a specification of logical structures and resources. In this sense and paraphrasing Élie Cartan, a category (of a certain type) contains all the logic of a specific mathematical theory. With these ideas and results in his pocket, Lawvere could see that a program of "functorizing" the study of mathematical concepts in general could be formulated.

The presentation of these fundamental facts and the program that ensued were made in the paper entitled *Adjointness in Foundations*, published in 1969. This paper contains the seeds of a categorical program in logic and the foundations of mathematics. Two other papers of that period also contain important parts of that program: first the paper on diagonal arguments presented in 1968 and published in 1969 ([172]) and the paper on quantifiers and the comprehension schema as adjoints also presented in 1968 and published in 1970 ([173]). In addition to these papers, the two abstracts mentioned earlier have to be mentioned, since they played a role in the development of categorical logic. The discovery of the notion of elementary topos in collaboration with Tierney in 1969–70, can be seen as a direct extension of this work, opening up vast horizons hitherto unseen.

We will now look more carefully at the details of this program. We will start with Lawvere's study of algebraic categories, then move on to the category of categories and to the elementary theory of the category of sets. The last two papers did not have the same fate as his work on universal algebra. Despite the fact that Lawvere's work on the category of categories suffered from a slight technical flaw, both it and his work on the category of sets were essentially metamathematical, and category theory was not yet seen as a potentially useful framework for metamathematics. Studies on the category of categories that followed Lawvere's pioneering work were mathematically motivated and I speculate that no one saw what to do with the category of sets. It simply did not have a clear function. His work on algebraic theories, however, inspired much of what was to follow in logic, including Lawvere's own work, and it still constitutes the starting point of what are now called "doctrines" in categorical logic and the categorical approach to universal algebra. We will therefore return to categorical logic afterwards.

6.1 Lawvere's Thesis

> Lawvere's imaginative thesis at Columbia University, 1963 contained his categorical description of algebraic theories, his proposal to treat sets without elements and a number of other ideas. I was stunned when I first saw it; in the spring of 1963, Sammy and I happened to get on the same airplane from Washington to New York. He handed me the just completed thesis, told me that I was the *reader*, and went to sleep. I didn't.
>
> ([196], 346.)
>
> Essentially, algebraic theories are an *invariant notion* of which the usual formalism with operations and equations may be regarded as "presentation".
>
> ([164], ii. My emphasis.)

The main concept of Lawvere's thesis is the notion of *algebraic category*. The main result of the thesis is a *categorical characterization* of algebraic categories. Together with algebraic categories, Lawvere also introduced *algebraic theories* and *algebraic functors*. The three notions are intimately connected to one another. As Lawvere himself points out, there is a strong analogy between the way his work is developed and the theory of sheaves that had been just introduced at that time: Grothendieck had provided an axiomatic characterization of categories of sheaves on topological spaces, and Lawvere's goal was to characterize algebraic categories in a similar manner. The main tool of the thesis — the key to the connections between these notions — is the adjoint functor. Adjoint functors constitute the methodological core of the thesis and of the whole approach. The framework is presented as a new foundation for universal algebra. In the very first chapter of the thesis, which gives the underlying context of the work, there is a sketch of a first-order theory of the category of categories. Within that context, sets are defined in categorical terms, the notion of equivalence of categories is given, as well as the category of small categories, the category of large categories, the category of finite sets, the category of small sets, and the category of large sets. A categorical version of the Peano postulates is also given.[1] But the bulk of the chapter is given to the presentation and development of the notions of adjoint functors and limits. In the second chapter, algebraic theories are introduced, the category of algebraic theories is defined and various properties of the category are proved, e.g., the existence of an adjoint that corresponds to the existence of free algebraic theories. Chapter three deals with algebraic categories (and here Lawvere explicitly exploits the analogy with sheaves). The notions of algebraic semantics and algebraic structure are defined and a categorical characterization of algebraic categories is given. Chapter four deals with algebraic functors and their adjoints. Finally, in chapter five, particular cases and extensions are considered.

Let us look at the central concepts and results of the thesis and see how the invariant content of universal algebra is analyzed.

---

[1] These axioms are sometimes called the Peano-Lawvere axioms. (See, for instance [199], 67.)

6.1 Lawvere's Thesis

First comes the notion of an algebraic theory. As we have seen in the foregoing chapters, a group is usually thought of as a set together with some specified operations, e.g., multiplication, inverse and unit, satisfying certain identities, e.g., associativity, unit law and inverses. Formally, this is encoded by a signature, e.g., $(\times, ()^{-1}, e)$ or some such, with the standard axioms. But it is clear that the signature can in general vary and so does the choice of axioms. These choices determine, though, the *same* theory, in the sense that all the definable operations and all the theorems are the same. Lawvere's idea is to define a category which will encode all the information at once, thus independently of the choice of signature and axioms and call *that* category the theory. Here is the general definition:[2]

Definition 6.1. An *algebraic theory* is a (small) category $\mathscr{A}$ such that:

1. The objects of $\mathscr{A}$ are the natural numbers $0, 1, 2, \ldots$;
2. Each object n is the product of 1 with itself n times; thus, for each n, the projection maps $\pi_{1n} \colon n \to 1, \ldots, \pi_{nn} \colon n \to 1$ exist;
3. A morphism $m \to n$ is an n-tuple $(\omega_1, \ldots, \omega_n)$ of m-ary operations, where an m-ary operation is any map $m \to 1$; in particular the map $(\pi_{1n}, \ldots, \pi_{nn}) \colon n \to n$ exists;
4. If $\omega = (\omega_1, \ldots, \omega_n) \colon m \to n$ and $\xi = (\xi_1, \ldots, \xi_k) \colon n \to k$, then $\xi \circ \omega = (\xi_1 \circ (\omega_1, \ldots, \omega_n) \ldots, \xi_k \circ (\omega_1, \ldots, \omega_n))$.

It is easy to verify that composition so defined is associative and that the maps $(\pi_{1n}, \ldots, \pi_{nn}) \colon n \to n$ are the identity maps.

The underlying motivation is very simple and makes perfect sense once the notion of an *algebra of type* $\mathscr{A}$, also called an $\mathscr{A}$-*algebra*, has been given: it is simply a product preserving functor from an algebraic theory to the category of sets, $F \colon \mathscr{A} \to \mathbf{Set}$. Thus, $F(1)$ picks a set, say A, and $F(n)$ is simply an n-fold product of A, i.e., $A \times \cdots \times A$ n-times. An operation $n \to 1$ becomes a standard set-theoretical operation $A \times \cdots \times A \to A$. Notice that an algebra of type $\mathscr{A}$ *is* a functor. It is also called a *model* of the theory $\mathscr{A}$. Thus, in particular, if $\mathscr{A}$ is the theory of groups, then each and every group *is* a functor.

Two categories can now be formed: (1) The category $\mathfrak{I}$ of algebraic theories whose objects are algebraic theories, morphisms functors preserving products and taking 1 to 1; and (2) The functor category $\mathbf{Set}^{(\mathscr{A})}$ of all *product preserving* functors $\mathscr{A} \to \mathbf{Set}$, which is the category of models of the theory $\mathscr{A}$. The latter category is called an *algebraic category*.

A morphism $f \colon \mathscr{A} \to \mathscr{B}$ of algebraic theories induces a functor $\mathbf{Set}^{(f)} \colon \mathbf{Set}^{(\mathscr{B})} \to \mathbf{Set}^{(\mathscr{A})}$ by composition. Such a functor is called an *algebraic functor*. Furthermore, for any algebraic category $\mathbf{Set}^{(\mathscr{A})}$, there is an obvious forgetful functor $U_{\mathscr{A}} \colon \mathbf{Set}^{(\mathscr{A})} \to \mathbf{Set}$ which sends to each object $F(n)$ the underlying set and to each

---

[2] I am presenting the definition given by Lawvere in his published papers. In his thesis, Lawvere defines the *category* of algebraic theories as a subcategory of the category of finite sets, which is itself a category in the category of categories. Pareigis' presentation is, in this respect, more faithful to Lawvere's original work. See [235].

morphism $F(n) \to F(m)$ the underlying set map. Notice that the forgetful functor is an algebraic functor.

Algebraic functors lead to another construction, named *algebraic semantics*: it assigns to each algebraic theory $\mathscr{A}$, the forgetful functor $U_{\mathscr{A}}$ and to each morphism $f\colon \mathscr{A} \to \mathscr{B}$ of algebraic theories the algebraic functor $Set^{(f)}\colon \mathbf{Set}^{(\mathscr{B})} \to \mathbf{Set}^{(\mathscr{A})}$. In fact this is itself a functor, sometimes called the *semantic functor*,

$$\mathfrak{S}\colon \mathfrak{T}^\circ \to \mathfrak{K}$$

where $\mathfrak{K}$ is the category of algebraic categories.

With these definitions, Lawvere's main results are:

1. Every algebraic functor has a (left) adjoint. This is the conceptual and unified formulation of various constructions in universal algebra, e.g., free algebras, tensor algebras, monoid rings, etc.
2. Algebraic semantics has a (left) adjoint, which can be called *algebraic structure*. This means that it is possible to recover an algebraic theory from the semantics, i.e., from the category of models.
3. The categorical characterization of algebraic categories: if $\mathscr{C}$ is a category with finite limits, has an abstractly finite regular projective generator G and every recongruence in $\mathscr{C}$ is a congruence, then there is an algebraic theory $\mathscr{A}$ and an equivalence $\Phi\colon \mathscr{C} \to \mathbf{Set}^{(\mathscr{A})}$. (There is no need to specify what the second and third conditions mean here. They are technical conditions that we do not have to look into.) Thus algebraic categories are characterized in an invariant manner.

These results were not only remarkable for what they accomplished, but also for the research avenues they opened. There were some obvious generalizations that were taken up rapidly by others and not so obvious generalizations that had to wait for other concepts to be fully worked out.

1. The obvious generalization was to consider infinitary operations and related work in universal algebra. This was done quickly by Linton. (See [182].)
2. One of the great advantages of the categorical language is that it is possible to replace the category **Set** of sets by an arbitrary category $\mathscr{C}$ with appropriate properties. The identification of the relevant properties of $\mathscr{C}$, expressed in categorical terms, leads to a *classification* of logical categories in *categorical* terms. The category **Set** of sets becomes a special, but very important, case of a type of category defined abstractly. Lawvere has given a characterization of algebraic categories. Further work led to characterizations of similar categories, i.e., categorical characterization of semantical frameworks.
3. An algebraic theory as defined by Lawvere can be thought of as a data type.[3] Lawvere's work shows how syntactical information of a specific kind can be encoded by categories. The search for a proper generalization to cover all types of logical theories, not only the algebraic or equational case, is irresistible. More

---

[3] This is more than a metaphor. There is indeed a formal connection with databases. See for instance [238].

specifically, the task is to find a general procedure to move from a theory written in a given formal system to a category that would be the invariant formulation of the latter. The notion of algebraic theory was specifically tailored for algebraic structures and it is not clear how one can go from there to other cases, e.g., cases with quantifiers and relations. In particular, Lawvere considered single-sorted theories and a generalization to many-sorted theories seems natural, although in practice we are used to the single-sorted case.

4. Once an element of one or all the previous points has been settled, the next task consists in looking at the various adjoint situations and seeing what one can get from them. For instance, Lawvere's work makes it clear that the adjoint situation is a special case of an algebraic duality and its importance is due to the fact that it is the very first case of such a duality where the category of sets appears as the dualizing object.

Mathematicians took up these tasks in the wake of Lawvere's thesis. Important results were obtained in the late sixties and early seventies by Lawvere himself, but also by Lambek, Freyd, Linton, Isbell and by Gabriel and Ulmer. (See [97].) Ehresmann and his students, most notably Bénabou, obtained similar results during that period. Ehresmann's motivation was different and was mainly oriented towards the foundations of differential geometry, but it led him to the notion of sketch which was recognized later as being significant. (See, for instance [64], [65], [21].)

We will come back to these developments later in this chapter. Before we do so, we will consider Lawvere's attempts at functorializing the foundations of mathematics via the category of categories and the category of sets.

6.2 The Category of Categories as a Foundational Framework

> In the mathematical development of recent decades one sees clearly the rise of the conviction that the relevant properties of mathematical objects are those which can be stated in terms of their abstract structure rather than in terms of the elements which the objects were thought to be made of. The question thus naturally arises whether one can give a foundation for mathematics which expresses wholeheartedly this conviction concerning what mathematics is about, and in particular in which classes and membership in classes do not play any role. Here by "foundation" we mean a single system of first-order axioms in which all usual mathematical objects can be defined and all their usual properties proved.
>
> ([167], 1.)

This was a fabulously bold claim to make in 1966. Lawvere had already made it in 1963. He was one of the few people to see "clearly the rise of the conviction that the relevant properties of mathematical objects are those which can be stated in terms of their abstract structure rather than in terms of the elements which the objects were thought to be made of". Bourbaki certainly can be said to have had similar views, but they were developed in a different and considerably less satisfactory framework.[4] Mac Lane, who was probably the first one to develop a part of mathematics without using set elements in his study of duality in groups, certainly did not see it *clearly*. Furthermore, Abelian categories and adjoints had only just been introduced and developed. It *was* clear that category theory was now migrating to other areas of mathematics (e.g., algebraic geometry and universal algebra), and that it was not confined to the simple role of being a useful language for algebraic topology and homological algebra, but it certainly was *not* obvious that it could be used as a language for the *whole* of mathematics, *including* set theory, or portions of it.

It should be said that Lawvere's first love was continuum mechanics, which he studied under Truesdell at Indiana and that he felt that set theory was simply *not* providing an adequate foundation for the understanding and development of that field. The study of continuum mechanics naturally led him to functional analysis and thence to category theory. He clearly saw how category theory could be used to clarify and develop functional analysis and his hope was that one could develop a categorical framework for continuum mechanics that would avoid what he took to be set-theoretical side-effects.

> My own motivation came from my earlier study of physics. The foundation of the continuum physics of general materials, in the spirit of Truesdell, Noll, and others, involves powerful and clear physical ideas which unfortunately have been submerged under a mathematical apparatus including not only Cauchy sequences and countably additive measures, but also ad hoc choices of charts for manifolds and of inverse limits of Sobolev Hilbert spaces, to get at the simple nuclear spaces of intensively and extensively variable quantities. But as Fichera lamented, all this apparatus gives often a very uncertain fit to the phenomena. This apparatus may well be helpful in the solution of certain problems, but can the problems themselves and the needed axioms be stated in a direct and clear manner? And might this not lead to a simpler, equally rigorous account? These were the questions to which I began to apply the topos method in my 1967 Chicago lectures. (...) I had spent 1961-62 with the Berkeley logicians, believing that listening to experts on foundations might be a road to clarifying foundational questions. (...) Though my belief became tempered, I learned about constructions such as Cohen forcing which also seemed in need of simplification if large numbers of people were to understand them well enough to advance further. ([177], 726.)

Lawvere had taken courses in logic both at Indiana and at Columbia, where Mendelson was teaching logic and the foundations of mathematics. He left Columbia for California, more specifically Palo Alto, and he attended seminars by Tarski and his colleagues. It is during that year that he wrote his thesis and that many ideas about a categorical analysis of logic took form, e.g., his analysis of quantifiers in terms of adjoints was announced at Berkeley during a seminar.[5]

---

[4] Grothendieck, who was a member of Bourbaki, had left the collective partly because influential members of the group were against category theory.

[5] Lawvere, personal communication 2003. The first announcement of this discovery appeared in an abstract in 1966. See [168]. As we have seen, Mac Lane was keenly interested in universal algebra and thought of using category theory to develop it. He was also interested in logic and the foundations of mathematics, but it never occurred to him to consider a categorical analysis of the concepts underlying these fields. In fact, as we will see, Mac Lane thought that it simply did not make sense to *try* to do such an analysis, at least for sets. Furthermore, it seems that Mac Lane

6.2 The Category of Categories as a Foundational Framework

As is seen from the introduction to his 1966 paper quoted above, Lawvere claims that a categorical foundation would be more faithful to the nature of mathematics, since it would only present or represent the *relevant* properties of mathematical objects. The relevant properties are stated in terms of abstract structures and, in turn, the latter are stated in the context of categories, not in terms of elements of sets. It is impossible not to think of Kreisel's comments (presented in the previous chapter) at this stage.[6] One could easily retort that sets are certainly abstract in some sense and that structures can be specified in a set-theoretical manner, e.g., in the manner specified above by Kreisel himself, where the notion of isomorphism plays a key role. So *what* precisely is the point here?

From the point of view of someone like Kreisel, it certainly cannot be the fact that a foundational framework has to be more readily usable in algebraic topology, functional analysis or any other such field, as Lawvere claims. This is, according to Kreisel, an aspect of organization, not foundations. For Kreisel, a foundational framework might be totally useless in various areas of "higher" mathematics, in other words, for the "working mathematician", in the same way that fundamental physics might be useless in various areas of engineering or "higher" physics, whatever that might be, or for the "working physicist", whatever that might be.

But the latter analogy is clearly misleading. Kreisel draws an analogy between foundational research and theoretical physics and argues that foundational concepts do not have to have an impact on the "working mathematician". The problem with the analogy is that concepts of theoretical physics *do* have an impact on *all* areas of the world, in one way or another. For example, physical properties are often relevant to biological phenomena. This was noticed explicitly, although with no reference whatsoever to Kreisel, by Taylor in the very first paragraph of the introduction of his book published in 1999:

> Foundations have acquired a bad name amongst mathematicians, because of the reductionist claim analogous to saying that the atomic chemistry of carbon, hydrogen, oxygen and nitrogen is enough to understand biology. Worse that this: whereas these elements are known with no question to be fundamental to life, the membership relation and the Sheffer stroke have no similar status in mathematics. ([258], viii.)

We can perhaps reformulate this point of view by saying that a foundational framework should exhibit the underlying unity of mathematics, in the sense that foundational concepts and results should be seen as being a part of mathematics, not something that is extraneous or on a different plane altogether. This is clear for the foregoing quote by Lawvere. More recently, in an appendix to a book on set theory written in collaboration with Robert Rosebrugh,[7] we read:

---

could not make the epistemological shift required to consider categories as data types or semantical universes. To do so, one had to *treat* categories in a different fashion, to see them as more than a useful language for algebraic topology and homological algebra.

[6] It should be mentioned that Kreisel and Lawvere had the occasion to meet and discussed these issues in the sixties. While Lawvere was in Palo Alto, he sometimes took the train with Kreisel to go to Berkeley. Furthermore, the two were in Zurich at the same time.

[7] Interestingly enough, the book is entitled *Sets for Mathematics*. I guess Kreisel could write a book entitled *Sets for Foundational Analysis*.

> A foundation makes explicit the essential general features, ingredients, and operations of a science as well as its origins and general laws of development. The purpose of making these explicit is to provide a guide to the learning, use, and further development of the science. A "pure" foundation that forgets this purpose and pursues a speculative "foundations" for its own sake is clearly a nonfoundation. ([179], 235.)

This statement can be reformulated as follows: a foundation should

1. Provide an analysis of the essential general features, ingredients and operations of mathematics *as it is*, thus it should be *a part* of mathematics and its role is to reveal the scaffolding of the discipline, not so much what it rests on, but rather how it all holds together as one discipline;
2. This analysis should make it possible to explain why previous developments were *successful*, i.e., it should provide the means to reconstruct the history of the field in such a way that one can *understand* why certain concepts were correct and fertile and others pointless;
3. Similarly, the analysis should suggest *new* ideas, concepts and developments of various mathematical fields, including, of course, foundations.[8]

In all three cases, adjoint functors play a key role. The analysis Lawvere has in mind should in fact be performed in terms of categories, functors and in particular, adjoint functors as long as they are relevant to the situation at hand; which is to say, for Lawvere, almost always. (Or, he might say, one should at least *try* to analyse any given situation in categorical terms, since, as he has shown, it is very often possible even though it might not seem so at first.) The presence and the role of adjoint functors in this analysis is precisely what makes it more that merely an organization of mathematics in Kreisel's sense. What Kreisel has in mind is probably an analysis performed on known mathematics expressed in set-theoretic language or a theory whose models are taken in a set-theoretical universe. Categories, functors, etc, might enter the picture, but they will be, in that context, categories of set-theoretical structures. I should nonetheless indicate that the analysis provided by the categorical point of view might very well be extremely insightful and in particular, the search for adjoints in such a case might also be extremely relevant. But this is certainly not what Lawvere has in mind in general. The whole analysis should be performed in a categorical framework at the level of categories and adjoints right from the beginning. The organization of mathematical disciplines thus obtained is structured and developed from the adjoints found in that context. The *nature* of the mathematical disciplines is also determined by the adjoints revealed in that context. Later we will see concrete examples of this organization and nature.

There is no doubt that Lawvere's early attempt at axiomatizing the category of categories was somewhat premature. For the original axiomatization of the category of categories presented by Lawvere was based on the properties of categories known

---

[8] Lawvere's work is an illustration of how this can be done. For the historical aspect, see for instance, [177]. Historians might disagree with the goals and the methods used by Lawvere in this enterprise. As to the direction of mathematics, his early proposal to develop differential geometry in a categorical setting is certainly telling. See [175].

6.2 The Category of Categories as a Foundational Framework

at that time. Indeed, we read in the first paragraph of the paper on the category of categories:

> The author believes, in fact, that the most reasonable way to arrive at a foundation meeting these requirements is simply to write down axioms descriptive of properties which the intuitively-conceived category of all categories has until an intuitively-adequate list is attained; that is essentially how the theory described below was arrived at. Various metatheorems should of course then be proved to help justify the feeling of adequacy. ([167], 1.)

Very little was known about the "intuitively-conceived category of all categories" and one could even argue that the proper language for such a description was not even available at the time. The requirements mentioned by Lawvere here are pragmatic:

> A foundation of the sort we have in mind would seemingly be much more natural and readily-useable than the classical one when developing such subjects as algebraic topology, functional analysis, model theory of general algebraic systems, etc. Clearly any such foundation would have to reckon with the Eilenberg-MacLane [*sic*] theory of categories and functors. ([167], 1.)

Again, as we have seen, Kreisel disagrees. According to him, a foundation for mathematics does not have to be readily-useable by working mathematicians. Lawvere's reply is that the set-theoretical foundations introduce extraneous elements that are confusing and that obscure the conceptual scaffolding of various theories, e.g., functional analysis. A foundational analysis should clarify the conceptual basis of a field, not obscure it. Notice that Lawvere's goal is *not* to get rid of sets altogether, it is clearly to keep the portions of set theory that do clarify things but to put them in a context, namely a categorical context, that allows one to get rid of the exogenous and irrelevant aspects of the theory.

Let us now turn to Lawvere's axiomatization of the category of categories as such. Lawvere works in a first-order language and the aim is to provide an axiomatization such that:

1. A model of the axioms should be a category;
2. The objects of the model should themselves be categories;
3. The axioms should be such that the basic properties and theorems of category theory can be proved; e.g., functor categories and adjoint functors should be definable, Yoneda's lemma and Freyd's special adjoint functor theorem should be provable, etc.
4. Sets should be definable within the model, and the relevant aspects of set theory should be definable and provable;
5. It should be possible to make a distinction between small and large categories, and Grothendieck universes should be models of the theory. (However, Lawvere claims explicitly that his theory is weaker than Grothendieck since only one inaccessible cardinal is required — which again shows where Mac Lane got his idea — and he also explicitly considers reflection principles, although not the one proposed by Feferman.)

The first requirement indicates how to begin: from the axioms of category theory themselves. However, a category cannot be defined as a set or a class, nor can

Hom-sets be used in the definition itself. But this was not insurmountable. The hard part seems to be to find a way to force the objects of a category to be categories themselves and arrows to be functors. This is precisely what most of the remaining axioms are meant to do. In modern terminology, the goal is to define categories internally, that is *within* a category.[9] Lawvere's strategy is to pick specific objects, basic "forms", that can be used to characterize categories in the sense that these objects can only be represented in categories, i.e., these objects encode the presence of morphisms and their properties, i.e., composition, associativity, and identity. They are then mapped into the objects of the model and the fact that they can be mapped tells us that the objects are indeed categories. Notice the deeply geometric method involved here, which I have already mentioned: the objects of the models are categories because certain specific forms can be mapped into them in the appropriate way.

Lawvere divided his presentation in two parts: a basic theory which occupies almost 70% of the paper, and a short and intricate presentation of what is called a strong theory. Both theories are preceded by an elementary theory of abstract categories, from now on called ETAC. The latter simply guarantees that the model of the category is a category. It is essentially a formalization of Eilenberg and Mac Lane's definition of a category. The theory is developed within first-order logic with equality.[10] There are three function symbols, two unary, $\Delta_0(x)$ and $\Delta_1(x)$, and one ternary $\Gamma(x,y;u)$, yielding the basic formulas

$$\Delta_0(x) = y, \Delta_1(x) = y \text{ and } \Gamma(x,y;u)$$

which should be read "y is the domain of x", "y is the codomain of x" and "u is the composition of x followed by y".[11] Formulas and sentences are defined as usual in first-order logic. ETAC is then given by the following axioms:

1. $\Delta_i(\Delta_j(x)) = \Delta_j(x)$ for $i, j = 0, 1$;
2. $(\Gamma(x,y;u) \wedge \Gamma(x,y;u')) \Rightarrow u = u'$;
3. $\exists u[\Gamma(x,y;u)] \Leftrightarrow \Delta_1(x) = \Delta_0(y)$;
4. $[\Gamma(x,y;u) \Rightarrow (\Delta_0(u) = \Delta_0(x) \wedge \Delta_1(u) = \Delta_1(y))]$;
5. $\Gamma(\Delta_0(x),x;x) \wedge \Gamma(x,\Delta_1(x);x)$;
6. $[\Gamma(x,y;u) \wedge \Gamma(y,z;w) \wedge \Gamma(x,w;f) \wedge \Gamma(u,z;g)] \Rightarrow f = g$.

The first four axioms are "bookkeeping" axioms. Their presence guarantees that composition and domain and codomain functions are coherent with one another. Axiom 5 is the identity axiom and axiom 6 is the associativity law. Thus, a model of the theory is a category.

---

[9] I should immediately point out that at about the same time, but with a very different motivation, Ehresmann was also trying to define categories internally. This led him to develop the notion of sketch.

[10] This was done, of course, before Freyd's paper presented in chapter 5 and Ehreshmann's work on sketches. Again, I should emphasize the fact that it is far from obvious that category theory requires a modification of the *syntax* of mathematics.

[11] Notice the order of composition. I follow Lawvere and write the order of the arrows from left to right in this section.

6.2 The Category of Categories as a Foundational Framework

One can introduce the usual abbreviations to depict arrows and commutative diagrams, as well as for objects that can be identified with the identity arrows. It is then possible to define monomorphisms, epimorphisms, isomorphisms, etc., as Lawvere did himself. Notice the real primitive notion here is the function $\Gamma(x,y;u)$. Thus, one could say, as apparently Lawvere did to Tarski after his talk on the matter, that whereas set theory is the formal expression of the binary relation "x belongs to A", category theory is the formal expression of the ternary function "u is the composition of x followed by y", where one could informally think of u, x and y as processes, or rules, etc.[12]

The next axioms are now intended to characterize the objects and the morphisms of the model, i.e., of a category and in fact to make sure that the morphisms are functors and objects are categories themselves. Formally, Lawvere added two constants ∂_0 and ∂_1 that are used to introduce five specific objects, denoted by **1, 2, 3, 4** and **E**. It is these objects that will be used to characterize the other objects of the model. The morphisms between these objects and the other objects of the underlying category will assure us that we are dealing with a category of categories.

We have already seen the first three: **1** is the category with exactly one morphism and it is defined by the abbreviated formula — we denote arbitrary objects by A, B, C, it being understood now that they are categories:

$$\exists \mathbf{1} \forall A \exists! x [A \xrightarrow{x} \mathbf{1}].$$

Needless to say, the formula simply says that **1** is a terminal object in the category. **2** should denote the category with two objects and one non-trivial morphism. Thus one can think of **2** as being the ordinal 2, considered as a category. It is defined by (any one of) the equations:

$$\Delta_i(\partial_j) = \mathbf{2}, i, j = 0, 1,$$

i.e. graphically

$$\mathbf{2} \underset{\partial_1}{\overset{\partial_0}{\rightrightarrows}} \mathbf{2}.$$

But **2** has to be characterized categorically, i.e., in the language of ETAC. Here is how Lawvere does it: ∂_0 and ∂_1 are constant, that is, they factor through **1**; this can be translated into the following commutative diagrams:

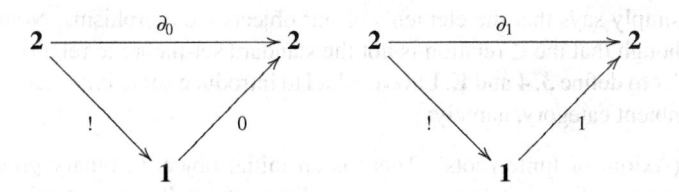

$$\mathbf{2} \xrightarrow{\partial_0} \mathbf{2} \qquad \mathbf{2} \xrightarrow{\partial_1} \mathbf{2}$$

(with factorizations through **1** via $!$ and 0; $!$ and 1 respectively)

---

[12] This is reported by Lambek. Lambek, personal communicaiton, 2003.

$\Gamma(\partial_i, \partial_j; \partial_j)$, $i, j = 0, 1$, which is the same as saying that the diagrams

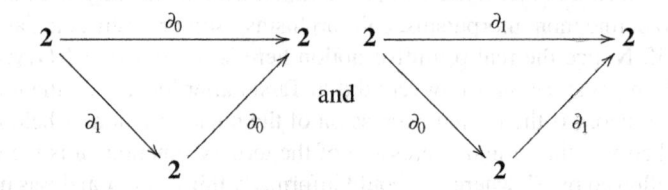

$$\begin{array}{ccc} \mathbf{2} & \xrightarrow{\partial_0} & \mathbf{2} \\ {\scriptstyle \partial_1} \searrow & \nearrow {\scriptstyle \partial_0} & \\ & \mathbf{2} & \end{array} \quad \text{and} \quad \begin{array}{ccc} \mathbf{2} & \xrightarrow{\partial_1} & \mathbf{2} \\ {\scriptstyle \partial_0} \searrow & \nearrow {\scriptstyle \partial_1} & \\ & \mathbf{2} & \end{array}$$

commute.

The next two axioms assure that ∂_0 and ∂_1 are distinct and are the only non-trivial endofunctors on **2**:

$$\partial_0 \neq \partial_1, \partial_i \neq 1_\mathbf{2}, i = 0, 1 \text{ and } \forall x(x\colon \mathbf{2} \to \mathbf{2} \Rightarrow x = \partial_0 \vee x = \partial_1 \vee x = 1_\mathbf{2}).$$

The last two axioms stipulate the role played by **2** in a model: **2** is a generator, i.e.,

$$\forall x \forall y [(\Delta_0(x) = \Delta_0(y) \wedge \Delta_1(x) = \Delta_1(y \wedge x \neq y) \Rightarrow \exists z(z\colon \mathbf{2} \to \Delta_0(x) \wedge zx \neq zy)].$$

If A is any generator, then $\exists x \exists y(x\colon \mathbf{2} \to A \wedge xy = 1_\mathbf{2})$, that is **2** is a retract of A.

This completes the characterization of **2**. The axioms yield an adequate characterization of **2**, since it can now be proved that for any generator A satisfying the axioms for **2**, then $A = \mathbf{2}$. The last equality follows from an axiom assumed by Lawvere but that I have not stated yet. It says that if the identity 1_A is the only automorphism of a category A, then A is the only category in its isomorphism class, i.e., if $A \approx B$, then $A = B$. Thus, the objects **1** and **2**, which clearly satisfy this condition, are unique in the sense of identity. Although the axiom might be convenient, it seems to introduce a certain amount of confusion within the theory. Indeed, it is not in the spirit of category theory to identify objects. As we have seen, the latter are usually characterized up to a unique isomorphism and the theory might be more faithful to category theory by characterizing objects in that way.

Now that the object **2** has been defined, it is possible to say what it is to be a morphism in an object A of the model, i.e., we are starting to look into the objects of the model to make sure that they are themselves categories.

Definition 6.2. $x \in A$ means $\mathbf{2} \xrightarrow{x} A$.

This simply says that the elements of our objects are morphisms. Notice immediately though that the $\in$ relation is not the standard set-theoretic relation.

In order to define **3**, **4** and **E**, Lawvere had to introduce some important properties of the ambient category, namely:

Axiom (Axiom of finite roots). There is an initial object **0**, binary products and coproducts as well as equalizers and coequalizers of parallel arrows exist.

It follows that finite limits and colimits exist.

6.2 The Category of Categories as a Foundational Framework

We will only show how **3** is defined and characterized. **3**, together with two arrows α, β, is first defined by the following pushout[13]:

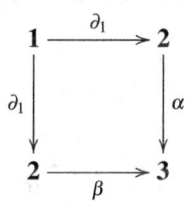

$$\begin{array}{ccc} \mathbf{1} & \xrightarrow{\partial_1} & \mathbf{2} \\ {\scriptstyle \partial_1} \downarrow & & \downarrow {\scriptstyle \alpha} \\ \mathbf{2} & \xrightarrow{\beta} & \mathbf{3} \end{array}$$

This already says that **3** has, together with the obvious three identity morphisms, two different morphisms, α and β such that the codomain of α is the domain of β. But the role of the object **3** is to depict commutative triangles in the objects of the model, that is any morphism $t: \mathbf{3} \to A$ to an object A of the category is a commutative diagram in A. Therefore, Lawvere postulates that **3** has exactly one morphism γ besides the five implied by the definition and it satisfies the equations

$$\partial_0 \gamma = \partial_0 \alpha \text{ and } \partial_1 \gamma = \partial_1 \beta.$$

3 allows one to say that a morphism h in an object A is the composition of two morphisms f and g of A.

4 and **E** are defined and characterized in similar fashions. **4** is required to state that composition of morphisms is associative and **E** to state that two morphisms are parallel.

With these five objects (together with a slight logical trick to assure that we can now talk about properties of the objects of the model), Lawvere claims that every object of the universe described is at least a category. To insure that the objects are no more than categories, more axioms are required.

The first axiom is the important statement that the category of categories is *Cartesian closed*, i.e., the claim that given two categories A, B, there is a category B^A and a functor

$$e: A \times B^A \to B$$

called *evaluation* and satisfying the already mentioned universal property. This axiom says that functor categories exist in the category of categories and there is no need to emphasize the importance of this fact at this point.

Lawvere introduced "sets" at this point, by the following

Definition 6.3. [14] The category A is said to be *discrete* (or to be a *set*) if and only if there is an isomorphism $A^!: A^2 \xrightarrow{\sim} A^1$.

This definition states that an object A is discrete if and only if the morphisms of A are in fact objects.

The last four axioms included by Lawvere can now be stated directly.

---

[13] A pushout is the dual construction of a pullback. The latter will be defined in section 6.6
[14] This definition is the basis of Bell's objection presented in chapter 2.

Axiom. For any category A, there is a discrete category A_c with a functor $A \to A_c$ satisfying the following universal property: for any functor $A \to B$ to a discrete category, there is exactly one functor such that the diagram

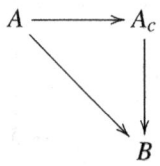

$$\begin{array}{ccc} A & \longrightarrow & A_c \\ & \searrow & \downarrow \\ & & B \end{array}$$

commutes. The object A_c is called the *set of components* of A.

Axiom (dual of the preceding). For any category A, there is a maximal discrete subcategory $|A|$ with a functor $|A| \to A$ such that for any functor $B \to A$ from a discrete category to A, there is a unique functor $B \to |A|$ such that the diagram

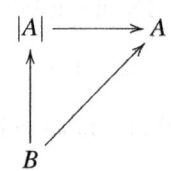

$$\begin{array}{ccc} |A| & \longrightarrow & A \\ \uparrow & \nearrow & \\ B & & \end{array}$$

commutes. The object $|A|$ is called the *set of objects* of A.

Definition 6.4. The *set of morphisms* of A is the discrete category $|A^2|$.

Axiom (Axiom of choice). For all $f\colon A \to B$ such that $A \not\cong \mathbf{0}$ and B is discrete, there is a $g\colon B \to A$ such that $fgf = f$.

Axiom. If $|A^2| \xrightarrow{\bar{f}} |B^2|$ and $[\forall t[t\colon \mathbf{3} \to A \Rightarrow \exists! u(u\colon \mathbf{3} \to B \wedge |u^2| = |t^2|\overline{f})]]$ then

$$\exists! f(f\colon A \to B \wedge \overline{f} = |f^2|).$$

This last axiom simply says that given a function from a set of morphisms into a set of morphisms satisfying the informal properties of a functor, there is an actual functor that corresponds to it in the model.

This constitutes the basic theory of the category of categories. Lawvere then states that the basic theory is sufficient to develop a large part of analysis, preuniversal algebra, category theory and universal algebra. Lawvere proceeds to prove a version of the general adjoint functor theorem in the basic theory.

Lawvere's axiomatization of the category of categories was bold in two important respects: (1) very little was known about the category of categories since pure category theory was still in its infancy — no one had looked into that structure carefully; (2) the motivation underlying such work might have seemed dubious at that time: what would one gain from such a concept? There were no clear applications, and the foundational benefits were far from obvious, despite Lawvere's claim to the

6.2 The Category of Categories as a Foundational Framework

contrary in the opening paragraph of his paper. But these elements were overshadowed by Isbell's review, published in 1967 ([114]), in which he showed that some claims made by Lawvere in his paper were in fact false.

The culprit is a statement, later called the *category description theorem*, or CDT, presented as a theorem by Lawvere towards the end of the exposition of the basic theory. Isbell showed in his review that there is a model C of the basic theory, namely the category of all (small) categories in which every endomorphism is an identity, such that CDT is false in that model. More specifically, the basic theory allows one to construct a certain category A which is not in the model (it has non-trivial endomorphisms). Therefore, CDT cannot be a theorem of the theory.

But this is not all. Lawvere also asserted "that the basic theory needs no explicit 'axiom of infinity'". ([167], 6.) This is of course a striking claim, because in set theory an axiom of infinity has to be assumed. According to Lawvere, no such assumption is required in the context of the category of categories. However, following Isbell's suggestion, it can be seen that there is a model in which all the categories are finite, thus showing that Lawvere's claim is wrong. So for Lawvere's original proposal at least, the situation is no different from the set-theoretical one.

From this last result, it also follows that Lawvere's claim that his own theory of the category of sets can be developed within the basic theory of the category of categories is also false. Overall, Lawvere's analysis was flawed.

But clearly it would be wrong to dismiss Lawvere's work on these grounds.

1. The problems simply show that the task is more difficult than one might have expected. As we have already said, little was known in pure category theory in the early and mid-sixties. Lawvere's work called for more research and it was soon to follow in various directions. (For instance, [22] and [100–102].) As to Lawvere's specific attempt, various proposals were made later to fix it. See [27] and [26], where an attempt is made at rehabilitating some of the original claims (e.g., the existence of an infinite object is claimed to be derived), and [222] for a development which follows Lawvere's own presentation very closely but avoids its pitfalls.
2. As more recent works have shown, a category of categories is a rich and intricate structure in itself and it seems reasonable to try to characterize a category of categories by grasping this structure instead of trying to "force" the objects to be categories; we will briefly come back to this point in the conclusion;
3. Be that as it may, the main point remains: what Lawvere was basically trying to do was to provide a proper setting for the categorical way of doing mathematics. More precisely, from his own research and the work done at that time, Lawvere understood that mathematical domains, for instance homological algebra or universal algebra, could be characterized in a purely categorical fashion, or developed in a categorical framework. In turn, adjoint functors constituted, according to Lawvere, the main methodological tool required for categorical work, since they yield an invariant presentation of given concepts. Thus, if category theory has to do with functors, and more specifically adjoint functors, one has to clarify to some extent what the underlying universe of these functors is. Instead of trying to force category theory into a set-theoretical universe, it seems

to make sense to describe the universe of categories and functors directly, which is what he did.

4. I can now reiterate more precisely the main connection to Klein's program. It is possible to characterize a system of mathematical concepts by categorical means, more precisely by adjoint functors and related properties, e.g., the so-called "exactness conditions" (the terminology comes directly from homological algebra and the characterization of Abelian categories in that context). This is now clearly in the spirit of the program and also coherent with Eilenberg and Mac Lane's original intention of using category theory to provide a classification of mathematical concepts in terms of invariants. Not only is a category itself a collection of "spaces" with their "transformations", but, going one level up, categories themselves are "spaces" with "transformations" between them (and, as we will see, "transformations of transformations", i.e., transformations similar to homotopies in topology).

As I have already mentioned, Lawvere meant to develop set theory within the category of categories, i.e., he thought it possible to characterize the concept of set without referring to the standard $\in$ relation and its associated axioms. The general strategy should now be obvious: identify enough categorical properties such that any category equivalent to a category satisfying these properties "is" the category of sets. This ought to yield an invariant presentation of the category of sets. Again, the connection with Klein's program is more than clear. To see how Lawvere thought this could be done and how it served as a springboard to his work on elementary toposes, we now turn to the elementary theory of the category of sets, ETCS for short.

6.3 The Elementary Theory of the Category of Sets

(...) and then [Lawvere] conceived the idea of giving a direct axiomatic description of the category of all categories. In particular, he proposed to do set theory without using the elements of a set. His attempt to explain this idea to Eilenberg did not succeed; I happened to be spending a semester in New York (at the Rockefeller University), so Sammy asked me to listen to Lawvere's idea. I did listen, and at the end I told him "Bill, you can't do that. Elements are absolutely essential to set theory." After that year, Lawvere went to California.

([196], 342.)

Thus we seem to have partially demonstrated that even in foundations, not Substance but *invariant Form* is the carrier of the relevant mathematical information.

([165], 1506. My emphasis.)

Some years ago I began an introductory course on Set Theory by attempting to explain the *invariant content* of the category of sets, for which I had formulated an axiomatic description.

([176], 5. My emphasis.)

6.3 The Elementary Theory of the Category of Sets

Let us briefly see how invariant Form is the carrier of the relevant information, even for the concept of set.

Lawvere assumes the standard axioms for a category and then postulates that the category of sets has a terminal object 1, an initial object 0, binary products and coproducts, equalizers and coequalizers, thus all finite limits and colimits (axiom 1). He also assumes that it is Cartesian closed, i.e., that the object B^A, together with the known morphism and universal property, exist for any A and B (axiom 2). These two axioms are "structural" and are satisfied by many categories, even, as we have seen, by the category of categories as axiomatized by Lawvere.

The next axiom is known as the Peano-Lawvere characterization of the natural numbers. It is therefore an axiom of infinity, but it contains more when it is assumed along with the previous axioms.

Axiom (3). There is an object N together with morphisms $1 \xrightarrow{z} N \xrightarrow{s} N$ such that given any object X together with mappings $1 \xrightarrow{x_0} X \xrightarrow{t} X$, there is a unique morphism $N \xrightarrow{x} X$ such that $x_0 = zx$ and $xt = sx$.

This is now the standard characterization of the natural number system by a universal property. A theorem on primitive recursion follows from it and the preceding axioms. Lawvere immediately points out that axiom (3) is satisfied, as well as the first two axioms, by the category **1**, i.e., the category with one object and one morphism. Thus, more is needed, both structurally and in terms of existence.

The next axiom is false in the category of categories, for it says that the terminal object 1 is a generator, i.e., if the morphisms $A \xrightarrow[g]{f} B$ are different, then there is a morphism $x\colon 1 \to A$ such that $fx \neq gx$. In more colloquial language, this axiom states that if f and g are different, then there is an element x, or a point, in A such that f and g are different on that element.

The fifth axiom is the axiom of choice as expressed in the category of categories. The last three axioms are *not* expressed in a categorical fashion. It is time I mentioned that Lawvere reintroduces the $\in$ notation in his axiomatization. Since there is a one to one correspondence between the morphisms $x\colon 1 \to A$ and the elements $x \in A$ in the universe of sets, it seems reasonable to say that x is an *element* of A if and only if $x\colon 1 \to A$. The remaining axioms are then expressed as follows:

Axiom (6). If A is not an initial object, then A has elements.

Axiom (7). An element of a sum is a member of one of the injections.

Axiom (8). There exists an object with more than one element.

Lawvere underlines the fact that the first seven axioms are satisfied by the category **1** with one morphism, thus the need for the eighth axiom. One could certainly object to the introduction of the $\in$ relation and the notion of element in a categorical framework and claim that this part of the axiomatization does not capture the invariant form of the category of sets.

Be that as it may, the claim that the invariant form of the concept of set is captured by the axioms is substantiated by a metatheorem and its corollary. Indeed, the metatheorem asserts that any two categories satisfying the eight axioms are equivalent, thus in particular any (complete) category satisfying the eight axioms is equivalent to the category of sets. We can now clearly see what was missing in Eilenberg and Mac Lane's approach in the development of invariance in a categorical setting: the notions of adjoint functors and of equivalence of categories. Lawvere is in a position to state precisely what it means to be "the" category of sets: the properties are (mainly) expressed in categorical terms and the invariance amounts to the claim that any other category satisfying these properties is equivalent to it. If this is not a clear generalization of Klein's program, then I do not know what it is.[15]

Surprisingly, perhaps, Lawvere's work did not lead to further investigation along similar lines. The category of sets was not taken as a foundational framework; it was not studied and explored. Although there is no clear explanation for this, Johnstone has suggested that the category of sets is simply too "rigid":

> In retrospect, the answer is that Lawvere's axioms were too specialized: the category of sets is an extremely useful object to have as a foundation for mathematics, but as a subject of axiomatic study it is not (*pace* the activity of Martin, Solovay *et al.*!) tremendously interesting — it is too "rigid" to have any internal structure. ([117], xiii.)

It is not exactly clear why it is such "an extremely useful object to have as a foundation for mathematics" if nothing is developed in it and no one proceeds to do more research within it. It is clear that Lawvere's axiomatization did not attract much interest for it was seen as being a simple translation of the standard axioms into the language of category theory and, as such, not providing any interesting novelty. Sociological factors must also probably be invoked. Lawvere's paper appeared in 1964, very shortly after Cohen's proof of the independence of the continuum hypothesis was published. (See [48].) Cohen's result, and probably most notably his method of forcing, deservedly attracted much attention. There is very little doubt that one of the key advantages of topos theory over ETCS is precisely that the former bridges the gap between a categorical description of sets and the method of forcing whereas the latter is a category of sets satisfying the axiom of choice.

6.4 Categorical Logic: the Program

> Categorical logic, in a very broad sense, can be seen to derive from the completeness and exactness properties of the category of sets, in a manner paralleling the earlier development of Abelian categories.
> ([207], 5.)

---

[15] And other similar claims for other concepts followed. For instance, D. Schlomiuk, following Lawvere's steps very closely, did the same for "the" category of topological spaces. See [247]. More recently, a different characterization of the category of sets has been given in terms of chains of adjoints. See [242].

6.4 Categorical Logic: the Program

After 1966, Lawvere's own published work goes back to connections between categories and logic. As I have said, three papers published in 1969 and 1970 are extremely important in many respects since they contain the seeds and the statement of a vast foundational program which was taken up and is still alive. Here are the basic methodological elements underlying this program.

First, the use of adjoint functors is emphasized both in practice and also from a more general point of view. In his paper on diagonal arguments and Cartesian closed categories, Lawvere defines a Cartesian closed category as a category $\mathscr{C}$ equipped with three kinds of right adjoints ([171].):

1. A right adjoint $\mathbf{1}$ to the unique $\mathscr{C} \to \mathbf{1}$;
2. A right adjoint $(- \times -)$ to the diagonal functor $\mathscr{C} \to \mathscr{C} \times \mathscr{C}$;
3. For each object A in $\mathscr{C}$, a right adjoint $(-)^A$ to the functor $A \times - : \mathscr{C} \to \mathscr{C}$.

Before that paper, adjoints were used by Lawvere to show that any category satisfying a certain specification was equivalent to a fixed category of interest or to establish certain properties of given functors in a context. In this Lawvere was following the examples of Abelian categories and sheaves. As we have seen, the axioms for the category of categories and the category of sets did not mention adjoints explicitly. Right from the beginning, Lawvere emphasized the fact that Cartesian closed categories are algebraic versions of type theories.

Second, in connection with the role of adjoint functors, there is an *explicit* recognition of the levels of abstraction introduced by category theory, more specifically the fact that category theory now allows for purely abstract characterizations of mathematical domains.

> More recently, the search for universals has also taken a conceptual turn in the form of Category Theory, which began with viewing as a new mathematical object the totality of all morphisms of the mathematical objects of a given species A, and then recognizing that these new mathematical objects all belong to a common non-trivial species C which is independent of A. ([171], 281.)

What is described by Lawvere should now be clear: in the beginning, mathematicians started with already defined mathematical objects and structure-preserving functions and moved to a new object, namely the *category* of these objects of species A. But it was soon realized that such a category participated in a different species, namely a category of type C, which can be described *independently* of A. The first example of this phenomenon is of course provided by Abelian categories: one started with Abelian groups (or modules over a commutative ring), the latter constituting the objects of species A, moved to the category of Abelian groups (or modules over a commutative ring) and then to an Abelian category, which is a category of type C. As we have seen, the important step is that the latter can be described independently of the former. Lawvere makes a bold generalization: he sees this case as a *general* phenomenon, even as a framework that should guide the development and analysis of mathematics. Of course, he had other examples at his disposal, e.g., algebraic theories and categories, Cartesian closed categories and hyperdoctrines, etc. Once more, the parallel with transformation groups in geometry is striking.

Third, Lawvere uses Cartesian closed categories to present an analysis of well-known diagonal arguments, i.e., those of Cantor, Russell, Gödel and Tarski. The motivation is similar to the one indicated in the foregoing paragraph: these diagonal arguments are similar and thus seem to form a *species* of argument. Lawvere hopes to be able to disclose the common abstract structure underlying them. This abstract structure takes the form of a fixed-point theorem based on the properties of Cartesian closed categories. In the process of his analysis, Lawvere introduces an object of truth values **2**, in fact the standard Boolean algebra, for this object appears in one way or another in all the arguments. As we will see, the object of truth-values, as well as the Cartesian closed structure, will become pivotal in his characterization of the notion of elementary topos in 1969.

Fourth, in all three papers, Lawvere suggests extensions of his earlier work on algebraic theories to theories written in higher-order type theories and, as a special case, first-order theories. I should point out immediately that he was no longer alone in looking for connections between categories and logic. Lambek's work on categorical analysis of deductive systems has to be mentioned at this stage. (See [155–157].) Lawvere's extensions are based on the following fundamental facts:

1. The logical quantifiers can be presented as adjoint functors to the simple and fundamental operation of substitution ([169], [171], [174]);
2. The comprehension principle can be presented as an adjoint functor in a proper context ([171], [173]);
3. Lawvere sketches how one can construct a category from a given theory formalized in higher-order logic ([171], [173], [174]).

I have to underline the fact once more that a categorical analysis of logical systems not only provided a novel and unifying understanding of logical operations and systems, but by the same token, it initiated a shift in the status of categories *themselves*. It is now possible to *identify* a type of category with a type of deductive system. The claim that category theory can be seen as a language can now be made more precise: category theory can be seen as a *formal* language for mathematics.

Fifth, all these constructions are incorporated in a general framework that constitutes nothing less than the frame of categorical logic as it developed afterwards.[16] This framework is presented in very broad strokes at the beginning and the end of the papers entitled "Adjointness in Foundations" and is itself of considerable philosophical interest.

Lawvere identified two aspects, which he qualifies as being "dual", since they appear to obey some sort of general duality or Galois connection inherent to mathematics, namely the Formal and the Conceptual. The Formall is more or less identified with the manipulation of symbols, either in deduction or calculations, whereas the Conceptual is identified with the content of these symbols, the subject matter of the Formal or what they refer to. Thus, at first sight, Lawvere's terminology coincides with the classical distinction between the syntax and the semantics of formal languages. However, Lawvere has the actual practice of mathematics in mind and

---

[16] At least before categorical logic became a standard tool in theoretical computer science. But even in this context, one can see the influence of Lawvere's suggestions.

6.4 Categorical Logic: the Program

therefore does not equate his distinction with the fundamental metamathematical distinction. In fact, he sees foundational research as being part of mathematics: "Being itself part of Mathematics, Foundations also partakes of the Formal-Conceptual duality." ([171], 281.) Thus, the syntax of a logical system is part of the Formal, whereas the semantics is part of the Conceptual. But Lawvere's presentation of the semantics is somewhat odd: "Naturally the formal tendency in Foundations can also deal with the conceptual aspect of mathematics, as when the semantics of a formalized theory T is viewed itself as another formalized theory T', or in a somewhat different way, as in attempts to formalize the study of the category of categories." ([171], 281.) Category theory is clearly put on the conceptual side of mathematics and, in fact, one can see that Lawvere sees his work on the foundations of universal algebra and the subsequent work on Cartesian closed categories and extensions thereof as being part of the *conceptualization* of the formal aspect of mathematics.[17] Indeed, he claims explicitly that "Foundations may conceptualize the formal aspect of mathematics, leading to Boolean algebras, cylindric and polyadic algebras, and certain of the structures discussed below."[18] ([171], 282.) At the center of this conceptualizations appear adjoint functors.

Adjoints are present in foundations in two senses. First, Lawvere introduces Cartesian closed categories and what he calls hyperdoctrines in that paper. The main property of these two concepts is that they are *entirely* given by adjoint functors. However, as we have already mentioned, Cartesian closed categories and hyperdoctrines correspond in a precise technical sense to logical frameworks. Thus, both Cartesian closed categories and hyperdoctrines are categorical codifications of logical structures, the algebraic counterpart of these structures. Adjoint functors are used to *define* the conceptual content of foundations. However, and this is the second point, adjoint functors also play a more general role. The Formal and the Conceptual mentioned above should be related by adjoint functors (but we are clearly at a programmatic stage here).

Lawvere is in fact more precise in the way these adjoints should show up and here we see him generalizing the work contained in his thesis. First, Lawvere suggests that one should consider *categories* of models of a theory, thus framing model theory in the context of category theory.[19] More specifically, a model in the standard model-theoretic sense can be described as a *functor* from a category T to the category Set of sets. The category of such models is then a subcategory of the functor category Set^T, or also written $Mod(T, Set)$ or $Mod_{Set}(T)$.

Second, the category T is the categorical encoding of a given formal theory. In Lawvere's own terms: "The *invariant* notion of theory here appropriate has, in all cases considered by the author, been expressed most naturally by identifying a

---

[17] Indeed, in the 1990's Lawvere and Steve Schanuel cowrote a book entitled *Conceptual Mathematics*. (See [180].)

[18] Notice here the reference to cylindric and polyadic algebras. Indeed, it is clear that Lawvere's first attempt at capturing logic by the means of category theory consisted in translating in the categorical language and spirit some of the results of algebraic logic of that period.

[19] I have to point out that Freyd had already made explicit connections between category theory and model theory before. See [90].

theory **T** itself with a category of a certain sort." ([171], 295.) I emphasize the fact that Lawvere is looking, once again, for an *invariant* notion of a theory and that this invariant notion is provided by a category. Thus, if a group of transformations is an invariant presentation of a theory of geometry, a category can be seen as an invariant presentation of a theory in general. Since a group is a special type of category, this is another illustration of the generalization of Klein's program at work.

Third, and the adjoint functors enter the picture explicitly at this stage, there ought to be an adjoint pair of functors encapsulating the general duality expressed above

$$\mathbf{T}^{\circ} \underset{\text{structure}}{\overset{\text{semantics}}{\rightleftarrows}} \text{Mod}(\mathbf{T}, \mathbf{Set}).$$

The Conceptual is identified with the category of models of **T**.[20] However, the Formal is *not* identified with the invariant formulation of the theory, since clearly there are aspects of the Formal, e.g., specific rules of computation or derivation, that are inherent to a formal framework. Therefore, Lawvere suggests that there is a further adjoint situation, left unspecified this time and simply written as:

$$\text{Formal} \rightleftarrows \text{Theories}.$$

This adjoint situation describes "the presentation of the invariant theories by means of the formalized languages appropriate to the species". ([171], 295.) Since adjoint functors compose, we get a family of adjoint functors

$$\text{Formal}^{\circ} \rightleftarrows \text{Conceptual}$$

that we started with.

As such, this description is almost completely programmatic. It is clearly a bold generalization of Lawvere's thesis. It is given as a completely general framework for foundational research. It is taken as being faithful to the essential elements of mathematical knowledge. Logicians who were about to enter the scene picked up that program and started to develop it systematically. It led to what were later called "categorical doctrines" and were presented as such by Kock and Reyes in their survey paper. (See [143].) Thus, I strongly disagree with Corry when he says:

> Lawvere himself proposed in an article of 1969 to connect the concept of duality, and other categorical concepts, with the epistemological issues related to the philosophy of mathematics. In order to do that, he identified two "dual aspects" of mathematical knowledge — the conceptual and the formal aspects — which appear in many domains of mathematics. (...) Now Lawvere proposed to dedicate efforts to develop the second aspect, the conceptual one, embodied in category theory. This proposal, however, remained at the programmatic level and no one seems to have developed it further. ([51], 388.)

It *is* true that different mathematicians may have interpreted the Formal and the Conceptual according to their own convictions, but the mathematical content of Lawvere's proposal, including the *mathematical* duality involved, led very quickly

---

[20] I am not entirely faithful to Lawvere here. He takes the Conceptual to be the functor category Mod(**CAT**, [**Set**^{I}]), which is of course slightly more general than what I have been describing.

6.4 Categorical Logic: the Program

to a host of important results. The fact is, no one had to quote Lawvere explicitly or say that their work was part of that program, for in a sense the program was already implicit in the *manner* Lawvere had set up his own work on universal algebra and that work, as we have seen, called for various generalizations and expansions.

Finally, one has to contrast categorical logic with other attempts at developing an algebraic framework for logic around the same time, e.g., Halmos's polyadic algebras and Tarski *et al.*'s cylindrical algebras. Joyal and Reyes described the advantages of the categorical approach in an unpublished paper from the mid-seventies:

> 1) The concept of category is used in all branches of mathematics, whereas the structure of polyadic algebra is exotic. Thus, we may hope to extend the application of logic in mathematics.
> 2) Certain categories, used in different fields of mathematics are in fact theories and is useful [sic] to consider them as such.
> 3) As we shall see in this paper, constructions which are actually used in model theory are specializations of general categorical constructions (hence logic is no exception to the generalized use of categories in mathematics). ([126], 5)

These considerations boil down to one fundamental fact: whereas cylindric algebras and polyadic algebras are isolated in the conceptual realm of mathematics, categories are omnipresent. The heuristic gain of using categories is therefore clear and powerful.

But there are indications that the gain is more than heuristic. As we have already seen in the section on adjoint functors, the algebraic expression of propositional logic is given by lattice theory and in the latter adjoint functors are usually called Galois connections. Disagreements appear when higher-order operations like quantifiers are considered, in other words, disagreements appear as to how to generalize the algebraic framework to higher-order logical operations. The fact that categories are a generalization of posets suggests that they might yield the correct generalization. Indeed:

1. Concepts and results about propositional logic are special cases of concepts and results in the categorical setting;
2. It is possible to naturally extend *proofs* in the propositional setting to the categorical setting;
3. It is possible to obtain *new* results in the categorical setting;
4. It is possible to make contacts with other areas of mathematics either by using results of different fields in the new context or by applying the new results in different fields.

It is a remarkable fact that the categorical machinery introduced for algebraic topology, homological algebra, homotopy theory and algebraic geometry constitutes at the same time the proper setting for an algebraic analysis of logic.

We now turn to the details of this program and its development. We will now leave the historical thread and concentrate on its conceptual aspects. Again, my goal is to emphasize how the introduction of category theory in logic is part of the extension of Klein's program.

6.5 An Adjoint Presentation of Propositional Logic

As we have seen in section 5.1, it is possible to introduce operations on a partially ordered set by adjoint functors (or Galois connections). Once this is seen, various systems of propositional logic can be presented by sets of rules that mimic the adjoint situations. Here are the details for classical propositional logic (CPL).

Let us recall some elementary conventions to fix the vocabulary. We denote propositional atoms by $p, q, r, \ldots$. A *conceptual frame* is an (arbitrary) set of atoms and it is denoted by $\mathcal{L}$. Thus a conceptual frame is nothing but a collection of propositions about something. The notion of well-formed formula is as usual. Formulas are denoted by φ, ψ, ξ, etc.

At the syntactic level, a logic $\mathbb{L}$ is characterized by a consequence relation, which we will denote by $\vdash$: and we write $\Gamma \vdash: \alpha$ to express the fact that α is derivable from Γ. A propositional logic is given by a set of connectives and a set of rules of inferences, which in turn determine the consequence relation. The set of connectives for CPL is the set $\{\top, \bot, \wedge, \vee, \rightarrow\}$. The rules are applied to entailments: an *entailment* is a formal entity of the form $\varphi \vdash \psi$, where φ and ψ are arbitrary formulas. A *rule of inference* is an n-ary relation on entailments, for n a positive integer. The first $n-1$ places of the relation are called the *premises*; the last place is the *conclusion* of the rule. When n is equal to 1, we have a *logical axiom scheme*. A rule $\mathfrak{R}$ has the form

$$\mathfrak{R}\frac{\alpha \quad \beta \quad \cdots}{\gamma}$$

where α, β, γ denote entailments.

The set of rules of CPL is given by:

Structural rules:

$$(\text{Taut}) \quad \varphi \vdash \varphi$$

$$(\text{Cut}) \quad \frac{\varphi \vdash \psi \quad \psi \vdash \theta}{\varphi \vdash \theta}$$

Logical rules:

$$(\text{True}) \quad \varphi \vdash \top \qquad (\text{False}) \quad \bot \vdash \varphi$$

$$(\wedge) \quad \frac{\varphi \vdash \psi \quad \varphi \vdash \theta}{\varphi \vdash \psi \wedge \theta} \qquad (\vee) \quad \frac{\psi \vdash \varphi \quad \theta \vdash \varphi}{\psi \vee \theta \vdash \varphi}$$

$$(\rightarrow) \quad \frac{\varphi \wedge \psi \vdash \theta}{\varphi \vdash \psi \rightarrow \theta} \qquad (\text{Bool}) \quad (\top \vdash \varphi \vee \neg\varphi)$$

where $\neg\varphi$ is defined as $\varphi \rightarrow \bot$. Some comments about the rules are in order. First, the structural rules can be thought of as defining a partial order on formulas, hence a category. Thus, one could define a deductive system as a category. Second, the logical rules are easily seen to be the syntactic expression of adjoint situations, except for (Bool). Indeed, if one reads the entailment relation as being a partial order, then

6.5 An Adjoint Presentation of Propositional Logic

the rules are nothing but the adjoint situations for partial orders. Third, if the rule (Bool) is removed, then one obtains intuitionistic propositional logic (IPL). Fourth, double lines indicate that it is possible to go either way: from top to bottom or from bottom to top. Thus they are abbreviations for two or three rules.

A *theory* **T** is a pair **T** = $(\mathfrak{L}, \mathfrak{U})$ where $\mathfrak{L}$ is a conceptual frame and $\mathfrak{U}$ is a set of entailments, thought of as the axioms of the theory **T**. Thus, we write $\mathfrak{U} \vdash \alpha$ to mean that α is derivable from the axioms $\mathfrak{U}$ of the theory (by the use of some specified rules, here the classical rules).

The foregoing syntax for CPL is such that some aspects of the semantics become obvious. Furthermore, and as we will see, it allows for the definition of basic notions that can be raised to higher-order contexts. In the context of categorical logic, semantics for propositional logics are given by algebras. In the case of CPL, it is of course given by Boolean algebras. We will be somewhat explicit in the presentation and discussion of the semantics, although the details are well-known, because the categorical framework allows much flexibility and generalization.

Let the category **Bool** be the category with objects $A, B, C, \ldots$, Boolean algebras and $f, g, h, \ldots$, Boolean homomorphisms. Let **T** = $(\mathfrak{L}, \mathfrak{U})$ be a fixed theory and A a fixed Boolean algebra. An *interpretation* $\mathfrak{I}$ of $\mathfrak{L}$ in A is any mapping $\mathfrak{I} \colon \mathfrak{L} \to A$. For any $\mathfrak{L}$-formula φ, that is any formula in which the only atoms occurring are from $\mathfrak{I}$, $\mathfrak{I}(\varphi)$ is defined by recursion as usual. A *model* of **T** in A is an interpretation $\mathfrak{I} \colon \mathfrak{L} \to A$ such that each entailment $\varphi \vdash \psi$ in $\mathfrak{U}$ is true in A, in the sense that $\mathfrak{I}(\varphi) \leq_A \mathfrak{I}(\psi)$ (recall that a Boolean algebra is also a poset). In this case, we write as usual $\mathfrak{I} \vDash \varphi \vdash \psi$. For CPL, it is a standard result that the two-elements Boolean algebra **2** is enough and can be used to prove soundness and completeness. In fact, in this context, verification of soundness is routine: it is simply a translation of the various rules of inference in the language of posets and they simply become the expression of the existence of adjoints for certain operations on the appropriate algebras. Completeness, although more complicated, also takes an algebraic dressing.

A somewhat more general framework is needed to be able to prove soundness and completeness in a uniform manner for other propositional logics. Recall, for instance, that for IPL, there cannot be a finite algebra that plays the same role that **2** plays for CPL. (This was shown by Gödel in 1932.) What we need in general is a specified collection of algebras for a given logic. Thus, to specify a *semantics* for a logic $\mathbb{L}$, a collection $\mathscr{S}$ of $\mathbb{L}$-algebras has to be given, where an $\mathbb{L}$-algebra is a partially ordered set L together with a set of operations, one for each logical connective and such that the rules of inference of $\mathbb{L}$ are satisfied by L with the obvious interpretation. Of course, as before, $\mathbb{L}$-algebras form a category. Thus, we have the category of distributive lattices, of Boolean algebras, of Heyting algebras, etc. We then say that an entailment α is a *semantic consequence* of a theory **T** *with respect to* $\mathscr{S}$ if for every $\mathbb{L}$-algebra S in $\mathscr{S}$, and every S-model $\mathfrak{I} \colon \mathfrak{L} \to S$ of **T**, we have $\mathfrak{I} \vDash \alpha$. This is denoted by $\mathbf{T} \vDash_{\mathscr{S}} \alpha$. We say that completeness holds with respect to $\mathscr{S}$, if $\mathbf{T} \vDash_{\mathscr{S}} \alpha$ implies $\mathbf{T} \vdash \alpha$ for any **T** and α. For instance, completeness holds with respect to $\mathscr{S} = \{\mathbf{2}\}$ for CPL.

In an algebraic context, the completeness theorem for a logic $\mathbb{L}$ is equivalent to a representation theorem, provided the logic $\mathbb{L}$ satisfies some very mild conditions, e.g., being invariant under substitution. More specifically, the representation theorem holds with respect to $\mathscr{S}$, if for every $\mathbb{L}$-algebra A, and any elements x, y in A, $x \leq_A y$ if and only if $h(x) \leq_{\mathscr{S}} h(y)$ for every S in $\mathscr{S}$ and every morphism $h \colon A \to S$. The Stone representation theorem for Boolean algebras is equivalent to the statement that the representation theorem holds with respect to the collection $\mathscr{S} = \mathbf{2}$. The equivalence between completeness and representation now takes the following form: for any propositional logic $\mathbb{L}$ satisfying mild conditions, and for any class $\mathscr{S}$ of $\mathbb{L}$-algebras, completeness with respect to $\mathscr{S}$ holds if and only if the representation theorem holds with respect to $\mathbb{L}$ in the category of $\mathbb{L}$-algebras.

Seeing that a completeness theorem for a propositional logic is equivalent to a representation theorem for a category of algebras might seem mildly interesting but of no real value, since completeness theorems can be proved directly. What are the advantages of proceeding via a representation theorem? We can at this point immediately underline two facts: (1) When moving to higher-order logics in a categorical context, we are simply moving to categorical algebra, and a uniform procedure to establish various representation theorems can be developed. In other words, in all cases, there is a part of the proof that relies on "general abstract nonsense", that is category theory, and another part that is specific to the logic under consideration; (2) The equivalence between completeness theorems and representation theorems links logical results to important and fundamental mathematical results. We will come back to these points in chapter 7.

Furthermore, in the algebraic setting, we can immediately introduce two important concepts which are more or less trivial at the propositional level but acquire more substance when we move to higher-order logics. As we will see, these constructions can be lifted directly into the categorical setting.

The first such concept is the notion of a *generic or universal model* of a theory. Let **T** be a theory in CPL. Then, as is well known, one can construct the associated *Lindenbaum-Tarski algebra* of **T** as follows: define a congruence relation $\sim_{\mathbf{T}}$ on the formulas of the underlying conceptual frame $\mathfrak{L}$ of **T** by

$$\varphi \sim_{\mathbf{T}} \psi \text{ if and only if } \mathbf{T} \vdash \colon \varphi \vdash \psi \text{ and } \mathbf{T} \vdash \colon \psi \vdash \varphi.$$

In words, two formulas are equivalent or logically indistinguishable if and only if they are deducible from one another in **T**, i.e., they have, so to speak, the same logical strength. It follows from the structural rules of CPL that $\sim_{\mathbf{T}}$ is an equivalence relation and the fact that it is a congruence relation follows from invariance under substitution.

Let us denote as usual the equivalence class of φ by $[\varphi]$ and the collection of all $\mathfrak{L}$-formulas by $\mathscr{F}_{\mathfrak{L}}$. Then the usual quotient $\mathscr{F}_{\mathfrak{L}}/\sim_{\mathbf{T}}$ can be seen to be a partially ordered set, the latter relation being defined by

$$[\varphi] \leq [\psi] \text{ if and only if } \mathbf{T} \vdash \colon \varphi \vdash \psi.$$

Furthermore, the usual algebraic operations can be defined in the usual manner on representatives of the equivalence classes. It can be shown that the resulting algebra, which we will denote by [**T**], is a Boolean algebra. There is an obvious interpretation

6.5 An Adjoint Presentation of Propositional Logic

$M_\mathbf{T}$ of $\mathbf{T}$ in $[\mathbf{T}]$: $M_\mathbf{T}(\varphi) = [\varphi]$ for all $\varphi \in \mathscr{F}_\mathscr{L}$ and it can be shown that $M_\mathbf{T}$ is a model of $\mathbf{T}$. It is called the *generic* or *universal* model of $\mathbf{T}$.

From a categorical point of view, the Lindenbaum-Tarski algebra is defined by a universal property. Indeed, the pair $\langle[\mathbf{T}], M_\mathbf{T}\rangle$ satisfies the following property: for any model $M\colon \mathbf{T} \to A$, there is a unique morphism $f\colon [\mathbf{T}] \to A$ of Boolean algebras such that the diagram

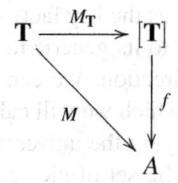

$$\begin{array}{ccc} \mathbf{T} & \xrightarrow{M_\mathbf{T}} & [\mathbf{T}] \\ & \searrow{\scriptstyle M} & \downarrow{\scriptstyle f} \\ & & A \end{array}$$

commutes. Thus, the Lindenbaum-Tarski algebra is another instance of a universal morphism. The *existence* of the Lindenbaum-Tarski algebra is established by the foregoing construction, but as usual in a categorical context, what matters is specified by the universal property it satisfies. Furthermore, the pair $\langle[\mathbf{T}], M_\mathbf{T}\rangle$ is determined by $\mathbf{T}$ up to isomorphism. What the diagram asserts is that there is a one-to-one correspondence between models $M\colon \mathbf{T} \to A$ and morphisms $f\colon [\mathbf{T}] \to A$. It is in this sense that $[\mathbf{T}]$ is generic or universal: every model of $\mathbf{T}$ is in fact an image of the generic model and any such model can be obtained in this manner. In particular, that is, in the case of Boolean algebras, it follows that any model $\mathbf{T} \to \mathbf{2}$ corresponds to a Boolean homomorphism $[\mathbf{T}] \to \mathbf{2}$.

The generic model is connected to deducibility in a very important way. Indeed, it follows that

$$\mathbf{T} \vdash \colon \varphi \vdash \psi \text{ if and only if } M_\mathbf{T} \vDash \varphi \vdash \psi \text{ if and only if } M_\mathbf{T}(\varphi) \leq_{[\mathbf{T}]} M_\mathbf{T}(\psi).$$

A few words of explanation about the generic model are probably needed at this stage. One has to be clear about what it amounts to. It is a quotient construction based on an equivalence relation. Thus, we start with a certain theory $\mathbf{T}$ in a given conceptual frame and we *ignore* the explicit content of the propositions to focus our attention on the *logical strength* of the formulas. In other words, when we move from the theory $\mathbf{T}$ to the generic model $[\mathbf{T}]$, we shift from the criterion of identity on *specific* formulas and the deducibility relation defined between them to the criterion of identity on *equivalence classes* of formulas and the order relation between them. But we have to be extremely clear about the nature of these equivalence classes. The fact that the quotient structure is an algebra, and not just a set, is of fundamental importance. The algebraic structure encodes in a condensed manner all possible logical paths between all possible formulas having the same logical strength. An equivalence class $[\varphi]$ stands for a certain property that is shared by all its representatives: given any representative φ of $[\mathbf{T}]$ and any other equivalence class $[\psi]$ with representative ψ such that $[\varphi] \leq_{[\mathbf{T}]} [\psi]$, there is a proof from φ to ψ in $\mathbf{T}$. The details of the proof, e.g., its length or complexity, are completely ignored. The partial order represents a basic logical connection between formulas; it ignores all the syntactical information contained in the proofs to keep the purely conceptual connection between logical types within the given theory. Thus, the identity condition used to identify the logical types is *not* that they are sets (i.e., it is not the axiom of extensionality), but rather it is the criterion of identity for the algebraic system they find themselves in. It is for this reason that we *can* treat the equivalence classes as (logical) types, for they simply yield a representation of an algebraic system, i.e., it is the fact that the construction of the Lindenbaum-Tarski algebra satisfies the foregoing universal property which constitutes the key fact.

We have moved from a theory to its generic model, to algebraic systems. There is a way to go in the opposite direction. We can start from a specified algebra A and try to construct a theory $\mathbf{T}_A$, which we will call the *internal theory* of A, and an interpretation $M_A \colon \mathbf{T}_A \to A$ satisfying the universal property of the generic model. The conceptual frame is simply the set of elements of A, i.e., $\mathfrak{L} = |A|$, where $|A|$ denotes the underlying set of A. The set of entailments Σ_A is the set of all entailments over $\mathfrak{L}_A$ that are true under the identity interpretation $1_{|A|} \colon \mathfrak{L}_A \to A$. Thus, the internal theory $\mathbf{T}_A$ of A is given by $\langle \mathfrak{L}_A, \Sigma_A \rangle$ and the generic model $M_A \colon \mathbf{T}_A \to A$ is the identity morphism. The universal property is in this case trivially verified.

Needless to say, the previous considerations do not apply exclusively to Boolean algebras and CPL. They cover a whole range of propositional logics. Furthermore and as we will see, the notions of generic model and of internal theory can be lifted into the categorical context and thus are also used for higher-order logics.

Let us recapitulate the use of categories and adjoints at the level of propositional logic.

1. At the syntactical level, logical systems can be presented in a language that mimic adjoint situations, as we have illustrated for classical logic;
2. At the semantic level, each specific algebraic structure, e.g., Boolean or Heyting algebras, used to interpret the syntax can be defined via adjoint connections on partial orders seen as categories;
3. The algebraic structures form categories in their own right and the structure of these categories is relevant from the logical point of view;
4. Given that adjoints are organized in a certain manner and that categories of algebraic structures are related to one another in systematic ways, various results connecting logical systems to one another are possible and can even be exploited. A striking example of this phenomenon will be given later when I present and discuss geometric or coherent logic.

6.6 Quantifiers as Adjoint Functors

One of the stunning discoveries made by Lawvere in the early sixties is the fact that quantifiers *can* be introduced as adjoint functors. This is an additional and very important indication that the slogan "algebraic logic is categorical logic" has a definite content.

To motivate the purely formal definitions, we will start with the case of sets. Let X be a set and $\wp(X)$ be the set of all subsets of X. $\wp(X)$ is a preorder, thus a category,

6.6 Quantifiers as Adjoint Functors

under the relation of inclusion of subsets. In fact, as is well known, it is a Boolean algebra. Each function $f: X \to Y$ induces two order-preserving functions between $\wp(X)$ and $\wp(Y)$. The first function is the direct image function $f_*: \wp(X) \to \wp(Y)$ defined by $f_*(S) = \{f(x) \in Y \mid x \in S\}$. The second is the inverse image function $f^*: \wp(Y) \to \wp(X)$ defined by $f^*(T) = \{x \in X \mid f(x) = y \text{ for some } y \in T\}$. This last object can be described in purely categorical terms. To see this, we need to introduce another universal construction, called the pullback.

Consider the following diagram in an arbitrary category $\mathscr{C}$.

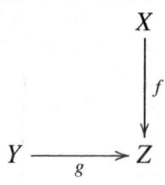

$$\begin{array}{ccc} & & X \\ & & \downarrow f \\ Y & \xrightarrow{g} & Z \end{array}$$

A *pullback* of this diagram, whenever it exists, is an object P together with arrows $p_X: P \to X$, $p_Y: P \to Y$ and $p_Z: P \to Z$ such that

1. $f \circ p_X = p_Z = g \circ p_Y$ and
2. For any object Q with arrows $h: Q \to Y$, $i: Q \to Z$ and $k: Q \to X$ such that $g \circ h = i = f \circ k$, there is a unique arrow $u: Q \to P$ such that $p_X \circ u = k$, $p_Z \circ u = i$ and $p_Y \circ u = h$.

In diagrammatic language, a pullback $\langle P, p_Y, p_Z, p_X \rangle$ for the arrows $f: X \to Z$ and $g: Y \to Z$ is a way to fill in the diagram in the following manner:

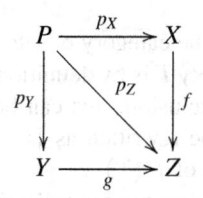

$$\begin{array}{ccc} P & \xrightarrow{p_X} & X \\ p_Y \downarrow & \searrow^{p_Z} & \downarrow f \\ Y & \xrightarrow{g} & Z \end{array}$$

such that the following universal property is satisfied:

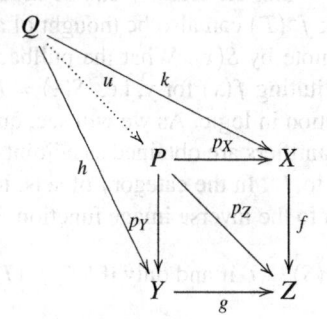

$$\begin{array}{c} Q \\ \end{array}$$

Notice that since $f \circ p_X = p_Z = g \circ p_Y$, the arrow p_Z can be dropped from the definition altogether (and so can the arrow $i\colon Q \to Z$, which we have not drawn). Such a P is sometimes denoted by $Y \times_Z X$ and is also known as the *fibered product* of Y, X over Z. Notice also that pullbacks are a generalization of the notion of products, for whenever Z is the terminal object 1, then a pullback is a product.

The category **Set** has pullbacks. A pullback of sets X and Y can be described in set-theoretical language as a subset of the Cartesian product of X and Y. Indeed, it is the set of all pairs (x, y) such that $f(x) = g(y)$, i.e., $\{(x,y) \in X \times Y \mid f(x) = g(y)\}$.

The inverse image function f^* can be described in terms of pullbacks. Indeed, consider the following diagram:

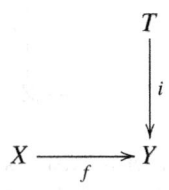

$$\begin{array}{ccc} & & T \\ & & \downarrow i \\ X & \xrightarrow{f} & Y \end{array}$$

Taking the pullback of the inclusion function $i\colon T \to Y$ along f yields the inverse image of f.

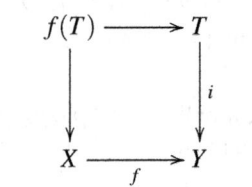

$$\begin{array}{ccc} f(T) & \longrightarrow & T \\ \downarrow & & \downarrow i \\ X & \xrightarrow{f} & Y \end{array}$$

Since we are by assumption in the category of sets, we can see directly why $f^*(T)$ is the pullback. The pullback $X \times_Y T$ is by definition the set $\{(x,y) \in X \times T \mid f(x) = f(y)\}$. Since $i\colon T \to Y$ is an inclusion, this can be rewritten as $\{(x,y) \in X \times T \mid f(x) = y\}$, which in turn can be rewritten as $\{x \in X \mid f(x) = y \text{ for some } y \in T\}$, which is precisely the definition of $f^*(T)$.

The inverse image of a given monic arrow i can be taken in any category with pullbacks. Think of the subset T of Y as being a unary predicate $T(y)$ over Y. What the above pullback shows is that the arrow f can be thought of as being "substitution". Indeed, the pullback $f^*(T)$ can also be thought of as being a unary predicate over X, which we can denote by $S(x)$. What the pullback says is that $S(x)$ is obtained from $T(y)$ by substituting $f(x)$ for y, i.e., $S(x) = T(f(x))$. It is in this sense that pullbacks are substitution in logic. As we will see, quantifiers can be described as adjoints to f^*. Thus quantifiers are obtained as adjoints to substitution.

Consider now adjoints to f^*. In the category of sets, it is easily verified that the direct image is left adjoint to the inverse image function, i.e.,

$$f_*(S) \subseteq T \text{ if and only if } S \subseteq f^*(T),$$

or, to use our previous notation

6.6 Quantifiers as Adjoint Functors

$$\frac{f_*(S) \subseteq T}{S \subseteq f^*(T)}$$

that is f_* is left adjoint to f^*. But by definition, $f_*(S)$ can be written in this particular case as the set of elements y of Y such that *there exists* an x in X which is in S such that $f(x) = y$. Thus, the left adjoint is directly connected to the *existence* of an element that can be substituted. It is the existential quantifier $f_*(S)$ and is written as $\exists_f S$. Thus,

$$\exists_f(S) = \{f(x) \mid x \in S\} = \{y \in Y \mid \exists x \in X (f(x) = y \wedge x \in S)\}.$$

A right adjoint to the inverse image function f^* can be defined and it exists in the category of sets. There is another function $\wp(X) \to \wp(Y)$, let us call it provisionally f^+, such that

$$f^*(T) \subseteq S \text{ if and only if } T \subseteq f^+(S).$$

The set $f^+(S)$ is defined in this context as

$$f^+(S) = \{y \in Y \mid \text{ for all } x, \text{ if } f(x) = y, \text{ then } x \in S\}.$$

It can be verified that f^+ is right adjoint to f^*, i.e.,

$$\frac{f^*(T) \subseteq S}{T \subseteq f^+(S)}$$

Notice the presence of a universal quantifier in the definition of this subset in the category of set. We write $f^+(S)$ as $\forall_f S$. Thus,

$$\forall_f S = \{y \in Y \mid f^{-1}(\{y\}) \subseteq S\} = \{y \in Y \mid \forall x \in X (f(x) = y \to x \in S)\}.$$

To summarize, we have:

$$\wp(X) \underset{\forall_f}{\overset{\exists_f}{\underset{\longleftarrow}{\rightrightarrows}}} \wp(Y), \quad \exists_f \dashv f^* \dashv \forall_f.$$

To see the connection with the standard quantifiers, it is enough to look at a particular case of the above. Let f be the first projection function $p \colon X \times Y \to X$. Hence $p^* \colon \wp(X) \to \wp(X \times Y)$ which to each subset S of X assigns the inverse image $p^*(S)$ in $X \times Y$. S ought to be thought of as a unary predicate $S(x)$ over X. Then $p^*(S)$ is also a predicate $p^*(S)(x,y)$, which we will simply write as $P(x,y)$, where y is a dummy variable. The pullback diagram in this case is

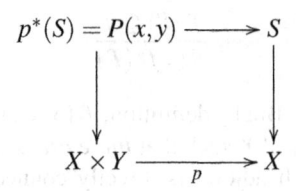

$$p^*(S) = P(x,y) \longrightarrow S$$
$$\downarrow \qquad\qquad \downarrow$$
$$X \times Y \xrightarrow{\ p\ } X$$

The pullback means that $P(p(x),y) = S(x)$, which is a special case of substitution. It is also called "weakening" in the literature since $p^*(S)$ yields a property of elements (x,y) of $X \times Y$ by ignoring y, that is by the weakening of the property S of elements of X. It can be verified that taking the pullback of a subset S of X in this case amounts to considering the cylinder over S in $X \times Y$.

The adjoint situation takes the following particular form:

$$\wp(X \times Y) \underset{\forall_p}{\overset{\exists_p}{\underset{\longleftarrow}{\overset{\longrightarrow}{\rightleftarrows p^* \rightleftarrows}}}} \wp(X), \quad \exists_p \dashv p^* \dashv \forall_p.$$

For $T \subseteq X \times Y$, $\exists_p(T) \subseteq X$: it is the set of elements of X such that there is a y and $\langle x,y \rangle \in X \times Y$, that is, it is the direct image p_*. We write $\exists_p T$ as $\exists y T(x,y)$. Similarly, $\forall_p T$ is the set of elements of x such that for all $\langle x,y \rangle$, if $p\langle x,y \rangle = x$, then $\langle x,y \rangle \in T$. We write $\forall_p T$ as $\forall y T(x,y)$. Symbolically,

$$\exists y T(x,y) = \exists_p T = \{x \in X \mid \exists y(x,y) \in T\}$$

and

$$\forall y T(x,y) = \forall_p T = \{x \in X \mid \forall y \in Y (x,y) \in T\}.$$

Rewriting the inclusions relations as implications, i.e., $\vdash_X$ for $\subseteq_X$ and $\vdash_{X,Y}$ for $\subseteq_{X \times Y}$, the adjoint situations become:

$$\frac{T(x,y) \vdash_{X,Y} p^*S(x)}{\exists y T(x,y) \vdash_X S(x)} \quad \text{and} \quad \frac{p^*S(x) \vdash_{X,Y} T(x,y)}{S(x) \vdash_X \forall y T(x,y)}.$$

All the foregoing was developed in the category of sets. It is easy to see that for any category $\mathscr{C}$ with finite products and pullbacks (and, of course, in which a notion corresponding to the lattice of subsets exists), it is possible to investigate whether quantifiers are definable in $\mathscr{C}$, i.e., if the relevant adjoints exist.

As Lawvere underlined in 1970, quantifiers are related to substitution in another way, at least in some important contexts, e.g., in the category of sets. Quantifiers are said to be *stable under substitution* whenever the pullback on the left implies that the square on the right is commutative:

$$\begin{array}{ccc} A \xrightarrow{g} B & & \wp(A) \xrightarrow{\exists g} \wp(B) \\ s \downarrow \quad \downarrow t & \Rightarrow & s^* \uparrow \quad \uparrow t^* \\ X \xrightarrow{f} Y & & \wp(X) \xrightarrow[\exists_f]{} \wp(Y) \end{array}$$

A similar condition can be stated for the universal quantifier $\forall_f$. These are called the *Beck-Chevalley conditions* (for the respective quantifier) in the literature. Thus, although quantifiers do satisfy this condition in the category of sets, it is not always the case, even in categories where the quantifiers exist.

6.7 Graphical Syntax: Sketches

> (...) categorical logic is, to a great degree, autonomous, even in matters *syntactical*.
>
> ([203], 54.) Makkai

As we have seen, category theory was at first considered to be a useful language. What was clearly useful was the use of diagrams to prove certain results either in algebraic topology, homological algebra or algebraic geometry. It is clear that doing category theory, or simply applying category theory, implies manipulating diagrams: constructing the relevant diagrams, chasing arrows by going via various paths in diagrams and showing they are equal, etc. This practice suggests that diagram manipulation, or more generally diagrams, constitutes the natural syntax of category theory and the category-theoretic way of thinking. Thus, if one could develop a formal language based on diagrams and diagrams manipulation, one would have a natural syntactical framework for category theory. However, moving from the informal language of categories which includes diagrams and diagrammatic manipulations to a *formal* language based on diagrams and diagrammatic manipulations is not entirely obvious. The gains obtained by the introduction of such a formal language are not at first clear. The motivation for such an enterprise is not a priori transparent. As we have seen in a previous chapter, Freyd was forced to consider some issues related to the language of categories by a metatheoretical preoccupation. It was not seen by category theorists as being a vital issue.

This idea of exploiting diagrams in a formal manner was first developed systematically by C. Ehresmann and his school in the sixties when they defined the notion of an *esquisse*, which was appropriately translated by "sketch" in English. However the notion was not taken up by many mathematicians and has only recently made a comeback. It has been generalized and used in various ways and certainly offers an interesting framework to clarify and again unify various important notions, even logical notions.

In order to see what a sketch is, I will first give a simple example to illustrate how the arrows of a given category can be used as a language to define various internal notions. Since groups have been the focus of our attention so far, we will once again take a look at them, but this time by giving the definition of a group internal to a category $\mathscr{C}$, that is given by the language of the category $\mathscr{C}$.

The data required to define a group G abstractly can be specified with the help of diagrams, *provided* the underlying category has a minimal amount of structure, in this case binary products and a terminal object must exist in the category (or equivalently, binary and empty products exist). Thus a group *in* a category $\mathscr{C}$ with products is an object G of $\mathscr{C}$ together with the following arrows:

$$m: G \times G \to G; \quad i: G \to G; \quad e: 1 \to G$$

such that the following diagrams are commutative in $\mathscr{C}$:

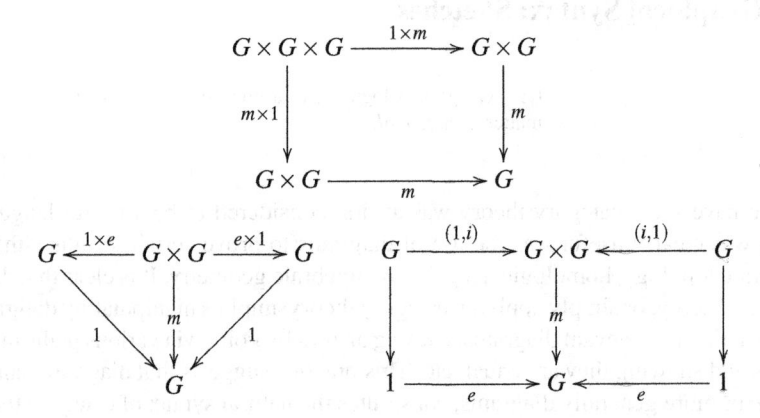

$$\begin{array}{ccc} G \times G \times G & \xrightarrow{1 \times m} & G \times G \\ {\scriptstyle m \times 1} \downarrow & & \downarrow {\scriptstyle m} \\ G \times G & \xrightarrow{m} & G \end{array}$$

$$\begin{array}{ccc} G \xleftarrow{1 \times e} G \times G \xrightarrow{e \times 1} G \\ {\scriptstyle 1} \searrow \quad \downarrow {\scriptstyle m} \quad \swarrow {\scriptstyle 1} \\ G \end{array} \qquad \begin{array}{ccccc} G & \xrightarrow{(1,i)} & G \times G & \xleftarrow{(i,1)} & G \\ \downarrow & & \downarrow {\scriptstyle m} & & \downarrow \\ 1 & \xrightarrow{e} & G & \xleftarrow{e} & 1 \end{array}$$

The first one states that the group operation is associative, the second that the element $e: 1 \to G$ is the identity element and the third one asserts that inverses exist. If $\mathscr{C}$ is the category of sets, we obtain the usual notion of group, if it is the category of topological spaces and continuous maps, G is a topological group, if $\mathscr{C}$ is the category of manifolds and smooth maps, we obtain a Lie group and if $\mathscr{C}$ is the category of groups and group homomorphisms, then G is an *Abelian* group. (This last statement ought to be surprising! One has to verify why it is so.)

The general idea is to construct an abstract graph which would simply contain the information in the above diagrams together with the information necessary for certain constructions to be of the right type, e.g., products and certain commutativity conditions. Thus, the concept of group can be presented by an abstract graph, and notice immediately that a group in the usual sense then becomes an interpretation of this abstract graph into a category, e.g., it is given by a functor which preserves the relevant structure from the graph to the underlying graph of a category, for instance as above the category of sets, or the category of topological spaces.

6.7 Graphical Syntax: Sketches

Thus, a specific *abstract graph* is used to characterize a certain concept, for instance the nodes and arrows above, in which certain diagrams will have to be considered as being limit or colimit diagrams and certain commutativity conditions are satisfied.[21] In the case of a group, we need a product node and a terminal object node in the data and certain commutativity expressing the equalities, hence very little data. These are the ingredients we need to specify for an abstract graph to become a definition or "axiomatization" of the group concept. This is precisely what the basic notion of a (finite product) sketch captures in general.

Before we give the definition of a sketch, recall the notion of a directed graph. A *directed graph* is given by a collection of arrows A and a collection of vertices or nodes O, together with two arrows $s\colon A \to O$ and $t\colon A \to O$, the source and target arrows, which associate to each arrow a in A, its source $s(a)$ and its target $t(a)$. We are now ready to specify what a sketch is. We will restrict ourselves to what is called a *finite limit sketch*, also called a left-exact sketch or LE-sketch, for it is a very natural notion and we can make all the points we are interested in with its help.

Definition 6.5. A *finite limit sketch* S or a *LE-sketch* is given by a quadruple $\langle S, I, D, C \rangle$ where:

1. S is a graph;
2. $I\colon O \to A$ is a map which assigns to each node g of S an arrow $g \to g$;
3. D is a collection of diagrams in S;
4. C is a collection of cones in S, where a *cone* with domain g in a graph S (respectively in a category) is a collection of arrows with domain g (such that, in a category, if $i\colon g_1 \to g_2$, $j\colon g_1 \to g_3$ and $k\colon g_2 \to g_3$ then $k \circ i = j$, i.e., the various diagrams commute). (Needless to say, these cones will be turned into limits in the interpretations or representations of the sketch in various categories.)

Let us consider the sketch for groups. This case, where only finite products are required, will again illustrate the nature of sketches.

First, let us see the graph that contains all the information. It is in fact quite involved, more than one might at first think. The collection O of objects contains four distinct objects, which we will denote by g_0, g_1, g_2 and g_3. For the arrows, we must have the arrows that will become the group operations and the arrows for the products, that is the projection arrows. Thus we have the following:

1. $1_i\colon g_i \to g_i$ ($i = 0, 1, 2, 3$ the identity arrows);
2. $p_1, p_2\colon g_2 \to g_1$ (projections of g_2 as a product of $g \times g$);
3. $q_1, q_2, q_3\colon g_3 \to g_1$ (projections of g_3 as a product of $g \times g \times g$);
4. $q_{12}, q_{23}\colon g_3 \to g_2$ (projections of g_3 on g_2 whose role will be clarified by diagrams);

---

[21] Of course, as soon as we have a graph, it is easy to pass to a category generated by this graph. Hence one could define a sketch as being a certain abstract category. The interesting point in this context is that an interpretation is a functor from this abstract category into a category, hence it is a representation. Recall that an abstract transformation group is simply a one-object category and a representation of a transformation group is a functor from this group into a category of vector spaces.

5. $m\colon g_2 \to g_1$ (operation of multiplication);
6. $e\colon g_0 \to g_1$ (operation of picking up the identity);
7. $i\colon g_1 \to g_1$ (send each element to its inverse element in g);
8. $m \times 1_1, 1_1 \times m\colon g_3 \to g_2$ (two morphisms whose name exhibit my intentions);
9. $!\colon g_1 \to g_0$;
10. $e \times 1_1, 1_1 \times e\colon g_1 \to g_2$ (two morphisms allowing us to pick appropriate pairs of elements);
11. $(i, 1_1), (1_1, i)\colon g_1 \to g_2$ (two morphisms allowing us to pick appropriate pairs of elements).

The collection of diagrams of the sketch contains all the foregoing diagrams describing the axioms of group theory together with the following diagrams which force the arrows given to be universal arrows and arrows of the proper type:

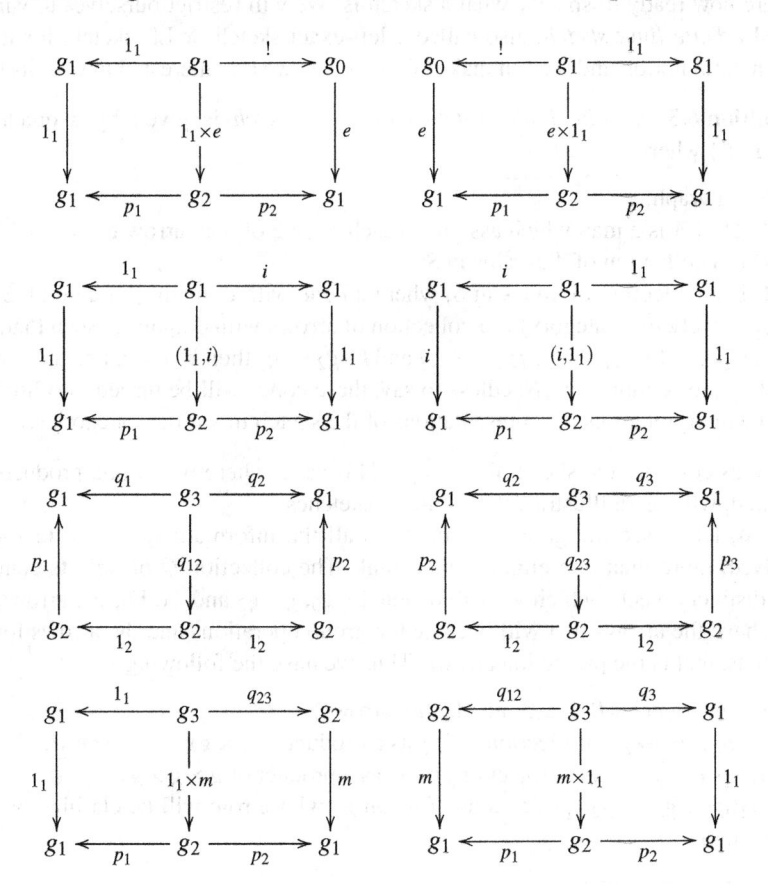

6.7 Graphical Syntax: Sketches

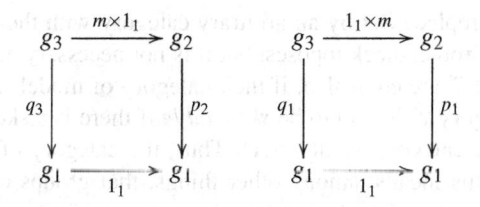

$$\begin{array}{ccc} g_3 & \xrightarrow{m \times 1_1} & g_2 \\ {\scriptstyle q_3}\downarrow & & \downarrow{\scriptstyle p_2} \\ g_1 & \xrightarrow{1_1} & g_1 \end{array} \qquad \begin{array}{ccc} g_3 & \xrightarrow{1_1 \times m} & g_2 \\ {\scriptstyle q_1}\downarrow & & \downarrow{\scriptstyle p_1} \\ g_1 & \xrightarrow{1_1} & g_1 \end{array}$$

The cones in this case are simple: we need the empty cone so that g_0 will become the terminal object and the cones given by 2, 3 and 4 above so that g_2 and g_3 become the appropriate product of g_1.

This is but *one* finite-product sketch. Notice that finite-product sketches are essentially finite-limit sketches in which diagrams are all discrete. A finite-limit sketch is a sketch in which all finite limits are allowed. Morphisms between sketches are defined as follows: given two sketches S and $S' = \langle S', I', D', C' \rangle$, a morphism $f: S \to S'$ is a graph homomorphism $f: S \to S'$ such that $f \circ I = I'$, every diagram in D is taken to a diagram in D' and every cone in C is taken to a cone in C'. Clearly, morphisms of sketches are simply special morphisms of graphs and it is therefore possible to consider the category of LE-sketches, which is a subcategory of the category of graphs. Notice that in this case, morphisms can be thought of as translations. The underlying groupoid of the category of sketches provide us with a criterion of identity for presentations of theories although we should immediately point out that it is not an *interesting* criterion of identity in this case, since our structures are already abstract. In view of their abstract nature, it is more interesting to introduce a criterion of identity which will rests on their representations, to which we now turn.

The foregoing sketch for groups constitutes a *presentation* of the concept of group by exhibiting, so to speak, the basic and generic structure of a group. Indeed, the sketch is the sketch of one object with certain transformations between it and certain of its products. Of course, the theory of groups has to do with all groups and we therefore have to find a way to move from this abstract generic group to various concrete groups. This is done by considering various representations, also called models in this context, of the given sketch in categories. In order to define models for sketches, observe that any category $\mathscr{C}$ has an underlying sketch whose graph is the underlying graph of $\mathscr{C}$, I picks out the identity maps of $\mathscr{C}$, D is the class of all commutative diagrams of $\mathscr{C}$ and C the class of all limit cones of $\mathscr{C}$. It is then easy to see how a representation or a model of a sketch can be defined. Define a map from a sketch S into a category $\mathscr{C}$, for instance the category **Set**, as a sketch morphism from S to the underlying graph of $\mathscr{C}$. In the category $\mathscr{C}$, diagrams become *commutative* diagrams, cones become *limit* cones in $\mathscr{C}$ (and so on for sketches with more structure). Given a sketch S and a category $\mathscr{C}$, consider the collection of all such representations or models $F: S \to \mathscr{C}$. They form a category whose objects are the models F and morphisms the natural transformations between such models. Notice that the collection of all graph morphisms $S \to \mathscr{C}$ is also a category and that the category of models of S in $\mathscr{C}$, denoted by $\mathrm{Mod}(S, \mathscr{C})$ is a subcategory of the functor category $\mathbf{Set}^S$. Given two sketches S and S', they are said to be *equivalent* or *similar* if the categories $\mathrm{Mod}(S, \mathbf{Set})$ and $\mathrm{Mod}(S', \mathbf{Set})$ are equivalent categories.

(In fact, we could replace **Set** by an arbitrary category with the appropriate structure, for instance Grothendieck toposes, but it is not necessary at this point.) Thus, two sketches S and S' are equivalent if their category of models are (categorically) equivalent. A category $\mathscr{C}$ is said to be *sketchable* if there is a sketch S such that $\mathscr{C}$ is equivalent to the category $\mathrm{Mod}(S, \mathbf{Set})$. Thus, the category of groups is sketchable. Notice that this means, among other things, that groups can be represented systematically as *functors* from a sketch in the category of sets.

Some important remarks can now be made.

1. There is a very natural way to think about the relationship between a sketch and its various representations. A sketch such as the above sketch for a group should be thought of as a *type*, in the standard philosophical sense of that expression, and its various representations or models as *tokens* of that type, again in the standard philosophical sense of that expression. But this shows how different from the standard approaches syntactical matters are in a categorical context. It is obvious that the relationship between the presentation of a theory in a standard syntactical framework with its models has nothing to do with a type/token relationship. The type/token relationship *is* the basic epistemological relationship in category theory.

2. Sketches are organized in a natural way. The basic idea is to follow the natural organization of limits and colimits and then additional structures that can be given on these. For instance, it is natural to start with finite-product sketches, then move to finite-limit sketches and to κ-infinite sketches, for various cardinals κ. One can then consider mixed sketches, that is, with limits and colimits, either finite or infinite. By playing with the sizes of limits and colimits and the presence and absence of some of them, it is possible to obtain various types of sketches. For instance, geometric sketches are sketches with finite products and no restriction on the size of coproducts.[22] There is therefore a natural classification of sketches that led to an intrinsic classification of categories. Furthermore since a sketch is fundamentally a geometric notion, it is possible that some geometric measure of complexity could be devised and would reflect a different kind of complexity than the standard measures. The idea is not to replace the standard measures, but to have a (potentially) different notion at hand. This brings us to my next point.

3. Since sketches are the categorical expression of syntax, it is natural to wonder how they correspond to the standard notions of logical syntax. There is indeed an interesting parallel as well as certain differences. Let us start with the case of finite-product sketches, the simplest case. Since a commutative diagram is basically the pictorial expression of an equation and since an arrow from a finite product object into an object, say $f: X \times \cdots \times X \to X$ is basically an operation defined on that object, it is easy to see that finite-product sketches are the graphical expressions of structures with operations whose basic properties are defined by universally quantified equations. Thus it is easy to construct, in addition to the sketch for groups, the sketch for semigroups, monoids, commutative

---

[22] The terminology still fluctuates somewhat at this stage.

6.7 Graphical Syntax: Sketches

monoids, Abelian groups, rings, commutative rings, semi-lattices, lattices, distributive lattices, etc. These are all cases of what are usually called *equational theories* or, equivalently, theories in *equational logic*. A model of an equational theory **T** is called a **T**-algebra. It is not possible to construct a finite-product sketch for the concept of a complete lattice nor is it possible for the concept of topological space, since both rest on an infinitary operation, nor is it possible to construct a finite product sketch for fields, simply because the axiom for multiplicative inverses is an implication whose antecedent stipulates that x is different from 0. For those, one has to introduce more operations and therefore more structure, hence more complicated sketches.

Finite limit sketches are particularly interesting with respect to traditional syntactical systems. From a categorical point of view, they constitute a very important and natural setting. They truly constitute the core of logic for they contain what could be considered the basic logical operations. Interestingly enough, although it is possible to precisely express the logic of finite limits, or finite-limit logics, in traditional syntactical terms, this characterization escaped the attention of logicians and model theorists. It is important to note that many important and natural mathematical notions can be expressed in this language, the very notion of a category being the most notable example. Here are some of the basic facts of the syntax of finite-limit sketches.

a. Pullback is a finite limit concept and a category having a terminal object, binary products and equalizers has pullbacks, and a category having pullbacks and a terminal objects has finite products and equalizers. In both cases, such a category has all finite limits. (See [220], 51.)

b. The concept of monomorphism is a finite limit concept, since a morphism $f: X \to Y$ is a monomorphism if and only if the following diagram is a pullback:

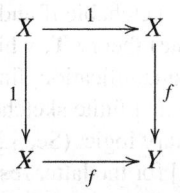

$$\begin{array}{ccc} X & \xrightarrow{1} & X \\ {\scriptstyle 1}\downarrow & & \downarrow{\scriptstyle f} \\ X & \xrightarrow{f} & Y \end{array}$$

As we have seen, a monomorphism $f: X \to Y$ can be thought of as being a predicate on Y, in the same way that a subset of a set X can be thought of as being a predicate on X. Given two such monomorphisms $f: X \to Y$ and $g: Z \to Y$, we say that f is included in g, $f \subseteq g$, if there is a (necessarily) unique $h: X \to Z$ such that $g \circ h = f$. The relation is easily shown to be a preorder on monomorphisms with common codomain. Furthermore, an equivalence relation can be defined on monomorphisms with common codomain thus: $f \approx g$ if and only if $f \subseteq g$ and $g \subseteq f$. The corresponding equivalence classes $[f]$ are called the *subobjects* of Y, where Y is the codomain of f. Thus, the notion of being a subobject of an object is essentially a finite limit notion. The poset of subobjects of Y is denoted by $S(Y)$.

Notice that it is a meet or inf semi-lattice, since we are in a category with finite limits.

c. Relations or predicates with several variables are essentially subobjects and are thus a finite limit notion. In particular, the diagonal monomorphism $\Delta_X \colon X \to X \times X$, given by $\langle 1_X, 1_X \rangle$, the equality predicate on X, is a finite limit notion.

d. The notion of *substitution* is also a finite limit notion, since it is a special case of pullbacks.

e. It is possible to give a precise definition of a finite-limit theory, the latter expression understood in its usual logical sense. It can also be shown that universal Horn logic can be interpreted in any category with finite limits. But finite-limit logic is richer than universal Horn logic. There are some notions which are definable within finite-limit logic but which are not in universal Horn logic. There is a special form of the unique existential quantifier hidden in finite-limit logic, which is just a consequence of the nature of what a finite limit is. This does not exist in universal Horn logic. The characterization of finite-limit logic in traditional quantification logic is somewhat delicate, but certainly not mysterious. (For a traditional syntactic presentation of finite-limit logic, called left-exact logic by the author, see [219]. See also [52] and [94] for other presentations.)

As the example shows, the correspondence between sketch-based logic and traditional syntax-based logic is somewhat delicate. But many important results are known. For instance (and glossing over important technical details regarding the notion of a (basic) theory in a certain formal system), it has been shown that a category is sketchable if and only if it is axiomatizable by a (basic) theory in $L_{\infty,\infty}$, the latter being full infinitary logic with no size restriction on quantifiers, on the one hand, and conjunction and disjunction on the other. (See [207], chap. 3.) A category is finitely sketchable if and only if it is axiomatizable in what is called a σ-coherent (basic) theory **T**, which is a theory between finitary logic and the logic with finite quantification, finite conjunctions and countably many disjunctions. This shows that finite sketches are somewhat "stronger" than theories axiomatized in finitary logic. (See [207] and [2] for more on these results in general and [3] and [1] for the latter result.[23])

4. There are two kinds of transfer of structures inherent to categorical logic. First, given a sketch morphism $t \colon S \to S'$, we automatically obtain a map $t^* \colon \mathrm{Mod}(S', \mathscr{C}) \to \mathrm{Mod}(S, \mathscr{C})$, simply by composing with t. This means that given a certain sketch S', any one of its models in a category $\mathscr{C}$ can be turned into a model of the sketch S. Second, given a functor $F \colon \mathscr{C} \to \mathscr{D}$ of the appropriate type, e.g., preserving finite limits, we get a functor $F^* \colon \mathrm{Mod}(S, \mathscr{C}) \to \mathrm{Mod}(S, \mathscr{D})$ by composing with F. This means that a model of a sketch S in $\mathscr{C}$

---

[23] Surprisingly, [1] also show that finite sketches are quite strong in the sense that for any geometric sketch, that is a sketch with finite limit-specifications and arbitrary colimit-specifications, there is a finite sketch such that their categories of models in **Set** are equivalent. There is only one restriction for this result to hold: the size of the colimit-specifications has to be less than or equal to the first measurable cardinal.

6.7 Graphical Syntax: Sketches

can be transfered to a model of S in $\mathscr{D}$. This is the basic setup for important results in categorical logic that will be expressible once we have introduced the intermediate level between the syntax, namely sketches, and the universes in which they are represented.

5. Given the notion of a sketch, it is natural to try to find out which categories are sketchable, i.e., which categories $\mathscr{C}$ are equivalent to categories of the form $\text{Mod}(S, \textbf{Set})$ for a sketch S, and what conditions $\mathscr{C}$ has to satisfy to be sketchable by a sketch S of a certain type — say, a finite-limit sketch. As we have already seen, the category of groups is sketchable, and we can say precisely what its sketch is: it is a *finite-product* sketch. Thus, the category **Grp** of groups and group homomorphisms is equivalent to $\text{Mod}(S, \textbf{Set})$, for the sketch S of groups presented. This is possible because the category **Grp** of groups has certain abstract categorical properties. Gabriel and Ulmer had already found many results in that area by in the late sixties, but a systematic investigation was only developed in the eighties, and research is ongoing. For instance, there is a complete abstract characterization of the categories that are sketchable and these categories are called *accessible* categories. These results are of course results about the *epistemic* accessibility of certain type of structures in the same way that results about the axiomatizability of a structure are results about the epistemic accessibility of certain type of structures. Notice, however, that there is here no restriction to the finite case. Sketches can be infinite too, and sketches of accessible categories certainly can be. But, as we have already said, one has a control on the upper bound of these sketches. (See [2], [3], [1], or [207].)

6. I have presented a simple and limited concept of sketch. There are various generalizations, which are needed to define, for instance, structured categories. [151] gives an historical survey and other results. [152], [153], [154] and [267] constitute generalizations of the notion in the spirit of the traditional concept introduced by Ehresmann. [203], [204] and [205] offer a generalization that goes in a somewhat different direction and with properties that are not quite the same as in the traditional framework. For one thing, in Makkai's framework, sketches are defined directly *within* a given category. Furthermore, there is no intrinsic hierarchy in his framework. One of the advantages of this approach is that it allows the definition of an entailment in a proof-theoretical sense and therefore a general framework for completeness proofs in the standard sense of that expression. Further work will show which of these notions turns out to be adequate.

6.8 Categorical Theories: Conceptual and Generic Structures

> A theory is a category with certain operations (defined up to isomorphism) (...) The notion of theory is thus "intrinsic", i.e., independent of a particular presentation via formal languages and axiomatic systems. In this sense, categorical logic may be viewed as "synthetic" or "intrinsic" logic by opposition to the usual "analytic", "formal" logic.
>
> ([126])

One of the most original aspects of categorical logic and categorical foundations of mathematics is the introduction of a presentation-invariant structure for theories. This is in itself a very interesting and conceptually significant contribution of category theory to logic and philosophy. Notice immediately the connection with the use and status of the group of transformations in geometry as a way of encoding the logical structure of a geometry. Unfortunately, it has escaped the attention of almost all logicians and certainly almost all philosophers. Furthermore, it illustrates once more in what manner category theory is essentially geometric.

Traditional logic divides things into two parts: syntax and semantics. This distinction is at the same time an epistemological and an ontological distinction, or at least it is motivated by epistemological and related ontological elements. On the one hand, syntax is suppose to represent the "concrete" facets of language and its epistemic accessibility is a crucial component: its basic constituents are an alphabet (a set of symbols), with rules of formation and rules of transformation defined on this alphabet and derived sets thereof. Formal systems and "languages" are defined in this context in the usual way. Syntax is usually finitistic: on the one hand, the alphabet can usually be surveyed and it is made up of distinguishable units; on the other hand, rules of formation and transformation are usually recursive. Semantics is the world of "entities" syntactical expressions are supposed to refer to. In mathematics, these are collections of mathematical entities with relations and functions defined on them. Semantics is the mathematical universe, what mathematicians are supposed to be thinking about or the content of their propositions. The whole point of model theory is to define the proper relations connecting syntax and semantics appropriately. So far we have seen how categorical logic treats syntactical matters, i.e., via the notion of sketch and sketch morphisms, and semantical matters, i.e., categories in which sketches are represented. Categorical logic introduces an intermediary step, which Makkai calls the "conceptual level"; I will adopt his terminology since it seems to me to be particularly apt.[24] The fundamental property of the conceptual level is that it shares a structure with the "world" of structures the formal system is referring to, and at the same time it shares with the syntax its epistemic accessibility. To motivate this intermediate level, I will start with a simple remark about traditional syntax and presentation of theories.

---

[24] As usual, the terminology is not adopted by everyone. What I call the category of concepts is sometimes called the syntactic category, which is also a reasonable terminology, or simply a theory, again a defendable choice. Another possibility would be to call it an intentional category associated to a theory.

6.8 Categorical Theories: Conceptual and Generic Structures

It is well known that many mathematical theories can be presented differently, either by different signatures and therefore, in this case, a different choice of primitive operations or simply by a different choice of primitive operations and axioms. For instance, the notion of group can be axiomatized in different manners. It is possible to take the inverse operation as primitive or to define it with the help of other operations. There are, therefore, some choices and some pragmatic elements involved in the presentation of a theory. This is particularly clear in the case of geometry, where many equivalent axiomatizations of the same theory exist. Although these choices might be very important pragmatically and pedagogically, when it is the theory as a whole that is the object of study, the presentation of a theory should not be of primary concern. For, the totality of relations and operations of the theory, given and defined, should be the same, no matter what the choice of primitive notions or the particular form of the axioms has been. In a sense then, the theory should be this totality of notions and theorems, independent of this or that presentation. This is precisely what a theory is defined to *be* in categorical logic. It is in fact the category of concepts of a particular field and it has some very interesting properties. Let us consider this in some detail.

From a purely conceptual point of view, the category of concepts of a sketch S can be characterized by the following universal property: given a sketch S, the *category of concepts* of S is a category $[\mathbf{T}]_S$ together with a functor $I: S \to [\mathbf{T}]_S$ such that for any category $\mathscr{C}$ and model $F: S \to \mathscr{C}$, there is a unique functor $F^*: [\mathbf{T}]_S \to \mathscr{C}$ such that $F^* \circ I = F$, i.e., such that the following diagram commutes:

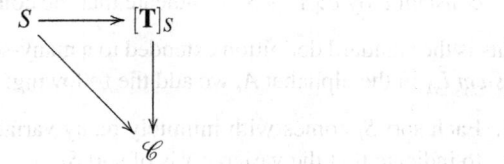

$$\begin{array}{ccc} S & \longrightarrow & [\mathbf{T}]_S \\ & \searrow & \downarrow \\ & & \mathscr{C} \end{array}$$

Thus, the category of concepts of a sketch S can be defined as a certain quotient of the free category generated by S, which makes sense since S is a graph. It can be constructed directly in various ways, depending on the context. Thus the conceptual category of a **T**-algebra can be given a very simple form which we have already seen: the conceptual category of a **T**-algebra is a category whose objects are the natural numbers $0, 1, 2, \ldots$ and in which every n is the n-fold product of 1, that is each n is equipped with an n-tuple of projection maps $p_i: n \to 1, i = 1, \ldots, n$. This is, of course, Lawvere's definition presented above. Notice that $[\mathbf{T}]_S$ is characterized up to *isomorphism* of categories. Furthermore, by definition, $\mathrm{Mod}(S, \mathscr{C})$ and $\mathrm{Mod}([\mathbf{T}]_S, \mathscr{C})$ will be equivalent categories. Notice also that the category of concepts $[\mathbf{T}]_S$ is already a model of the sketch S and that since the category of concepts $[\mathbf{T}]_S$ is a category, we can now consider the category $\mathrm{Mod}([\mathbf{T}]_S, [\mathbf{T}]_S)$ of models of the category of concepts into itself. Among the latter, there is a privileged, albeit trivial, model, given by the identity functor. We immediately recognize the constructions already presented at the propositional level. Although these considerations might still look ridiculous, they in fact lead to an important concept that we will see in a short while.

It is important to note that the category of concepts can be defined even when theories are defined in the traditional syntactical manner. The main difference with the case of sketches is that the relationship between a given theory **T** and its category of concepts is not a functor anymore, but an interpretation in the standard model-theoretic sense. It is probably worthwhile to look at the construction of the category of concepts of a theory in this context, for it may shed a different light on the reasons underlying the choice of terminology.

In a categorical framework, it is very natural to consider formal systems with many sorts. A *similarity type* or *alphabet A*, often called a language in the literature, is given by:

1. A collection of sorts $S_1, S_2, S_3, \ldots$;
2. A collection of relation symbols $R_1, R_2, R_3, \ldots$, each of which is given with the sorts of its arguments; we denote symbolically, in view of the intended interpretations, a relation R as $R(x_1, \ldots, x_n) \rightarrowtail S_1 \times \cdots \times S_n$; this is a purely notational device with absolutely no meaning;
3. A collection of function symbols $f_1, f_2, f_3, \ldots$ each of which is given with the sorts of its arguments and the sort of its target; we denote a function symbol f as $f: S_1 \times \cdots \times S_n \to S$ if f takes n arguments of sorts $S_1, \ldots, S_n$ respectively to a value of sort S.
4. A collection of constants $c_1, c_2, c_3, \ldots$ each with a specified sort; we denote a constant c by $c: 1 \to S_i$ to indicate that the constant c is of sort S_i.

This is the standard definition extended to a many-sorted context. To obtain a *formal system L_A* in the alphabet A, we add the following:

1. Each sort S_i comes with infinitely many variables $x_1, x_2, x_3, \ldots$; we write $x: S_i$ to indicate that the variable x is of sort S_i;
2. Each sort has an equality relation $=_S$; notice immediately that this means that equality is not treated as a universal or purely logical relation and that in the interpretation, whatever will correspond to a sort will have to come equipped with a criterion of identity or equality for *its* objects;
3. The usual logical symbols and two propositional constants, $\top$ and $\bot$;
4. The usual deductive machinery of intuitionistic predicate logic (if any other deductive procedures are assumed, they are made explicit).

Terms (of a given sort) and atomic formulas are defined as usual. We will denote an arbitrary term by the letter t. For the collection of formulas, one can restrict or extend the possible logical operations depending on the type of categories in which these formulas will be interpreted. For instance, a formula φ is said to be *coherent* if it is obtained from atomic formulas by applying finite conjunction, disjunction and existential quantification. More formally, the collection of coherent formulas is the smallest collection of formulas such that:

1. The atomic formulas $R(t_1, \ldots, t_n)$, $t = t'$, $\top$, $\bot$ are coherent formulas;
2. If φ and ψ are coherent formulas, then so are $\varphi \vee \psi$ and $\varphi \wedge \psi$;

6.8 Categorical Theories: Conceptual and Generic Structures

3. If $\varphi(x_1,\ldots,x_n)$ is a coherent formula, then so is $\exists x\colon S\varphi(x_1,\ldots,x_n)$ where S is any sort and x is a variable of that sort.

An *implication* of coherent formulas φ and ψ has the form

$$\forall x_1\ldots\forall x_n(\varphi(x_1,\ldots,x_n) \Rightarrow \psi(x_1,\ldots,x_n))$$

where $\varphi(x_1,\ldots,x_n)$ and $\psi(x_1,\ldots,x_n)$ are coherent formulas. Notice that instead of implications, we could have resorted to sequents; they are essentially equivalent in most contexts, although the formulation in terms of sequents is more flexible. A theory **T** in the given language L is said to be a *coherent theory* if all its axioms are implications of coherent formulas. Many mathematical theories can be expressed in this form. In particular, every equational theory can be presented this way. For instance, the theory of semigroups, the theory of monoids, the theory of groups, the theory of Abelian groups, the theory of R-modules for a fixed ring R, the theory of chain complexes, etc. are all coherent theories. (See for instance [200], chap. X, §3. However what we call *coherent*, they call *geometric*. The terminology is not entirely settled. See also [122], D.1) We will now give a specific example to which we will come back in the next chapter.

We can define linear orders with bottom and top elements in a language with one sort I, one relation symbol $\leq$, and two constants b and t. The following axioms for the theory of linear orders show that it is a coherent theory:

$$\forall x\colon I(\top \Rightarrow x \leq x),$$
$$\forall x,y,z\colon I(x \leq y \wedge y \leq z \Rightarrow x \leq z),$$
$$\forall x,y\colon I(x \leq y \wedge y \leq x \Rightarrow x = y),$$
$$\forall x\colon I(\top \Rightarrow b \leq x \wedge x \leq t),$$
$$(b = t) \Rightarrow \bot,$$
$$\forall x,y\colon I(\top \Rightarrow x \leq y \vee y \leq x).$$

This theory is intimately connected to the simplicial category and, more specifically, to simplicial sets. We will also come back to this point in the next chapter.

A different example of a class of significant formulas is provided by the notion of a *geometric* formula: a formula is said to be *geometric* if it is obtained from atomic formulas by applying *finite* conjunction, *finite* existential quantification and *infinite* disjunction. (Warning: some authors call a formula *geometric* if both conjunctions and disjunctions are allowed to be infinite.) As in the previous case, we can define an implication of geometric formulas and stipulate that a *geometric theory* **T** is a theory in which all axioms are geometric axioms.

The concepts of generic model of a theory **T** and of the internal theory of a category seen above can now be considered. Given, say, a coherent theory **T**, the category of concepts [**T**] of **T** is constructed from the language and the axioms of **T** as follows. (See also [208], chap. 8 or [200], chap. X, §5 for more details and proofs or again [122], D.) To get the objects of this category, first consider what is called a *formal set* $[\mathbf{x}; \varphi(\mathbf{x})]$, where $\mathbf{x}$ denotes a n-tuple of distinct variables containing all free variables of φ and φ is a formula of the underlying formal system L. Two such formal sets, $[\mathbf{x}; \varphi(\mathbf{x})]$ and $[\mathbf{y}; \varphi(\mathbf{y})]$ are equivalent if one is the alphabetic variant of the other, that is if $\mathbf{x}$ and $\mathbf{y}$ have the same length and sorts and $\varphi(\mathbf{y})$ is obtained from $\varphi(\mathbf{x})$ by substituting $\mathbf{y}$ for $\mathbf{x}$ (and changing bound variables if necessary). This is clearly an equivalence relation and it is therefore possible to consider equivalence classes of such formal sets. An object of the category of concepts $[\mathbf{T}]$ is such an equivalence class of formal sets $[\mathbf{x}; \varphi(\mathbf{x})]$, where φ is a formula of the formal system L. The objects of $[\mathbf{T}]$ are the equivalence classes of these formal sets, for all formulas of L. Notice this last important point: we take *all* formulas of the language, not only those which appear in $\mathbf{T}$. Thus, in a sense, the space of objects is the collection of all possible properties and sentences expressible in that language, thus all possible theories in the given formal system. No logical relationship is considered at this stage, we have only identified our points. The next step will introduce the structure corresponding to the structure of that particular theory $\mathbf{T}$ and it is this step that will capture the particular features of $\mathbf{T}$. This is just as one would expect in a categorical framework: the structure of $\mathbf{T}$ is captured by the morphisms we will define and the properties resulting therefrom.

It is easier to motivate the definition of morphism with an eye on the semantics, although the properties of the morphisms, e.g., that they form a category, have to be proved with the syntactical features of the theory (unless one has a completeness theorem at hand). The basic idea is this: a functor from $[\mathbf{T}]$ to **Set** should transform the objects of $[\mathbf{T}]$ into genuine sets and the morphisms of $[\mathbf{T}]$ into genuine functions automatically, and these functions should be functions that are definable in $\mathbf{T}$ — i.e., for which we can prove in $\mathbf{T}$ that they are indeed functions. Furthermore, $[\mathbf{T}]$ should contain all of them. By sending a formal set $[\mathbf{x}; \varphi(\mathbf{x})]$ to the set of n-tuples satisfying the formula, i.e. $\{(x_1, \ldots, x_n) \mid \varphi(\mathbf{x})\}$, a morphism from $[\mathbf{x}; \varphi(\mathbf{x})]$ to $[\mathbf{y}; \psi(\mathbf{y})]$ should become a genuine function between genuine sets $\{(x_1, \ldots, x_n) \mid \varphi(\mathbf{x})\}$ and $\{(y_1, \ldots, y_m) \mid \psi(\mathbf{y})\}$ respectively. Such a morphism should simply be given by a formula of the theory $\mathbf{T}$ that defines such a function, that is a formula $\theta(\mathbf{x}, \mathbf{y})$ of $\mathbf{T}$ that is provably functional. The only trick in the construction is to construct a morphism between two (equivalence classes of) formal sets $[\mathbf{x}; \varphi(\mathbf{x})]$ and $[\mathbf{y}; \psi(\mathbf{y})]$ in such a way that, when interpreted, it yields the *graph* of the function, in the standard set-theoretical sense of that expression, between the actual sets $\{(x_1, \ldots, x_n) \mid \varphi(\mathbf{x})\}$ and $\{(y_1, \ldots, y_m) \mid \psi(\mathbf{y})\}$. Thus, all definable functions in $\mathbf{T}$ will be represented by a morphism in $[\mathbf{T}]$.

Formally, consider a triple $(\mathbf{x}, \mathbf{y}, \gamma)$, where $\mathbf{x}$ and $\mathbf{y}$ are disjoint tuples of distinct variables and γ is a formula with free variables possibly among $\mathbf{x}$ and $\mathbf{y}$. Such a triple defines a formal function if the following formulas are provable:

$$\mathbf{T} \vdash \forall \mathbf{x} \forall \mathbf{y} (\gamma(\mathbf{x}, \mathbf{y}) \Rightarrow (\varphi(\mathbf{x}) \wedge \psi(\mathbf{y})));$$
$$\mathbf{T} \vdash \forall \mathbf{x} (\varphi(\mathbf{x}) \Rightarrow \exists \mathbf{y} (\gamma(\mathbf{x}, \mathbf{y})));$$
$$\mathbf{T} \vdash \forall \mathbf{x} \forall \mathbf{y} \forall \mathbf{y}' (\gamma(\mathbf{x}, \mathbf{y}) \wedge \gamma(\mathbf{x}, \mathbf{y}') \Rightarrow \mathbf{y} = \mathbf{y}');$$

6.8 Categorical Theories: Conceptual and Generic Structures

where we have used some obvious abbreviations. The underlying motivation should be clear: these formulas will be true in any interpretation of **T** in which γ is indeed a morphism.

We now define an equivalence relation $(\mathbf{x},\mathbf{y},\gamma) \sim (\mathbf{u},\mathbf{v},\eta)$ if

$$\mathbf{T} \vdash \forall \mathbf{x} \forall \mathbf{y}(\gamma \Leftrightarrow (\eta(\mathbf{x}/\mathbf{u},\mathbf{y}/\mathbf{v}))).$$

The equivalence relation guarantees that for every model M of **T**, the functions corresponding to γ and to η will coincide. We can now stipulate that a formal function is an equivalence class of the foregoing equivalence relation. Given a representative $(\mathbf{x},\mathbf{y},\gamma)$ of such an equivalence class, we denote the equivalence class containing it by $\langle \mathbf{x} \mapsto \mathbf{y} \colon \gamma \rangle$. Thus, a *formal morphism* in [**T**] is denoted by:

$$\langle \mathbf{x} \mapsto \mathbf{y} \rangle \colon [\mathbf{x} \colon \varphi] \to [\mathbf{y} \colon \psi]$$

For each formal set $[\mathbf{x} \colon \varphi(\mathbf{x})]$, the identity morphism is provided by the formal morphism $\langle \mathbf{x} \mapsto \mathbf{y} \colon (\mathbf{x}=\mathbf{y}) \wedge \varphi \rangle$. Given two formal morphisms $\langle \mathbf{x} \mapsto \mathbf{y} \colon \gamma \rangle \colon [\mathbf{x} \colon \varphi] \to [\mathbf{y} \colon \psi]$ and $\langle \mathbf{y} \mapsto \mathbf{z} \colon \eta \rangle \colon [\mathbf{y} \colon \psi] \to [\mathbf{z} \colon \zeta]$, their composition is defined by the formal morphism $\langle \mathbf{x} \mapsto \mathbf{z} \colon \mu \rangle \colon [\mathbf{x} \colon \varphi] \to [\mathbf{z} \colon \zeta]$ where $\mu = \exists \mathbf{y}(\gamma \wedge \eta)$. These two definitions satisfy the usual requirements of a category. Thus, [**T**] is a category.

Notice that [**T**] is *not* a category of structured sets and structure-preserving functions! A lot of information about **T** is lost when all we have at our disposal is [**T**]. It is, for instance, impossible to know which atomic formulas are involved in specific formal sets or what were the primitive symbols of the language $L_\mathbf{T}$. Furthermore, two different theories **T** and **T**′ can very well yield isomorphic categories of concepts, thus essentially the same category. As we have seen, the syntactic logical operations, i.e., quantifiers and connectives, become categorical operations in the category and this part of the structure is not lost. Again, moving from a theory **T** to its category of concepts [**T**] is an *abstraction*. The category of concepts [**T**] is the algebraic encoding of the logical theory. We should immediately point out that although the foregoing presentation was restricted to coherent theories, it does not need to be.

However, when restricted to coherent theories, the foregoing construction has additional interesting features. For instance, one can show that it is a category with finite limits. This means that the category of concepts has pullbacks and a terminal object. To show this, one has to show that given three formal sets related to one another in the appropriate manner, there is a formal set with formal projections satisfying the universal property of pullbacks. In fact, whenever **T** is a coherent theory, then [**T**] *as a category* is what is called a *coherent* category (see below for the main elements of the definition; these categories also go by the name of *logical* categories or *logos* (See [208], chap. 8)). By the foregoing construction, it can be seen that the category of concepts is the category of all definable sets and functions of a theory **T**. Thus, in a sense, it contains all the formally expressible concepts of **T**, whence its name. Furthermore, it should be clear that under any reasonable notion of interpretation, there is a *canonical* or *generic* or *universal* model, denoted by G, of **T**, in fact based on the interpretation of the whole underlying language of **T**, in its category of concepts [**T**]. In fact, when **T** is coherent more is true. Given any model M of **T** in a coherent category $\mathscr{C}$, there is a (coherent) functor $I\colon [\mathbf{T}] \to \mathscr{C}$ which extends the canonical model and which is uniquely determined up to a unique isomorphism. This can be rephrased as follows: for any coherent category $\mathscr{C}$ and model M of **T** in $\mathscr{C}$, there is a (coherent) functor $I\colon [\mathbf{T}] \to \mathscr{C}$ and an isomorphism of **T**-models $M \to I(G)$; if I_1 and $I_2\colon [\mathbf{T}] \to \mathscr{C}$ are such that $f\colon I_1(G) \to I_2(G)$, then there is a unique natural isomorphism $\eta\colon I_1 \to I_2$ such that $f = \eta(G)$.

In a sense, **T** and [**T**] are interchangeable: from a practical point of view it means that proof-theoretical means can be used to study [**T**] and that categorical methods can be used to study **T**, depending on the context and the methods available.

This last statement can be strengthened and one can see more precisely to what extent one can replace a theory by a category or replace a category by a logical theory. For given a (small) category $\mathscr{C}$, at least with finite limits, it is possible to associate or construct the language $L_\mathscr{C}$ of $\mathscr{C}$ as follows. We first have to identify the alphabet of $L_\mathscr{C}$. The sorts are given by the objects $X, Y, Z, \ldots$ of $\mathscr{C}$. Every morphism $f\colon X \to Y$ of $\mathscr{C}$ becomes a function symbol of $L_\mathscr{C}$. (In particular, a constant $c\colon 1 \to X$ is seen as 0-ary function symbols.) This is called the *canonical language* of $\mathscr{C}$. Notice that $L_\mathscr{C}$ is obtained as if we had taken $\mathscr{C}$ and destroyed its categorical structure, retaining only the symbols, and keeping in mind that function symbols are sorted. It is possible to extend this language to reflect the structure of $\mathscr{C}$ more closely. Although subobjects of $\mathscr{C}$ can be denoted naturally by formulas of $L_\mathscr{C}$, it is possible to introduce relation symbols for each subobject $R(x_1, \ldots, x_n) \rightarrowtail X_1 \times \cdots \times X_n$ and n-ary function symbol for morphisms $f\colon X_1 \times \cdots \times X_n \to X$. This is called the *extended* canonical language of $\mathscr{C}$. (See [208], chap. 2, sec. 4.) In order to get the *internal theory* $\mathbf{T}_\mathscr{C}$ of $\mathscr{C}$ in its canonical language, $\mathscr{C}$ has to have more structure than just finite limits. Once again, it has to be what is called a *coherent* category. In this case, it is possible to give a list of *coherent* axioms $\Sigma_\mathscr{C}$, that is a set of coherent formulas, and prove that $\mathbf{T}_\mathscr{C}$ is sound in $\mathscr{C}$. (See [208], chap. 3.) The internal theory $\mathbf{T}_\mathscr{C}$ is related to $\mathscr{C}$ by two expected properties:

1. There is of course a canonical interpretation G of $\mathbf{T}_\mathscr{C}$ in $\mathscr{C}$;
2. For any model M of $\mathbf{T}_\mathscr{C}$ in a coherent category $\mathscr{D}$, there is a unique (coherent) functor $I\colon \mathscr{C} \to \mathscr{D}$ such that I applied to G is equal to M.

It is of course possible to complete the circle: starting with a coherent category $\mathscr{C}$, construct its internal theory $\mathbf{T}_\mathscr{C}$ and then move to its category of concepts $[\mathbf{T}_\mathscr{C}]$. How are $\mathscr{C}$ and $[\mathbf{T}_\mathscr{C}]$ related? They are in fact *equivalent as categories*, i.e., $\mathscr{C} \to [\mathbf{T}_\mathscr{C}]$. This can be seen by the following (standard) argument. As we have seen, there is a canonical interpretation from $\mathbf{T}_\mathscr{C}$ into $[\mathbf{T}_\mathscr{C}]$. Since $\mathbf{T}_\mathscr{C}$ is the internal language of $\mathscr{C}$, there is a unique coherent functor $I\colon \mathscr{C} \to [\mathbf{T}_\mathscr{C}]$. Similarly, since $[\mathbf{T}_\mathscr{C}]$ is the category of concepts of $\mathbf{T}_\mathscr{C}$, there is a coherent functor $I'\colon \mathscr{C} \to [\mathbf{T}_\mathscr{C}]$, which is unique up to a unique isomorphism. Composing I and I' in both directions, we get functors which are isomorphic to the identity functors on the respective categories, i.e., an equivalence, as required. This yields the very important result that every (small) coherent category is equivalent to a category of concepts for some theory **T**.

6.8 Categorical Theories: Conceptual and Generic Structures

Given the foregoing equivalence between categories and logical theories, it makes perfect sense to say that a category of concepts in finite limit logic, say, *is* simply a (small) category with finite limits. This is what opens the doors to the direct and abstract definition of various kinds of conceptual categories, also called *categorical doctrines* or simply *doctrines* in the literature, which are basically logical. Thus a category $\mathscr{C}$ with all finite limits is called *Cartesian* or *Left-Exact*. As we have already mentioned, the property of being Cartesian is completely characterized by the fact that every diagonal functor $I\colon \mathscr{C} \to \mathscr{C}^{\mathscr{D}}$ has a right adjoint, for $\mathscr{D}$ an arbitrary finite diagram (category). We can now define coherent categories.

A *coherent* category $\mathscr{C}$ is a Cartesian category such that

1. Every subobject meet semilattice $S(X)$ is a lattice;
2. Each $f^*\colon S(Y) \to S(X)$ is a lattice homomorphism;
3. f^* has a left adjoint, denoted by $\exists_f$ and which is in the category of sets the existential quantifier and
4. $\exists_f$ satisfies the Beck-Chevalley condition stated previously.

As we have seen, any coherent theory **T** yields a coherent category, and conversely there is a coherent category corresponding to every coherent theory.

A *Heyting category* $\mathscr{C}$ is a coherent category in which each $f^*\colon S(Y) \to S(X)$ has a right adjoint, denoted by $\forall_f$. The last condition is sufficient to entail that each $S(X)$ is a Heyting algebra, that f is a homomorphism of Heyting algebras and that the right adjoint is also stable under substitution. Heyting categories are common: for any small category $\mathscr{P}$, the functor category $\mathbf{Set}^{\mathscr{P}}$ is a Heyting category. They correspond to theories in intuitionistic predicate logic.

A *Boolean category* $\mathscr{C}$ is a coherent category such that every $S(X)$ is a Boolean algebra, i.e., every subobject has a complement. Whenever the category $\mathscr{P}$ is a groupoid, the category $\mathbf{Set}^{\mathscr{P}}$ is a Boolean category. Thus, in particular, when $\mathscr{P}$ is a group, the category $\mathbf{Set}^{\mathscr{P}}$ is a Boolean category.

A *pretopos* $\mathscr{C}$ is a coherent category having (1) quotients of equivalence relations and (2) finite disjoint sums. The category **Set** suffers, or one should say in this case benefits, from a multiple-personality syndrome — some say that such an object is schizophrenic in the literature — for it is a coherent category, a Heyting category, a Boolean category and a pretopos (in fact, it is also other things...).

The terminology (coherent, Heyting, Boolean, etc.) reflects the fact that starting from a theory **T** given in a certain language L as above, the resulting category of concepts [**T**] of **T** is a category of the corresponding kind, e.g., a Heyting category. These categories correspond to logical theories and we now see how the categorical perspective introduces an organization of the logical landscape.

The transfer of structures we have discussed above for sketches takes a new twist for categories of concepts. Indeed, the foregoing transfers for sketches automatically extend to categories of concepts but in this new context it is reasonable to look for adjoints to these functors and, as usual, finding and determining the properties of these adjoints is an important aspect of the categorical investigation. We use the same terminology as for sketches: a structure-preserving functor $I\colon \mathbf{T} \to \mathbf{T}'$ between (small) categories of concepts is called an *interpretation* of **T** in **T**'. (When **T** and **T**' have been constructed from theories, one can verify that it is a legitimate notion of interpretation. See [208], chap. 7, 196.) A structure preserving functor $M: \mathbf{T} \to \mathbf{Set}$ is called a (set-)*model* of **T**. A natural transformation between models $\eta: M_1 \to M_2$ is a *homomorphism* of models of **T** and they are the traditional model-theoretic structure-preserving functions between models. The functor category $\mathrm{Mod}(\mathbf{T}, \mathbf{Set})$ is the *category of (set-)models* of **T**. More generally, for a category $\mathscr{C}$ with the appropriate structure, one can consider the category $\mathrm{Mod}(\mathbf{T}, \mathscr{C})$ of models of **T** in $\mathscr{C}$. Transfers of structure are as for sketches:

1. Given an interpretation $I: \mathbf{T} \to \mathbf{T}'$, one can transfer models of $\mathbf{T}'$ to models of **T** by composing with I, that is given a model $M: \mathbf{T}' \to \mathbf{Set}$, we get by composition with I a model $MI: \mathbf{T} \to \mathbf{Set}$. Hence, there is a functor $I^*: \mathrm{Mod}(\mathbf{T}', \mathbf{Set}) \to \mathrm{Mod}(\mathbf{T}, \mathbf{Set})$;
2. Given a functor $F: \mathscr{C} \to \mathscr{D}$ of the right type (that preserves the right kind of structure in each case), we get a functor $F^*: \mathrm{Mod}(\mathbf{T}, \mathscr{C}) \to \mathrm{Mod}(\mathbf{T}, \mathscr{D})$ by composing models M with F.

This is the appropriate set-up to develop some important ideas and significant results of categorical logic.

1. As we have already seen, as soon as **T** and $\mathbf{T}'$ are equational theories defined as above, the functor I^* has a left adjoint. This is simply a reformulation of Lawvere's original result for universal algebra. This situation can be thought about in the following way. The functor I provides us with an interpretation of the concepts of **T** into the concepts of $\mathbf{T}'$. We can think of $\mathbf{T}'$ as extending **T**, as a way of adding concepts to **T**. Then the functor I^* can be thought of as "restricting" the models of $\mathbf{T}'$ to models of **T** is a systematic way. The existence of the adjoint means that the models of **T** can be extended in a systematic way to models of $\mathbf{T}'$. In other words, the category $\mathrm{Mod}(\mathbf{T}', \mathbf{Set})$ has the means to represent within itself how one goes from a model of **T** to a model of $\mathbf{T}'$. Consider, for instance, the case when **T** is the theory of groups and $\mathbf{T}'$ is the theory of Abelian groups, as categories of concepts. There is an obvious product-preserving functor $I: \mathbf{T} \to \mathbf{T}'$ from the category of concepts of the theory of groups to the category of concepts of the theory of Abelian groups. The induced functor I^* takes a model of the theory of Abelian groups and sends it to a model of the theory of groups in an obvious way (since every Abelian group is a group). The left adjoint to I^* takes a group and sends it to its Abelianization, a standard construction of group theory.
2. An interesting case is the case when we end up with an equivalence of categories between $\mathrm{Mod}(\mathbf{T}', \mathbf{Set})$ and $\mathrm{Mod}(\mathbf{T}, \mathbf{Set})$. This means that from a categorical point of view, the category of models of $\mathbf{T}'$ is indistinguishable from the category of models of **T**. In a sense, then, adding new concepts to **T** simply does not modify in any essential way what it can express. This means that **T** has some sort of completeness and in this context it makes perfect sense to say that it is *conceptually complete*. More precisely, we say that **T** is *conceptually complete* whenever the following is satisfied: if the functor

I^*: $\mathrm{Mod}(\mathbf{T}', \mathbf{Set}) \to \mathrm{Mod}(\mathbf{T}, \mathbf{Set})$ is an equivalence of categories, then the functor $I: \mathbf{T} \to \mathbf{T}'$ was one already. This literally means that by moving to $\mathbf{T}'$, we did not add anything essentially new to $\mathbf{T}$, although we might have thought we had. Conceptual completeness is in fact equivalent to a standard result of model theory, namely Beth definability theorem. However, one of the advantages of working in the categorical framework is that categorical methods make it possible to prove results which might not be accessible otherwise, for instance, a constructive proof of this result for intuitionistic logic. (See [237] for a categorical proof of conceptual completeness of intuitionistic first-order logic.)

I should point out that conceptual completeness is *analogous* to a general phenomenon in mathematics, a particularly interesting case of which is the so-called "Morita equivalence". In the case of conceptual completeness, an equivalence between categories of models induced by a given functor I yields an equivalence on the underlying structures involved, namely the theories. This result can be formulated in this way because the theories are taken to be categories. If they were not, we could instead stipulate that two theories, presented in the traditional manner, are equivalent if and only if their category of (set-)models are categorically equivalent. (This has been done in practice for theories as well as for sketches. See, for instance, [1] for sketches and [38] for theories.) This is obviously an equivalence relation and in this form, which is not a form of conceptual completeness as above, we have a direct parallel with Morita equivalence. Very often in practice, a certain category $\mathscr{C}$ of structures is given whose objects "depend" in one way or another on a certain underlying structure. Categories of models are but one particular case. But there are many others. Consider, for instance, the category of (left-)modules M over a ring R. The underlying ring influences the structure of the whole space of modules. Changing the ring can certainly modify the structure of the category of modules. There are some very strange and interesting relations between such categories of modules and the category of rings, in particular a strange circularity. For given a module M, one can consider the *ring* of endomorphisms $\mathrm{End}(M)_R$ of M and then consider modules over *that* ring. Then, given the later category of modules, we can once more consider rings of endomorphisms of the latter. We could clearly go on and on like that. But more to the point, two rings R and S can be very different and yet yield *equivalent* categories of modules. For instance, a ring R and the ring of all $(n \times n)$-matrices over R *are* equivalent in this sense. But these two rings are not isomorphic in general. (Consider finite rings!) Morita has provided a ring-theoretical criterion when two rings R and S are equivalent in this sense: it is necessary and sufficient that there exists, in the category of R-modules, what is called a finitely-generated projective generator U, that is an R-module U with specific properties, such that the ring of endomorphisms of U is isomorphic to S. Furthermore, one can define the notion of Morita invariance for properties of rings: a ring-theoretical property P is said to be *Morita invariant* if a ring R has the property P if and only if every ring Morita equivalent to R has the property P. Many important ring-theoretical properties turn out to be Morita equivalent. Notice that a property P is Morita invariant if it is defined by means of a categorical property of the category of modules over R. Exactly the same definition could be given for logical theories.

In fact, the property of Morita equivalence can itself be expressed in purely categorical language, in terms of certain properties of a functor, i.e., a transformation between certain categories. Once more, we have an example of properties of certain structures that are captured by global transformations of the appropriate spaces. (See [17], chap. 2).

3. From a categorical point of view, the category of (set-)models of a theory **T** should capture the essential properties of **T**. In a sense, conceptual completeness is already an indication of this fact. But in principle it could be stronger: is it possible to recover a category of concepts **T** from its category of models Mod(**T**, **Set**) in a uniform manner and up to equivalence of categories? This would mean that the latter category can be enriched in such a way that the resulting category contains all the information necessary to recover the category of concepts in an essential manner. A theory for which this is possible is said to be *strongly conceptually complete*. A different way to formulate this result is to say that if Mod(**T**, **Set**) and Mod(**T**′, **Set**) are equivalent, then **T** and **T**′ are equivalent too. Notice that it is not assumed that the first equivalence is induced by an interpretation in this case. Whereas conceptual completeness is a local phenomenon, since it depends on the interpretation I, strong conceptual completeness is a global phenomenon, since there is no underlying interpretation at hand. The construction of **T** can be thought of as a case of abstracting certain data out of another, more "concrete", situation. Finite limit categories of concepts are strongly conceptually complete.[25] If the category of models is adequately enriched in a precise technical sense, then in these circumstances first-order classical logic is strongly conceptually complete. (See [201].)

In the same way that conceptual completeness is analogous to other common cases in mathematics, the same can be said about strong conceptual completeness. Indeed, strong conceptual completeness is directly tied to duality theorems and many such theorems in mathematics can be interpreted informally as saying that a certain abstract structure can be reconstructed from a structured collection of more concrete structures. Stone duality itself can be stated in this form. Pontrjagin duality asserts that certain groups can be reconstructed from their group of characters. The latter result is extended to more general groups via the so-called Tanaka duality. In this case, one starts with a *category* of representations of a (certain type of) group and the question is whether the given group can be recovered from its category of representations. Various groups can indeed be so recovered in a uniform manner. Of course, the methods used in logic are totally different from the methods used in representation theory. (For a presentation of Tannaka duality, see, for instance, [127].)

4. How can we not say that in the case of conceptual and strong conceptual completeness it is the whole *logical structure* of a given mathematical theory which is contained in the category of models and even, in a more precise manner, in the

---

[25] Thus they are the so-called Barr-exact categories

6.8 Categorical Theories: Conceptual and Generic Structures

law according to which operations of that category compose with one another, independently of the nature of the objects on which these transformations act (to paraphrase Cartan's claim about the relation between the structure of a Lie group and the logic of the corresponding geometry)? The fact is, in a categorical framework, theories are replaced by categories of concepts and the latter are characterized in a purely abstract way, i.e., as categories that are Cartesian, regular, coherent, Heyting, Boolean, pretoposes, toposes, etc. In a sense, the next step corresponds to the step in geometry where one shifts from transformation groups to transformation groups given abstractly and the various representations of these abstract groups.

5. As we have seen, completeness results for various propositional logics are equivalent to representation theorems for various algebras, e.g., in the case of classical propositional logic, the completeness theorem is equivalent to Stone's representation theorem for Boolean algebras. It has been shown that the classical (Gödel) completeness theorem is equivalent to a representation theorem for coherent categories, which can be stated thus: for any small coherent category C, there is a (small) set I and a conservative coherent functor $F: \mathscr{C} \to \mathbf{Set}^I$. A functor $F: \mathscr{C} \to \mathscr{D}$ is said to be *conservative* if it reflects isomorphisms, i.e., if $F(f)$ is an isomorphism in $\mathscr{D}$, then f was already an isomorphism in $\mathscr{C}$. Needless to say, the key property is precisely that of being conservative. For what it amounts to is the fact that for any diagram in $\mathscr{C}$ such that its image under F in $\mathscr{D}$ is a diagram of a universal morphism, then the original diagram was already a diagram of a universal morphism in $\mathscr{C}$. For instance, given the diagram

$$X \xleftarrow{p_X} Z \xrightarrow{p_Y} Y$$

in $\mathscr{C}$, such that

$$F(X) \xleftarrow{F(p_X)} F(Z) \xrightarrow{F(p_Y)} F(Y)$$

is a *product* diagram in $\mathscr{D}$, then the original diagram was already a product diagram in $\mathscr{C}$. As it can be verified, the category **Set** is coherent and so is the functor category $\mathbf{Set}^I$. Since the functor $F: \mathscr{C} \to \mathbf{Set}^I$ is conservative, it follows that $\mathscr{C}$ shares all the coherent properties of $\mathbf{Set}^I$, and in fact of **Set**. The equivalence between the representation theorem and the completeness theorem can be established as follows. Assuming the representation theorem, we start with a coherent theory **T** and construct the category of concepts [**T**] of **T**, which is a coherent category. Applying the representation theorem to [**T**], we obtain the completeness theorem. To prove the other direction, we assume the completeness theorem and start with a coherent category $\mathscr{C}$. Using the internal language of $\mathscr{C}$, one constructs as above the coherent theory $\mathbf{T}_\mathscr{C}$ of $\mathscr{C}$. The models of $\mathbf{T}_\mathscr{C}$ are then constructed so that they are identical with functors $\mathscr{C} \to \mathbf{Set}$. The representation theorem then follows from the completeness theorem for $\mathbf{T}_\mathscr{C}$.

Two important elements have to be added to the picture. First, the representation theorem for coherent categories is but one representation theorem for a whole collection of relevant categories: exact categories, Heyting categories and Boolean categories. Second, these results in fact follow a general pattern. Indeed, the foregoing representation theorem takes a general, purely categorical form. Given any categories $\mathscr{S}$ and $\mathscr{C}$, we can always consider the repeated functor category $\mathscr{S}^{(\mathscr{S}^\mathscr{C})}$. In this situation, there is a canonical functor, the evaluation functor

$$e\colon \mathscr{C} \to \mathscr{S}^{(\mathscr{S}^\mathscr{C})}$$

for which, given any object X of $\mathscr{C}$, and any functor $F\colon \mathscr{C} \to \mathscr{S}$, $e(X)(F)$ is simply $F(X)$, the evaluation of F at X. For any subcategory $\mathscr{D}$ of $\mathscr{S}^\mathscr{C}$, the same functor $e\colon \mathscr{C} \to \mathscr{S}^\mathscr{D}$ can be defined. It is then easy to show that the representation theorem for coherent categories is equivalent to the claim that the functor $e\colon \mathscr{C} \to \mathscr{S}^{\mathrm{Mod}(\mathscr{C})}$ is conservative. The fact that the evaluation functor is coherent holds on purely general grounds. We therefore have a purely categorical description of the representation theorem. Moreover, in the early seventies Joyal demonstrated that the functor e preserves all existing instances of the Heyting structure in $\mathscr{C}$. This automatically yields a representation theorem for Heyting categories and, in turn, a canonical completeness theorem for intuitionistic logic.

6.9 Summing Up

When Lawvere presented his program of using category theory to develop logic algebraically, very few people thought it possible or useful. The first developments were however promising — so much so that it seemed reasonable to investigate the matter further. We have seen that at the center of his program was the idea of invariance: once a theory has a categorical characterization, it is possible to say precisely what it means to have an invariant presentation of it. This invariance is captured by the categorical properties themselves, more often than not by specific adjoints. There is a deep geometric flavor to the whole enterprise, directly in line with Klein's program. However, at the end of the sixties, Grothendieck and his students had developed various important notions in their search for a proof of the Weil conjectures. Among the fruits of their labor, especially Grothendieck's work, was the concept of a topos. The latter notion was about to give to logic and to the foundations of mathematics a definitive and profound geometric twist, still in the spirit of Klein's program.

Chapter 7
Invariants in Foundations: Geometric Logic

> By an additional stage of abstraction Grothendieck, followed by Lawvere and Tierney, proposed an abstract concept of "topos" that was for him the ultimate generalization of the concept of space.
> ([44], 7.) *Cartier*

> We consider forcing over categories as a way of constructing objects by geometric approximation, including a construction of a generic model of a geometric theory as its special case.
> ([246], 1.) *Sčedrov*

Topos theory is a world in itself. It is mathematically rich, intricate and multifaceted. Its history is extremely complex, still quite recent and would deserve a whole book. Its philosophical relevance is unquestionable.

One of the important features of the history I have sketched so far is the unifying power of the concepts that appeared in it. The group concept had such unifying power as well, and that power certainly motivated Klein, Lie and others in geometry, elementary or not. But it is also true of the concept of category—and as we have seen this was already recognized by Eilenberg and Mac Lane right from the start—and the concept of adjoint functors. The axiomatic characterization of various types of categories also constitutes another mode of unification. The concept of Abelian category, for instance, allowed the presentation and development of various cohomology theories and related methods. What is striking about the concept of topos is that, like the concept of group, it has a tremendous unifying power: it is at the same time topological, geometrical, arithmetical and logical. From a mathematical point of view, what makes this diversity interesting is the fact that it is possible to use means, ideas, intuitions, from one domain inherent to toposes and transfer them to another inherent domain, e.g., from the geometrical to the arithmetical, from the logical to the topological. This in itself would warrant careful examination.

In this chapter, we will first take a quick look at Grothendieck toposes and some of their most important features. We will then move to elementary toposes, introduced by Lawvere and Tierney in the early seventies, and look at their most important properties. All the previous considerations surrounding Klein's program and its categorical generalization apply directly to the concept of topos. First, an (elementary) topos is characterized by the existence of certain adjoints to given elementary functors. Second, it can also be presented in a purely logical fashion, as a higher order type theory and thus, in a certain sense, it contains all the logic of a situation. Third, and perhaps most important for us, a topos *is* a space and we are right back in geometry.

7.1 Grothendieck Toposes: Generalized Spaces

> Such was the case, for instance, with the crucial unifying notion of topos, at the very heart of the new geometry – the one that provides the common geometric intuition for topology, algebraic geometry and arithmetic – the one also that allowed me to unveil the étale and *l*-adic cohomological tool and the main ideas (more or less forgotten since then, it is true...) of crystalline cohomology.
>
> ([106], 11, My translation.[1])

As the foregoing quotes indicate, Grothendieck saw toposes in two complementary ways: first, the concept itself provides the common geometric intuition for topology, algebraic geometry and arithmetic; second, one of its function is to allow for the definition and development of various cohomology theories. We will not look at the historical development of the concept of topos, although it will certainly make a fascinating story when all is told. (See, however [221] for a preliminary sketch and [224].) We will rather concentrate on some of its most important conceptual elements.

The motivation underlying topos theory can be summarized as follows:[2] one can study, say, a compact Hausdorff space X by studying the associated ring $C(X)$ of complex-valued functions on X. (Notice once again how the study of an object is transferred to the study of how this object is transformed or mapped into another, usually well-known, object.) The maps between spaces become maps between the associated rings and the space X can be completely recovered (up to homeomorphism) from $C(X)$. This is what Gelfand duality more or less says. Considering an arbitrary topological space X, one can, instead of looking at the *ring* of complex-valued functions on X, consider the "continuous set-valued functions" on X, that is, sheaves of sets on X. As in the previous case, continuous mappings between spaces can be described in terms of their associated sheaves and the space X *can be* fully recovered from the sheaves. Furthermore, and this is an important feature, *all* the (cohomological) invariants of the space X are transferred to the category of sheaves on X. Thus, one might as well replace or even *identify* the space X with the category

---

[1] Tel a été le cas, notamment, de la notion unificatrice cruciale de topos, au cœur même de la géométrie nouvelle—celle-là même qui fournit l'intuition géométrique commune pour la topologie, la géométrie algébrique et l'arithmétique—celle aussi qui m'a permis de dégager aussi bien l'outil cohomologique étale et *l*-adique, que les idées maîtresses (plus ou moins oubliées depuis, il est vrai...) de la cohomologie cristalline.

[2] I owe this way of presenting the motivation to [230].

7.1 Grothendieck Toposes: Generalized Spaces

of sheaves on X. From a categorical point of view, the latter is characterized by the fact that it is a (Grothendieck) topos. Hence, a space *is* a topos and a topos *is* a space.

Notice the shift from Eilenberg and Mac Lane's methodological standpoint. We are not looking at how the space X is transformed into other spaces or how other spaces are transformed into it, but rather we are looking at all the possible "continuous" transformations of X into a "known space", namely the category of sets! From an abstract point of view, we are looking at how X "acts" continuously on sets. The sheaves of sets on a topological space X form a topos. Thus studying the latter topos amounts to studying X. One of the advantages of looking at the topos instead of the space itself is that the "points" of a topos (the term has a precise technical definition), have a lot of structure and, in fact, in many concrete cases where X is constructed out of other structures, e.g., when X is the spectrum of a commutative ring Y, these 'points' of the topos are parts of what one wants to know. Furthermore, it is possible to generalize to more abstract spaces: sheaves of sets on a locale and sheaves of sets on a site are also toposes.

In the same way that transformation groups in the hands of Klein-Lie-Killing-Cartan were a tool to study and understand geometry better, toposes were first and foremost a tool to study and understand algebraic geometry, in particular étale cohomology theory. In this context, in the same way that one keeps an eye on the geometry at hand when using the group, one usually keeps an eye on the objects one starts with when using a topos. But, as in the case of groups in geometry, toposes acquired an autonomous standing and became an object of study with various applications.

In order to understand better in what sense a topos is a generalized topological space, let us start by observing what the fundamental property of a topological space is: a topology is essentially a structure which allows mathematicians to define things locally and study things locally, things that are transformed *continuously* into one another. A topological structure is a structure whose essence is to make sense of (1) local phenomena and (2) continuity. A topos is a generalized topological space more or less because it provides the means to express and study local phenomena and continuity. It possesses an intrinsic concept of localization and naturally extends the notion of continuity. This comes from Grothendieck's notion of a *site*, on the one hand, and the notion of a *geometric morphism*, on the other. We will briefly look at these two notions.

A site is more or less a generalized topology on a category in the sense that it provides a notion of localization for that category. Thus, a category $\mathscr{C}$ is thought of as the underlying space and a site is a family of "neighborhoods" in a generalized sense. Informally, the idea is this. What should it mean for a category to have something "locally defined" or a notion of localization? In the case of a topological space, "locally" means "in a neighborhood" or, what is equivalent, "in an open set". Of course, this is not enough. What this really means is that these open sets satisfy a certain structure, that they are related to one another in a specified manner. It is this structure (or a logically equivalent one) that constitutes the essence of "being open".

These open sets form, from the algebraic point of view, a lattice. In fact, they form a complete Heyting algebra and can therefore be considered as a category: its objects are the open sets and its morphisms are inclusion maps between open sets. We can concentrate on a topology from this point of view and find the essential properties of these mappings from a categorical point of view. Keeping the arrow-theoretic properties, we should obtain a purely categorical definition of "localizing" which should be applicable to arbitrary categories. Thus, the notion of a localization in a category, or of a "topology", is an abstraction from a certain notion of localization in the topological case and in this sense it is already a generalization of that notion. However, as usual, this abstraction is not trivial and not unique. The "right" properties of these open sets do not impose themselves and various abstractions are possible. The trick here is to focus on the *process*, so to speak, of localization; that is, the ability one has, in a topological space, of moving from an open set U to a family of open sets V_i that cover U. Informally, "localizing" is the ability to move from any given open set U to any open set V contained in U. But to make sure that this process is well defined and "stable", certain conditions have to be imposed on it. Roughly the conditions taken can be the following. The idea is to concentrate on covers of U, for any open set U of a topological space. A *cover* is a collection of open sets that cover U in the sense that U is contained in the union of these open sets. Such covers are the backbone of the process of localization. Intuitively, these covers should satisfy some simple properties: the collection of all open sets included in a given U should constitute a cover of U; a cover should be "stable" in the sense that for any open set V element of a cover of U, if we localize U via V further, that is by replacing V by a collection of open sets W whose union contain V, then the latter collection of Ws should be a cover of V; and finally localizing should be "transitive", that is a cover of a cover of U should be a cover of U too.

Formally, this leads us to the following considerations and definition. Given a topological space X and its category (lattice) of open sets $O(X)$, we consider an open set U of $O(X)$ together with its "covers", that is families of open sets $\{V_i\}_{i \in I}$ such that $U \subset \bigcup_i V_i$. We say that such a family $S = \{V_i\}_{i \in I}$ covers U. An axiomatization on covers has to be an axiomatization on families of covers for all open sets. Formally:

1. For every open set U, the collection of all open subsets of U is a cover of U;
2. If $S = \{V_i\}_{i \in I}$ is a cover of U, and given open sets $W_j \subset V_i$ for all $j \in J$, then $\{W_j\}_{j \in J}$ is a cover of V_i;
3. If $S = \{V_i\}_{i \in I}$ is a cover of U and if for every V_i, $\{W_j\}_{j \in J}$ is a cover of V_i, then $\bigcup_i \{\bigcup_j W_j\}$ is a cover of U.

These properties are obvious properties of covers of a topology and a choice of such covers corresponds to a notion of localization in a topological space. This is what localization is taken to mean. But it is now easy to translate these properties in the context of an arbitrary category $\mathscr{C}$ and define the cover of an object X of $\mathscr{C}$. To do this, we need a simple preliminary notion, for we will talk about covers of objects instead of covers of open sets. A *sieve* S—a *"crible"* in French—on an object X of a category $\mathscr{C}$ is a family of morphisms with codomain X in $\mathscr{C}$ such that

if $f: Y \to X$ is in S and $g: Z \to Y$, then $f \circ g: Z \to X$ is also in S.

7.1 Grothendieck Toposes: Generalized Spaces

Thus a sieve on an object X is a collection of morphisms with codomain X that is closed under right-composition. Informally, we can think of this situation as follows and this partially explains the choice of terminology. If the image $f[Y]$ is thought as being a (structured or coherent) part of X, that is a part of X that is still of the same type as X, and if $g[Z]$ is a part of Y, then $fg[Z]$ is a part of X. Thus, a sieve S on X is a way to "decompose" X. The name "sieve" is quite apt.

A sieve S on X can also be thought of as a subfunctor of the functor $\mathrm{Hom}(-,X)\colon \mathscr{C}^\circ \to \mathbf{Set}$. This simply means that for each Y, $S(Y)$ is a subset of $\mathrm{Hom}(Y,X)$ and for each morphism $f\colon Z \to Y$, $S(f)$ is simply a restriction of

$$\mathrm{Hom}(f,X)\colon H(Y,X) \to \mathrm{Hom}(Z,X).$$

If S is a sieve on X and $g\colon Y \to X$ is any arrow with codomain X, not necessarily in S, then $S^g = \{h \mid \mathrm{cod}(h) = Y \wedge g \circ h \in S\}$ is a sieve on Y.

Definition 7.1. [3] A *Grothendieck topology* or a *covering system* (or a *localization system*) on a category $\mathscr{C}$ is a function J which assigns to each object X of $\mathscr{C}$ a collection $J(X)$ of sieves on X, such that:

1. The maximal sieve $\{f \mid \mathrm{cod}(f) = X\}$ is in $J(X)$;
2. If $S(X) \in J(X)$, then $S^g \in J(Y)$ for any arrow $g\colon Y \to X$;
3. If $S(X) \in J(X)$ and R is any sieve on X such that $R^g \in J(Y)$ for all $g\colon Y \to X$ in S, then $R \in J(X)$.

Thus, a localization system is a collection of ways of decomposing an object X, ways of moving from the object X to its parts.

A *site* is a pair $(\mathscr{C},J)$ where $\mathscr{C}$ is a (small) category equipped with a Grothendieck topology J. Notice that in practice one has to *choose* a topology J on a category $\mathscr{C}$ to obtain the corresponding site $(\mathscr{C},J)$. It should be clear that a site is taken to be what one means by localizing in a category $\mathscr{C}$. There are of course many possible choices of localizations in a given category.

We now have a categorical characterization of the process of localization. Notice immediately that it *is* an adequate generalization of the topological notion, at least in the following sense. If X is a topological space, the partial order $\mathscr{O}(X)$ on open subsets $U \subseteq X$ is, as we have seen, a category: there is at most one arrow $U \to V$ whenever $U \subseteq V$. A covering in the usual topological sense, i.e., for any U, a collection of open sets $\{U_i\}_{i \in I}$ such that $U = \bigcup_i U_i$, yields a covering in the categorical sense, simply by letting $\{U_i \to U\} \in J(U)$ for each object U of $\mathscr{O}(X)$. It is easy to verify that this is a Grothendieck topology J on $\mathscr{O}(X)$.

Clearly, the notion of Grothendieck topology is more flexible than the usual notion of a topology and this flexibility was required by the context of algebraic geometry when one works with algebraic varieties in an affine space k^n where k is an arbitrary commutative ring. In this context, Grothendieck topologies are needed. But one could go much further in the heuristic direction. 20th century mathematics

---

[3] There are of course alternative definitions, in particular definitions that do not rely on the notion of sieve. See for instance [117] or [208].

has clearly shown that it was more than useful to try to find a topology in a given situation and to use topological ideas to clarify, solve and understand various mathematical contexts. If toposes are generalized topological spaces, it would seem to be a useful idea to try to import such a space into various situations. One can then use what Grothendieck called "topological intuition" in contexts where there is no obvious topology involved. Grothendieck clearly believed that various problems could be solved easily if only one could find the appropriate topos for them.

But the story does not end with sites. Although we have a categorical notion of localization, we do not have a space yet. To understand why, we have to go back to topological spaces. We have seen in an earlier chapter how the category of sheaves on a topological space X is defined. The basic point is that one can define the category of sheaves over a site. It is this latter category that should be thought of as a space. But one might wonder why this is so. In fact, a site is already an interesting structure, much like a space, which contains useful mathematical information. One would want to associate invariants to sites but it is precisely at this point that toposes appear to be useful. As Grothendieck and his colleagues observed, the important invariants associated to a site are definable in the associated topos. This led Grothendieck and his colleagues to the important observation that two sites can be considered to be the same whenever their associated toposes are equivalent categories (Morita again...).

Given a site $(\mathscr{C}, J)$, a functor $F: \mathscr{C}^\circ \to \mathbf{Set}$ is called a *presheaf* on $\mathscr{C}$. The category of presheaves on $\mathscr{C}$ is simply the functor category $\mathbf{Set}^{\mathscr{C}^\circ}$. Sheaves on a site can be defined in various manners. I will give a purely diagrammatic definition. A presheaf $F: \mathscr{C}^\circ \to \mathbf{Set}$ is a *sheaf* if and only if for all covering sieves S of objects X, any natural transformation $\eta: S \to F$ has a unique extension to $\mathrm{Hom}(-, X)$, that is there is an isomorphism between the set of natural transformations induced by the inclusion $S \rightarrowtail \mathrm{Hom}(-, X)$,

$$\mathrm{Hom}(S, F) \approx \mathrm{Hom}(\mathrm{Hom}(-, X), F)$$

or, equivalently, the following diagram

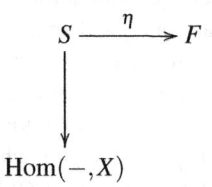

$$\begin{array}{ccc} S & \xrightarrow{\eta} & F \\ \downarrow & & \\ \mathrm{Hom}(-, X) & & \end{array}$$

can always be extended into the following commutative diagram:

7.1 Grothendieck Toposes: Generalized Spaces

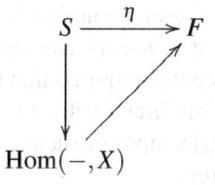

$$S \xrightarrow{\eta} F$$
$$\downarrow \nearrow$$
$$\text{Hom}(-,X)$$

Thus, whereas a presheaf associates a set $F(X)$ to an object X of $\mathscr{C}$ and a function $F(f)$ to a morphism $f: X \to Y$, a sheaf is systematically connected to the covering sieves of the objects of $\mathscr{C}$. In a sense, a sheaf systematically encodes the localization systems of the site $(\mathscr{C}, J)$ and thus each sheaf has a localization system. Needless to say, this is a generalization of the notion of sheaf over a topological space that we saw in Chapter 5. The category of sheaves over a site, with natural transformations as morphisms, is denoted by $\text{Sh}(\mathscr{C}, J)$. It is this latter category that ought to be thought of as a generalized space.

A *Grothendieck topos* is defined to be a category $\mathscr{E}$ equivalent to a category $\text{Sh}(\mathscr{C}, J)$ of sheaves over a site.

This is a rather strange definition, but of great interest from a philosophical point of view. For although there is a uniform description of categories of the form $\text{Sh}(\mathscr{C}, J)$, the definition says that any category equivalent to a category of this form is a Grothendieck topos. The definition is nothing less than an *abstraction* of the notion of a category of sheaves on a site. The reason why it is a *legitimate* abstraction is that it gives an identity criterion for Grothendieck toposes. It does not characterize Grothendieck toposes intrinsically but it says when two such toposes are *identical*, i.e., equivalent. Thus, given a Grothendieck topos $\mathscr{E}$, there are many different sites that yield equivalent categories, thus basically the same topos, although it is impossible to tell which of the sites is preferable. It is an extrinsic characterization that nonetheless provides us with a *type* of mathematical entity. It says absolutely nothing about the structure of toposes, their intrinsic properties. Of course, one would want a characterization of toposes independent of sites and similar to the characterization of Abelian categories for instance. This characterization was provided by Giraud, one of Grothendieck's students, in the form of the following theorem.

Theorem 7.1. *A category $\mathscr{E}$ with small Hom-sets and all finite limits is a Grothendieck topos if and only if it has the following properties:*

1. *$\mathscr{E}$ has all small coproducts, and they are disjoint and stable under pullback;*
2. *Every epimorphism in $\mathscr{E}$ is a coequalizer;*
3. *Every equivalence relation $R \rightrightarrows X$ in $\mathscr{E}$ is a kernel pair and has a quotient;*
4. *Every exact diagram $R \rightrightarrows X \to Q$ is stably exact;*
5. *There is small set of objects of $\mathscr{E}$ which generate $\mathscr{E}$.*

I will not give technical details about these conditions. Suffice it to say that the conditions given by the theorem do *not* refer to a site, only to the capacity of a category to represent in a universal way, certain properties, and the existence and size of a collection of certain specific morphisms. It is therefore an *intrinsic* characterization and it provides an abstract characterization of Grothendieck toposes.

Hence, the theorem provides a "presentation-free" characterization of Grothendieck toposes. Notice, however, that it is *not* an elementary characterization (that is, a first-order characterization) since it is required that the category $\mathscr{E}$ has *all small* coproducts. Of course, the *proof* of Giraud's theorem does more: it tells you *how* to construct the given category $\mathscr{E}$ as a topos of sheaves on a site. (See [45] for more on sheaf theory and Giraud's theorem.)

As I have already mentioned, Grothendieck observed that the choice of a site was arbitrary in a very specific sense. Two non-isomorphic sites and even two nonequivalent sites $(\mathscr{C}, J)$ and $(\mathscr{C}', J')$ can yield equivalent toposes. (See, for example, the so-called *comparison lemma* in [200], 588, or [121], C2.2, 547.) In this specific case, $(\mathscr{C}, J)$ and $(\mathscr{C}', J')$ can be considered to be essentially the same. This is by now a familiar situation. Thus two algebraic geometers could choose two different sites for specific purposes and work on sheaves over these sites without noticing that they are doing the "same" thing. As Grothendieck emphasized, it is the topos of sheaves on these sites that is really significant and that topos is determined up to equivalence of categories. This situation can again be compared with the case of elementary geometry that I have illustrated above, where the geometers A, B and C chose various presentations, in particular axiomatic presentations, of one and the same geometry. In the present case, it is as if A, B and C had chosen different sites to start with. From a global point of view, different presentations will yield the same (or rather isomorphic) transformation groups and one is led to conclude that it is the transformation group which really "matters": likewise, two sites can be said to be equivalent if their respective toposes of sheaves are equivalent. (Morita equivalence again.) [4] As we have already seen, this is a recurrent theme in category theory and categorical logic. Interestingly enough, sites and their associated toposes of sheaves can be used in logic. This is a direct *geometrization* of logic. The first steps are simple applications of what we have already seen.

Here are some standard and important examples of Grothendieck toposes.

Obviously, any category of sheaves over a topological space X is a Grothendieck topos. This already provides a large class of examples. In particular, the category **Set** of sets is a Grothendieck topos: simply let the space X be the one-point space **1** with the obvious trivial topology (the category **1** has only one morphism). Thus, one can think of the category of sets as a generalized space, although as a limiting case.

More generally, given any category $\mathscr{C}$, one can always consider the so-called *trivial topology*: the only covering families are the one-elements families of isomorphisms $\{f\colon Y \xrightarrow{\sim} X\}$. With this site, every presheaf is a sheaf and therefore every category of presheaves $\mathbf{Set}^{\mathscr{C}^\circ}$ is a Grothendieck topos.

---

[4] I should also point out that Grothendieck very often used the existence of an isomorphism between induced structures to define the equivalence of original structures. For instance in his work with Dieudonné on the foundations of algebraic geometry, it is the isomorphism between two varieties which allows one to say that two polynomial rings are equivalent. Once again, polynomial rings can look different because of the choice of variables or other contingent elements of the same type. (See [60].)

7.1 Grothendieck Toposes: Generalized Spaces

Two specific examples of toposes of sheaves, which were introduced later, after the discovery of elementary toposes, are worth mentioning, if only because they are directly connected to foundational studies. Let B be a complete Boolean algebra, considered as a category. There is an obvious Grothendieck topology definable on B: a family $\{b_i \to b\}$ is a covering if and only if $b = \vee b_i$ in B. We thus get a site and a Grothendieck topos $\text{Sh}(B)$ of Boolean sheaves. These toposes are closely related to Boolean-valued models of set theory.

The second example is related to Cohen forcing. Let P be a partially ordered set and $p \in P$. A subset $D \subseteq \{q \in P \mid q \leq p\}$ is said to be *dense below* p if $\forall r \leq p \exists q \leq r$ ($q \in D$). A sieve S on an element p of P in this case is simply a subset S of elements $q \leq p$ such that $r \leq q \in S$ implies $r \in S$. It is possible to define a topology J on P by

$$J(p) = \{D \mid (\forall q \in D)(q \leq p) \wedge D \text{ is a sieve below } p\}.$$

Although we have defined this topology on a partially ordered set, it can easily be defined on an arbitrary category. Given this site, we can consider the topos $\text{Sh}(P)$ of sheaves.

But there is another important and different class of examples. Let G be a topological group and X a set with discrete topology. We have already seen the category of continuous G-sets. It can be shown that the category of continuous G-sets is a Grothendieck topos, denoted by $\mathbf{B}G$. Notice how different these examples are from the examples of sheaves over a topological space, at least at first sight. But the notion of Grothendieck toposes covers both. In fact, a slightly more general class of examples can be included.

Indeed the last example can be generalized in the following way. Let G be a topological group, X a space and $\alpha \colon G \times X \to X$ an action of G on X. A G-*space over* X is a space $p \colon E \to X$ over X with an action $\beta \colon G \times E \to E$ such that p respects this action. In other words, the following diagram commutes:

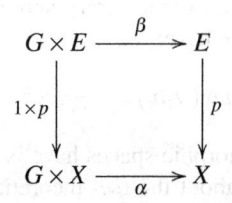

$$\begin{array}{ccc} G \times E & \xrightarrow{\beta} & E \\ {\scriptstyle 1 \times p} \downarrow & & \downarrow {\scriptstyle p} \\ G \times X & \xrightarrow{\alpha} & X \end{array}$$

A morphism of G-spaces over X is a morphism of spaces over X

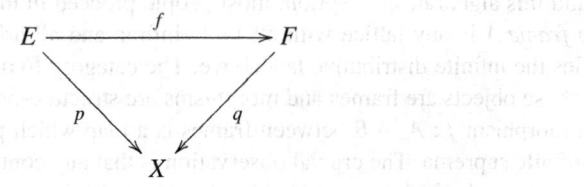

$$\begin{array}{ccc} E & \xrightarrow{f} & F \\ & {\scriptstyle p} \searrow \quad \swarrow {\scriptstyle q} & \\ & X & \end{array}$$

that respects the G-action. This constitutes the category of G-spaces over X. If, furthermore, the morphism $p\colon E \to X$ is *étale* (see Chapter 5 for the definition), we say that the G-space over X is *étale*. We can therefore consider the subcategory of étale G-spaces over X and this category is in fact a Grothendieck topos, it is called the topos of *G-equivariant sheaves* on X, denoted by $\mathrm{Sh}_G(X)$.

In fact, Joyal and Tierney showed in the early 1980s that the notion of Grothendieck topos is in a sense the precise generalization of these two classes of examples, sheaves over a topological space, on the one hand, and continuous actions of a group on a space, on the other. This is an extremely interesting and important result. For, it is clear that the notion of sheaves on a site is a vast generalization of the notion of a sheaf on a topological space and one is left wondering whether there is anything spatial left in the general notion. What Joyal and Tierney have shown is that Grothendieck's generalization is not as extreme as one might think and it is still essentially spatial or geometrical. More specifically Joyal and Tierney proved a theorem that asserts that for any Grothendieck topos $\mathscr{E}$, it is possible to find a localic or continuous groupoid G such that $\mathscr{E}$ is equivalent to the category of étale G-spaces. Since a locale can be seen as a rather straightforward generalization of the notion of topological space and a groupoid action is essentially a geometric notion intrinsic, for instance, to differential geometry, Grothendieck toposes are still fundamentally spatial. (See [128], and [229], where these results are extended, [125] and [124] for a slightly different approach and [121], C5 for a systematic and comprehensive exposition.)

Let us try to unpack the conceptual significance of this fundamental result. A localic or continuous groupoid G is a groupoid *in* the category of locales. This is probably not very helpful. The category of locales can be thought as the category of "formal" or "pointless" spaces in the following sense. A topology can be looked at from the algebraic point of view: the collection of open sets of a space X form a complete distributive lattice, that is, a lattice with finite infima, arbitrary suprema and for which infima distribute over arbitrary suprema i.e., for all a, $\{b_i\}_{i \in I}$ in a complete distributive lattice, we have:

$$a \wedge (\vee_i b_i) = \vee_i (a \wedge b_i).$$

It is obvious that two homeomorphic spaces have isomorphic lattices of open sets. Thus, it is possible to forget about the (set-theoretical) points of a space and define a topological space directly from the lattice-theoretical point of view. In order to keep a distinction between the set-theoretical conception of a topological space and this algebraic conception, most people proceed in the following manner. First, a *frame A* is any lattice with all finite infima and all arbitrary suprema that satisfies the infinite distributive law above. The category **Frm** of frames is the category whose objects are frames and morphisms are structure-preserving mappings, that is a morphism $f\colon A \to B$ between frames is a map which preserves finite infima and infinite suprema. The crucial observation is that any continuous map $f\colon X \to Y$ between topological spaces induces a morphism of frames $f^{-1}\colon \mathscr{O}(Y) \to \mathscr{O}(X)$ from the frame of open sets of Y into the frame of open sets of X.

7.1 Grothendieck Toposes: Generalized Spaces

The category of frames is a category of algebraic objects. But notice the connection between frames and topological spaces: given a continuous map between topological spaces, the induced morphism of frames is a map going in the *opposite* direction. It is tempting to *start* with the category of frames and *define* a category of spaces as the dual of the latter, e.g., the category with the same objects but the morphisms going in the opposite direction. By reversing the morphisms, we end up with the continuous maps between the spaces. Hence, dualizing the category of frames, one obtains what is called the category **Loc** of *locales*, i.e., **Loc** = **Frm**°. (A warning: what I have been calling the category of *frames*, Joyal and Tierney call the category of *locales* and what I have been calling the category of *locales*, they call the category of *spaces*, but as far as I know, no one has endorsed that terminology.) Thus, the objects of the category of locales are the same as the objects of the category of frames. However, a morphism of locales is the dual of a morphism of frames, i.e., $f\colon X \to Y$ is a morphism of locales if and only if $f^\circ\colon Y \to X$ is a morphism of frames.

The category of locales is naturally connected to the category of topological spaces. There is a functor $L\colon \mathbf{Top} \to \mathbf{Loc}$ from the category of topological spaces to the category of locales defined on objects by $L(X) = \mathcal{O}(X)$, associating to each space X, the frame of open sets of X, and to any continuous map $f\colon X \to Y$, the morphism of locale $L(f)\colon L(X) \to L(Y)$, given by the dual of the frame morphism $f^{-1}\colon \mathcal{O}(Y) \to \mathcal{O}(X)$. However, the notion of locale *is* a generalization of the notion of topological space. It should be pointed out that the idea of characterizing a space by its algebra of parts goes back at least to Wallman, who made the explicit connection between topological spaces and lattices in 1938. ([265])

A groupoid G in the category of locales, or a *localic* or *continuous groupoid*, is a groupoid $s,t\colon G_1 \to G_0$ such that G_0 and G_1 are locales and the source and the target maps are morphisms of locales. (To be totally rigorous here, I should add that the identity map associating to each object of G the identity arrow for that object is a morphism of locales and so are the morphisms defining composition in the groupoid and projections. I should also point out that these morphisms also have to be *open* morphisms in the appropriate localic sense. But these are important technical details that I can leave out for present purposes.)

Given a continuous groupoid G, a *G-space* E over G_0 is a locale (a "space") together with maps $p\colon E \to G_0$ and $E \times_{G_0} G_1 \xrightarrow{\bullet} E$, where the domain of the action $\bullet$ is the "set" of pairs (e,g) with $t(g) = p(e)$, where $t\colon G_1 \to G_0$ is the target map, satisfying the following conditions:

1. $p(e \bullet g) = s(g)$, where $s\colon G_1 \to G_0$ is the source map;
2. $e \bullet i(p(e)) = e$, where $i\colon G_0 \to G_1$ assigns to each element of G_0, its identity arrow;
3. $(e \bullet g) \bullet h = e \bullet (g \bullet h)$, where $G_1 \times_{G_0} G_1 \xrightarrow{\circ} G_1$ is the composition of morphisms.

Notice that these conditions could be expressed directly in terms of commutative diagrams in the category of locales. Informally a G-space is a space over G_0 with an (contravariant) action of G defined on it. As we have seen, this is a fundamental geometric notion. A *morphism* of G-spaces is a map of spaces $E_1 \to E_2$ over G_0 which preserves the action. We thus get a category of G-spaces.

An *étale* G-space E is a G-space such that $p\colon E \to G_0$ is a local homeomorphism. The word "étale" here means that the "topology" of G_0 is uniformly "spread out", that is "horizontally", in E. (Keep in mind that G is a continuous groupoid and E is a locale. So this is a more general definition.) As usual, this means that E is a sheaf on G_0 and as usual, this means that any such E can be reconstructed by the collection of local sections from G_0 into E. Furthermore, the morphisms of G, that is elements of G_1, act on the fibers of E; that is, if $f\colon x \to y$ is a point of G_1, the action defines a map $f^*\colon E_y \to E_x$ by $f^*(e) = e \bullet g$. A *morphism* of étale G-spaces is a morphism of G-spaces. The fact that each map over G_0 is a local homeomorphism implies that morphisms of G-spaces are local homeomorphisms too. The category of étale G-spaces is denoted by $\mathbf{B}G$ and it is a topos. Notice that when G_0 is a one-element space and G_1 is a discrete space, then the groupoid G is a group and the topos $\mathbf{B}G$ is the category of G-sets: that is, sets equipped with an action. At the other end of the spectrum, so to speak, when G is a trivial groupoid, namely when $G_1 = G_0$ (that is, when the only morphisms are the identities) then $\mathbf{B}G$ is the topos of sheaves on G_0, i.e., $\mathrm{Sh}(G_0)$. Thus, for any space X, the topos $\mathrm{Sh}(X)$ of sheaves on X is of the form $\mathbf{B}X$.

Joyal and Tierney's representation theorem states that any Grothendieck topos $\mathscr{E}$ is equivalent to a topos of the form $\mathbf{B}G$, for an appropriate open continuous groupoid G. We can now make more precise comments on the geometric content of this theorem.

First, everything is going on in the category of locales, thus of *spaces* in a generalized sense of the term. The objects of the topos $\mathbf{B}G$ are spaces over G_0, which is itself, of course, a space. Second, for each of these spaces over G_0, the groupoid G acts on them. To understand the meaning of this action in general, we can take a closer look at the situation. The action is defined on certain pairs (not on all pairs) of the product $E \times_{G_0} G_1$ (as the notation indicates, it is a fibered product, i.e., a pullback, over G_0), and thus the groupoid acts only "locally", for it acts on the points e of E for which $p(e) = t(g)$. A possible picture of the situation is to think of G_0 as the "base" space, that is the geometry one investigates, about which G_1 gives the structure of local "homeomorphisms", and a space E can be thought of as a space of "moving frames" over G_0. Then the groupoid will send moving frames to moving frames according to the underlying structure of local homeomorphisms of G_0. The conditions 1, 2 and 3 simply make sure that this process is done in a coherent fashion. Hence in general, a Grothendieck topos can be thought of as a framework in which one examines a given space, namely G_0, together with a basic equivalence relation defined on that space, namely by G_1, where the space is examined by considering the étale spaces over G_0 which "respect", in a certain sense, the geometry of the latter. The two ends of the spectrum, namely the toposes of G-sets and the toposes of sheaves over a topological space constitute degenerate cases where the "geometry" is "simplified" in a sense. For in the first case, the underlying space G_0 is trivial, whereas in the second, the groupoid action G_1 is trivial.

7.1 Grothendieck Toposes: Generalized Spaces

However, the foregoing examples and Joyal and Tierney's result obscure an important aspect of Grothendieck toposes, already clear to Grothendieck himself. In the examples, we have considered toposes of sheaves over a topological space X or, slightly more generally, over a locale X, or toposes of G-sets, for a group (or a groupoid) G. But Grothendieck's notion of a generalized space covers a much wider range of possibilities: in certain applications, one can and should consider a topos of sheaves over a site $\mathscr{C}$, where $\mathscr{C}$ is *itself* a category of spaces, e.g., the *category* **sSet** of simplicial sets, the *category* **Man** of smooth manifolds or the *category* **Top** of topological spaces or any other category of spaces of a certain kind. Of course, these toposes, being Grothendieck toposes, are also covered by Joyal and Tierney's representation theorem, but they might have specific properties of intrinsic interest, as Lawvere has emphasized later. (See [178], for instance.)

A topos can thus be thought of as a generalized topological space. If the latter notion was conceived to clarify the notion of continuity, since it leads directly to the definition of a continuous function, it seems reasonable to believe that morphisms between toposes would generalize that concept too.

First, recall the notion of a continuous function $f\colon X \to Y$ between topological spaces. The function f is *continuous* if for every open U of Y, $f^{-1}(U)$ is an open of X. Moving to sheaves over these topological spaces, namely $\mathrm{Sh}(X)$ and $\mathrm{Sh}(Y)$, it can be shown that a continuous function $f\colon X \to Y$ induces *two* functors

$$\mathrm{Sh}(X) \underset{f^*}{\overset{f_*}{\rightleftarrows}} \mathrm{Sh}(Y).$$

The first functor $f_*\colon \mathrm{Sh}(X) \to \mathrm{Sh}(Y)$, called the *direct image functor*, is defined by composing with f^{-1} thus: given a sheaf on X, $F\colon \mathscr{O}(X)^\circ \to \mathbf{Set}$ and an open set U in $\mathscr{O}(Y)$, define $f_*\colon (F)(U) = F(f^{-1}(U))$. This automatically yields a sheaf on Y. It is easier to define the other functor, the so-called *inverse image functor* $f^*\colon \mathrm{Sh}(Y) \to \mathrm{Sh}(X)$ by looking at the étale spaces corresponding to sheaves. Thus, given an étale space $E \xrightarrow{p} Y$ over Y, one defines $f^*(E \xrightarrow{p} Y)$ by taking the following pullback along f:

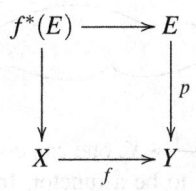

$$\begin{array}{ccc} f^*(E) & \longrightarrow & E \\ \downarrow & & \downarrow p \\ X & \underset{f}{\longrightarrow} & Y \end{array}$$

It can be shown that, at least when Y is Hausdorff (a slightly weaker condition will do, e.g., being sober), then any such pair of functors $\mathrm{Sh}(X) \underset{f^*}{\overset{f_*}{\rightleftarrows}} \mathrm{Sh}(Y)$ necessarily comes from a unique continuous function $f\colon X \to Y$. It can also be verified that in this case $f^* \dashv f_*$ and it follows from the construction of f^* that it preserves finite limits. (See, for instance, [200], 348–349, or [121], A4.1, 166.) We are thus led to the following definition of morphisms between toposes: a *geometric morphism* $f: \mathscr{F} \to \mathscr{E}$ between toposes is a pair of functors $f^*: \mathscr{E} \to \mathscr{F}$ and $f_*: \mathscr{F} \to \mathscr{E}$ such that f^* is left adjoint to f_* and f^* is left exact. Of course, it would be enough to say that a morphism $f_*: \mathscr{F} \to \mathscr{E}$ is a *geometric morphism* if it has a left exact left adjoint functor f^* (necessarily unique up to isomorphism). It is entirely reasonable to think of f_* as being the geometric part of the morphism and to think of f^* as being the algebraic part. A nice illustration of the fact that f_* should be thought of as the geometric part is offered by the functor of global sections $\Gamma: \text{Sh}(X) \to \textbf{Set}$. If we think of the sheaves over X as an étale space $E \xrightarrow{p} X$, then a *global section* of p is a continuous map $s: X \to E$ such that the diagram

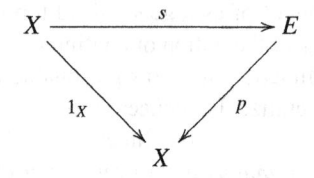

$$\begin{array}{ccc} X & \xrightarrow{s} & E \\ & \searrow{1_X} \swarrow{p} & \\ & X & \end{array}$$

commutes. This is clearly a geometrical concept. We can think of s as projecting X in E evenly. An informal picture of such a global section might look like this:

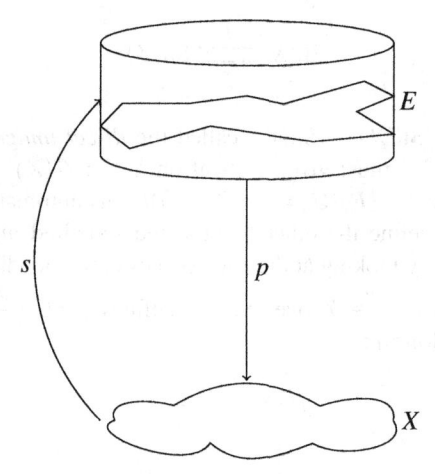

Thus, for each étale space $E \xrightarrow{p} X$, one can consider the *set* $\Gamma(E)$ of its global sections and it is easily verified to be a functor. In general, given a Grothendieck topos $\mathscr{E}$, one can define the functor $\Gamma: \mathscr{E} \to \textbf{Set}$ by $\Gamma(E) = \text{Hom}_{\textbf{Set}}(1, E)$. The latter functor is the geometric part of a geometric morphism. Its algebraic part, that is its left adjoint, is the functor $\Delta: \textbf{Set} \to \mathscr{E}$ defined by $\Delta(S) = \coprod_{s \in S} 1$, the co-product of S-many copies of the terminal object of $\mathscr{E}$, clearly an algebraic construction. It can be shown that $\Delta \dashv \Gamma$ and that Δ is left-exact. This is an entirely general situation: given a Grothendieck topos $\mathscr{E}$, there is one geometric morphism $\gamma: \mathscr{E} \to \textbf{Set}$ such that $\gamma_* = \Gamma$ and $\gamma^* = \Delta$, and it can be shown that there can be only one (up to natural isomorphism).

I should point out immediately that Grothendieck's reasons for considering a topos as a generalized space were somewhat different and more technical. To put it simply, Grothendieck's motivations rested on the fact that the cohomological properties of a space are completely determined by the topos of sheaves over it, or in the words of Johnstone: "topos cohomology is a minimal common generalization of the (sheaf) cohomology of spaces and of the (Galois) cohomology of groups" ([119], 104). Again, an analogy with the role of groups in geometry seems natural. In the same way that a transformation group captures all the essential properties of a geometry, a topos of sheaves over a topological space captures all the essential properties of a topology, even in the generalized sense Grothendieck gave to the word "topology".

One last element has to be mentioned: if the notion of Grothendieck topos can be thought of as a legitimate generalization of the notion of topological space, it seems reasonable to expect that various concepts arising in the context of topology could be lifted to the context of toposes, and indeed they can. Thus, it is possible to define various notions, like open maps of toposes, connected, locally connected, hyperconnected, etc. for toposes.

But, as I have said, the spatial facet of a topos is but one of its aspects.

7.2 Elementary Toposes

Grothendieck toposes are not all of topos theory. In the late sixties, Lawvere and Tierney provided elementary (that is, first-order) axioms for toposes. An elementary topos is an abstract category that captures by global properties of a theory of *abstract* sets, that is, sets that are *not* defined by their elements, or in other words, for which nothing is known about the individual and intrinsic properties of their elements. What is known is known via universal arrows and other arrow-theoretic properties. Notice that this does not mean that absolutely nothing is known about the "elements" or the "parts" of these abstract sets. Quite the contrary is in fact true. As above, since a topos is a special type of category, it can be thought of as a space and its objects can be thought of as the definable "figures" in it. As in the geometric case, two figures are equal if there is a transformation of the appropriate sort between them, which in this case means that there is an isomorphism between them. The identity of a geometric form *as a form* should not depend upon specific, "contingent" factors, like its place relative to a chosen "system of coordinates" or its actual "parts" or "points". It surely depends on these points in a certain manner, but it is also independent of them. The crucial fact about abstract sets within a topos is that there is a uniform part-whole relation within that topos. In other words, a topos has the means to represent within itself the property "being a part of", seen as an abstract property. This can be seen directly by looking at the axioms for elementary toposes.

Definition 7.2. A category $\mathscr{E}$ is an *elementary topos* if it satisfies the following conditions:

1. It has all finite limits or, equivalently, it has pullbacks and a terminal object 1;
2. There is an object Ω, called the *subobject classifier*, together with a monic arrow $\top\colon 1 \to \Omega$ such that for any monic $m\colon Y \to X$, there is a unique arrow $\chi_m\colon X \to \Omega$ in $\mathscr{E}$ such that the following square is a pullback:

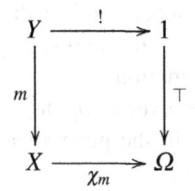

$$\begin{array}{ccc} Y & \xrightarrow{!} & 1 \\ {\scriptstyle m}\downarrow & & \downarrow{\scriptstyle \top} \\ X & \xrightarrow[\chi_m]{} & \Omega \end{array}$$

3. It has power-objects, that is for each object X, there is an object PX, called the *power object* of X, and an arrow $\in_X\colon X \times PX \to \Omega$ such that for every arrow $f\colon X \times Y \to \Omega$ there is a unique arrow $g\colon Y \to PX$ such that the following diagram commutes:

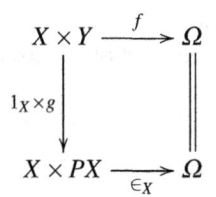

$$\begin{array}{ccc} X \times Y & \xrightarrow{f} & \Omega \\ {\scriptstyle 1_X \times g}\downarrow & & \| \\ X \times PX & \xrightarrow[\in_X]{} & \Omega \end{array}$$

Let us briefly examine these axioms one by one.

The first axiom asserts that elementary constructions are possible in any elementary topos. These constructions are finite products, equalizers, pullbacks and combinations thereof, in particular monomorphisms. Notice that these can be given, as usual, by asserting the existence of adjoint functors to elementary functors.

The second axiom is crucial from the logical point of view *and* the geometric point of view. The terminology itself, *subobject classifier*, comes directly from topology: Tierney suggested the name when he noticed the parallel between the topos-theoretical notion and the concept of a *classifying space* in homotopy theory. The latter concept comes from a fundamental paper of E. H. Brown Jr. published in 1962. ([33].) In a nutshell and very roughly, Brown proved that cohomology theories, as functors, are representable. More precisely, given a contravariant functor $H\colon \mathbf{hTop}_\bullet^\circ \to \mathbf{Set}_\bullet$ from the pointed homotopy category to the category of pointed sets and pointed functions satisfying certain mild conditions, there is a space Y, unique up to homotopy, and a natural isomorphism $\phi\colon [-, Y]_\bullet \to H$. To see the similarity with the notion of a subobject classifier, one has to look at it from a slightly different point of view, namely the point of view of representable functors.

First, given any category $\mathscr{C}$ with finite limits, there is a functor $\mathrm{Sub}\colon \mathscr{C}^\circ \to \mathbf{Set}$, called the *subobject functor*, constructed as follows. For an object X of $\mathscr{C}$, $\mathrm{Sub}(X)$ is the set of (equivalence classes) of subobjects of X, i.e., the set of (equivalence classes) of monomorphisms $m\colon A \to X$. Given a morphism $f\colon Y \to X$ and

7.2 Elementary Toposes

a monomorphism $m\colon A \to X$, $\mathrm{Sub}(f)\colon \mathrm{Sub}(X) \to \mathrm{Sub}(Y)$ is obtained by pulling back m along f as in the following diagram[5]:

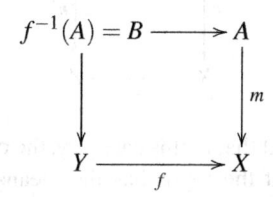

$$\begin{array}{ccc} f^{-1}(A) = B & \longrightarrow & A \\ \downarrow & & \downarrow m \\ Y & \xrightarrow{f} & X \end{array}$$

The important point here is that the existence of the subobject classifier Ω amounts to the claim that the functor Sub *is* representable and is represented by Ω. This simply means that there is a natural isomorphism $\varphi\colon \mathrm{Hom}_{\mathscr{E}}(-,\Omega) \to \mathrm{Sub}(-)$. One subtle technical point has to be emphasized in order to understand how close the analogy is between the subobject classifier and the concept of classifying space. It is not only the fact that, in both cases, a certain functor is representable that underlies the analogy, it is also the fact that in both cases, a certain object and a morphism are obtained by *pulling back* along the representable object. Thus, in the case of the classifying space, every n-dimensional cohomology class in any space X arises by pulling back along a morphism in the universal or representing space Y. The similarity with the notion of subobject classifier Ω is now immediate.

From the logical point of view, Ω can be thought of as an object of truth-values. Its presence in an elementary topos amounts to the possibility of representing internally, that is with the means of the arrows of the category $\mathscr{E}$, "properties" of its objects, since we can think of a subobject Y of an object X, that is a monomorphism $m\colon Y \to X$, as a predicate over X. Each such predicate m corresponds to a characteristic map $\chi_m\colon X \to \Omega$. This means that there is a bijection between predicates $m\colon Y \to X$ over X and characteristic maps $\chi_m\colon X \to \Omega$. A subobject classifier can be thought of as an object of "truth values", since what the foregoing pullback defining it says is that every predicate over X corresponds to a characteristic map which sends this predicate to "true", that is composing to predicate m with the characteristic map is equal to the component $\top$ of Ω.

Given the equivalence between the notion of representable functor and the existence of an adjunction, it is no surprise that the existence of a subobject classifier amounts to the existence of an adjoint functor to an elementary functor. It is certainly worth making the construction explicit. The subobject classifier can be exhibited as a terminal object in an appropriate category. Let $\mathbf{Mon}(\mathscr{E})$ be the category whose objects are monomorphisms $m\colon Y \to X$ of $\mathscr{E}$ and given two objects $m\colon Y \to X$ and $n\colon U \to Z$, a morphism $m \to n$ is a pair of morphisms (f,g) such that $f\colon Y \to U$ and $g\colon X \to Z$ and the following square

---

[5] This is, of course, a special case of the inverse image of f.

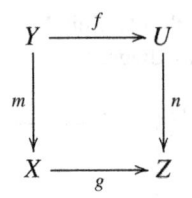

$$\begin{array}{ccc} Y & \xrightarrow{f} & U \\ {\scriptstyle m}\downarrow & & \downarrow{\scriptstyle n} \\ X & \xrightarrow{g} & Z \end{array}$$

is a pullback. It can be verified that in this category, the object $\top: 1 \to \Omega$ is terminal.

The third axiom says that the topos has the means to represent internally the lattice of subobject of an object X. This is exactly what PX is. In the case of the topos of sets, PX is simply the power set of X. In other words, there is an object in the topos $\mathscr{E}$ which represents faithfully the parts of X. Of course PX will be closely connected to the subobject classifier as it should be. In particular, Ω *is* $P1$ and PX is the same as Ω^X. Notice that, as usual in category theory, a subobject classifier and power objects are determined up to (a unique) isomorphism. The basic claim here is that these are the fundamental operations that any theory of abstract sets should satisfy. Thus a category of sets is thought of as being a category that admits certain transformations, namely those which amount to the representation of finite limits and, essentially, the part-whole relation.

Two stunning facts are direct consequences of Lawvere and Tierney's definition of an elementary topos. First, it follows from the fact that Ω represents the functor Sub that is has in general the structure of a Heyting algebra. Thus, it is possible to define operations $\Omega \times \Omega \xrightarrow{\wedge} \Omega$, $\Omega \times \Omega \xrightarrow{\vee} \Omega$ and $\Omega \times \Omega \xrightarrow{\to} \Omega$ satisfying the usual identities of a Heyting algebra. Furthermore, quantifiers can also be defined. Given a morphism $f: X \to Y$, there is always a morphism $f^*: \Omega^Y \to \Omega^X$, namely, once again, by pulling back along f. The quantifiers are then morphisms $\exists_f: \Omega^X \to \Omega^Y$ and $\forall_f: \Omega^X \to \Omega^Y$ satisfying the expected conditions. Notice that the definition of quantifiers is simply obtained by using the fact that quantifiers are adjoints to substitution, as we have seen previously. Thus, from the simple constraint that Ω is the representing object of a representable functor, the whole logical structure of the space follows. Of course, it is not *that* surprising, since Ω represents the functor Sub, which gives all possible decompositions of an object into its parts. We can therefore *expect* to get a lattice-like structure. But the fact that it is *in general* a Heyting algebra was not expected.

The second fundamental fact brings us back to Grothendieck topologies and Grothendieck toposes. One of Lawvere and Tierney's goals was to provide an elementary axiomatic approach to sheaves. Their axiomatization of an elementary topos constituted a preliminary step in that respect, a characterization of an abstract type of category in which it is possible to define a *topology* by the means given by the definition. Specifically, given a topos $\mathscr{E}$ with its subobject classifier Ω, a *Lawvere-Tierney topology* on $\mathscr{E}$ is a morphism $j: \Omega \to \Omega$ satisfying the following properties:

1. $j \circ \top = \top$;
2. $j \circ j = j$;
3. $j \circ \wedge = \wedge \circ (j \times j)$.

7.2 Elementary Toposes

These identities can be pictured thus:

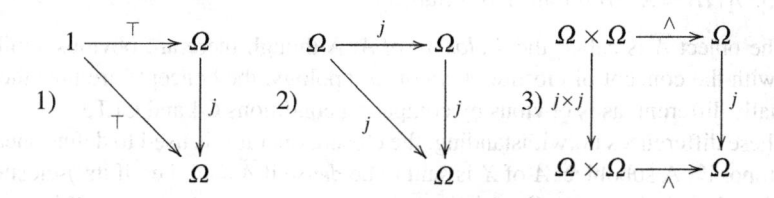

1) $$\begin{array}{ccc} 1 & \xrightarrow{\top} & \Omega \\ & \searrow_{\top} & \downarrow j \\ & & \Omega \end{array}$$

2) $$\begin{array}{ccc} \Omega & \xrightarrow{j} & \Omega \\ & \searrow_{j} & \downarrow j \\ & & \Omega \end{array}$$

3) $$\begin{array}{ccc} \Omega \times \Omega & \xrightarrow{\wedge} & \Omega \\ j \times j \downarrow & & \downarrow j \\ \Omega \times \Omega & \xrightarrow{\wedge} & \Omega \end{array}$$

I should immediately point out that Johnstone calls these operators *local operators* instead of *topologies*. I will stick to the original terminology for now. Since Ω is the subobject classifier and $j\colon \Omega \to \Omega$ is an endomorphism, it determines up to isomorphism a unique subobject $J \to \Omega$ obtained from the following pullback

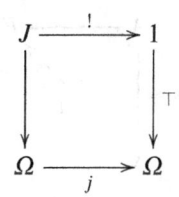

$$\begin{array}{ccc} J & \xrightarrow{!} & 1 \\ \downarrow & & \downarrow \top \\ \Omega & \xrightarrow{j} & \Omega \end{array}$$

A Lawvere-Tierney topology is given in algebraic terms: it is a modal operator on the object of truth-values of the topos $\mathscr{E}$. Informally, it should be thought of as a "closure" operator, with two caveats. First, recall that a topology, in the standard sense of that expression, can be given by a closure operator on the collection of all subsets of a set X, namely by a map $\overline{}\colon \wp(X) \to \wp(X)$ such that:

C1. $\overline{\emptyset} = \emptyset$;
C2. $A \subset \overline{A}$ for all $A \subset X$;
C3. $\overline{\overline{A}} = \overline{A}$ for all $A \subset X$;
C4. $\overline{A \cup B} = \overline{A} \cup \overline{B}$ for all $A, B \subset X$.

Given a Lawvere-Tierney topology $j\colon \Omega \to \Omega$ in a topos $\mathscr{E}$ and a characteristic map $\chi_A\colon X \to \Omega$, the composition $j \circ \chi_A\colon X \to \Omega$ yields a subobject of X by considering the following pullback:

$$\begin{array}{ccc} \overline{A} & \xrightarrow{!} & 1 \\ \downarrow & & \downarrow \top \\ X & \xrightarrow{\chi_A} \Omega \xrightarrow{j} & \Omega \end{array}$$

This construction obviously yields a map $\overline{}\colon \mathrm{Sub}(X) \to \mathrm{Sub}(X)$ and it can be shown that j is a Lawvere-Tierney topology if and only if this operator satisfies the following properties:

CLT1. $A \subset \overline{A}$ for all $A \in \mathrm{Sub}(X)$;

CLT2. $\overline{\overline{A}} = \overline{A}$ for all $A \in \mathrm{Sub}(X)$;
CLT3. $\overline{A \cap B} = \overline{A} \cap \overline{B}$ for all $A, B \in \mathrm{Sub}(X)$.

The object $\overline{A}$ is called the *j-closure of A*. Although there are obvious similarities with the concept of closure operator in topology, the concepts are nonetheless formally different, as is obvious by comparing conditions C4 and CLT3.

These differences notwithstanding, the closure operator is used to define sheaves in a topos $\mathscr{E}$. A subobject A of X is said to be *dense* if $\overline{A} = X$, i.e., if its j-closure is equal to the whole object. One also says that the monomorphism $A \to X$ is a *dense monomorphism* whenever the foregoing condition is satisfied. An object F of $\mathscr{E}$ is a *sheaf* (or sometimes a *j-sheaf*) if any morphism $A \to F$ from a dense object can be uniquely extended to a morphism on the whole of X. Thus, an object F is a sheaf of $\mathscr{E}$ if the following diagram can be filled uniquely as shown:

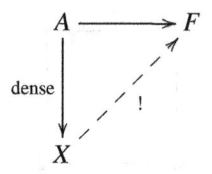

$$\begin{array}{ccc} A & \longrightarrow & F \\ \text{dense} \downarrow & \nearrow_{!} & \\ X & & \end{array}$$

The full subcategory of $\mathscr{E}$ of j-sheaves, usually denoted by $\mathrm{Sh}_j \mathscr{E}$, can be shown to be an elementary topos.

It is certainly not obvious at first sight how this characterization captures Grothendieck's notion of a sheaf. Informally, if one moves from an object X to its dense subobjects, and then from a dense subobject A to *its* dense subobjects, once starts to understand how this definition captures a process of moving from the local to the global, which is what sheaves are all about in the first place. Formally, it is possible to show that every Grothendieck topology J on a small category $\mathscr{C}$ determines a Lawvere-Tierney topology j on the presheaf topos $\mathbf{Set}^{\mathscr{C}^\circ}$. Conversely, assuming that $\mathscr{C}$ is a small category, every Lawvere-Tierney topology j on the presheaf topos $\mathbf{Set}^{\mathscr{C}^\circ}$ determines a Grothendieck topology J on $\mathscr{C}$. (See [200], 222–234.) However, it is clear that the notion of a Lawvere-Tierney topology is more general than Grothendieck's notion, since, for instance, the double negation operator $\neg\neg : \Omega \to \Omega$ is a Lawvere-Tierney topology in any elementary topos $\mathscr{E}$. It is of course this correspondence between Grothendieck toposes defined via Grothendieck topologies and elementary toposes in which sheaves can be defined via Lawvere-Tierney topologies that allows us to see why and how elementary toposes can still be thought of as generalized spaces.

It is certainly worth comparing how categories are used in both characterizations of the notion of sheaf. Whereas in the case of a Grothendieck topos, we start with a category $\mathscr{C}$ with a Grothendieck topology on it, that is a site, then consider a functor category of the form $\mathbf{Set}^{\mathscr{C}^\circ}$ and finally move to any category equivalent to the category of sheaves over that site. The language of categories is used in the notion of a Grothendieck topology and in the definition of a site. But there is a sense in which one can say that we are still very much in the ambient universe of sets, and in both definitions. That is, we are still considering functor categories into $\mathbf{Set}$ and sheaves are still defined relative to certain collections. In the case of an elementary topos, we start with a topos $\mathcal{E}$, define a topology by using the structure of $\mathcal{E}$ itself and characterize sheaves as being those objects of $\mathcal{E}$ into which a certain type of morphisms can be mapped. In this case, the whole characterization is intrinsic to categories, so to speak. Once again, one could say that the characterization of an elementary topos and of the topos of j-sheaves both lead to an *invariant* characterization, for the underlying categorical structure is made explicit.

I have not mentioned morphisms between toposes. There are, in fact, two types of morphisms of toposes: geometric morphisms and *logical* morphisms. We will first consider the category of toposes with geometric morphisms between them. In this case, one usually considers toposes *defined over a base topos* $\mathcal{S}$, which means that an object of that category is a topos $\mathcal{E}$ together with a geometric morphism $f \colon \mathcal{E} \to \mathcal{S}$ and a morphism from an object $g \colon \mathcal{F} \to \mathcal{S}$ to an object $f \colon \mathcal{E} \to \mathcal{S}$ is a geometric morphism $p \colon \mathcal{F} \to \mathcal{E}$ satisfying a certain condition. There is slightly more structure involved here, a 2-categorical structure, but we will ignore it. See [121], B3. We will restrict ourselves to the case when the base topos is the topos **Set**.

7.3 Invariants Under Geometric Transformations

Geometric morphisms are defined between elementary toposes exactly as they are for Grothendieck toposes. Geometric morphisms provide some of the most important links between topos theory and categorical logic. For one thing, the notion of a geometric theory, a concept that I have already introduced in Chapter 6, was in fact extracted from geometry. The reason for this is simple: it follows from the definition of a geometric morphism $f \colon \mathcal{F} \to \mathcal{E}$ that the inverse image functor f^* preserves finite limits, colimits and the existential quantifier, i.e., it preserves geometric logic or, put differently, any model of a geometric theory **T** in $\mathcal{E}$. (Recall that the toposes $\mathcal{E}$ and $\mathcal{F}$ are defined over **Set**.) In other words, if M is a model of a geometric theory **T** in $\mathcal{E}$, then $f^*(M)$, the inverse image of M in $\mathcal{F}$ is also a model of the geometric theory **T**, but now in $\mathcal{F}$. In more categorical terms, this means that for a geometric theory **T**, any geometric morphism $f \colon \mathcal{F} \to \mathcal{E}$ induces a functor $f^* \colon \mathrm{Mod}(\mathbf{T}, \mathcal{E}) \to \mathrm{Mod}(\mathbf{T}, \mathcal{F})$ from the category of models of **T** in $\mathcal{E}$ into the category of models of **T** in $\mathcal{F}$. In other words, geometric logic is the *invariant* part of first-order logic under geometric morphisms.

This brings us to one of the most important and original aspects of topostheoretical model theory. Given any geometric theory **T**, we have seen in Chapter 6 that it is possible to construct the category of concepts [**T**] of **T**. So far, we are still in algebra. But we can now move to geometry. Indeed, the latter category of concepts has a natural Grothendieck topology definable on it and is therefore a site. It is thus possible to construct the topos of sheaves over that site, which we will denote by **B(T)**, and the latter topos is called the *classifying topos* of the theory **T**. We have moved from a theory **T** in a certain language, to its category of concepts [**T**], its algebraic counterpart, ending up in a generalized space, a topos of sheaves over [**T**]. It is therefore possible to investigate the geometric properties of the classifying topos and apply geometrical methods in order to learn and understand properties of **T** itself. In other words, classifying toposes introduces geometrical features into logic and does so in several ways.

Once more, the term "classifying topos" comes directly from the concept of a classifying space in geometry and topology. (We should point out that the first instances of classifying toposes appeared in the sixties, in the work of Monique Hakim, then a student of Grothendieck. The general notion and its relations to logic were formulated around 1973–74 by various people, notably Joyal, Reyes, Bénabou and Tierney.) The fact is that the classifying topos **B**(**T**) of a geometric theory has exactly the same properties as the classifying topos of G-torsors. (See [200], chap. VIII, §2, for the notion of G-torsors and topos theory.) More specifically, for any cocomplete topos $\mathscr{E}$ (a cocomplete topos has all colimits), there is an equivalence of categories

$$\mathrm{Hom}(\mathscr{E}, \mathbf{B}(\mathbf{T})) \to \mathrm{Mod}(\mathbf{T}, \mathscr{E}),$$

between the category of models of **T** in $\mathscr{E}$ and the category of geometric morphisms between $\mathscr{E}$ and **B**(**T**). This equivalence is natural in $\mathscr{E}$: for any geometric morphism $f\colon \mathscr{F} \to \mathscr{E}$, the following square commutes:

$$\begin{array}{ccc}
\mathrm{Hom}(\mathscr{E}, \mathbf{B}(\mathbf{T})) & \longrightarrow & \mathrm{Mod}(\mathbf{T}, \mathscr{E}) \\
{\scriptstyle \mathrm{Hom}(f, \mathbf{B}(\mathbf{T}))} \downarrow & & \downarrow {\scriptstyle f^*} \\
\mathrm{Hom}(\mathscr{F}, \mathbf{B}(\mathbf{T})) & \longrightarrow & \mathrm{Mod}(\mathbf{T}, \mathscr{F})
\end{array}$$

It follows that there is a *universal* or *generic* model $G_\mathbf{T}$ of **T** in **B**(**T**): any model M of **T** in any topos $\mathscr{E}$ is, up to isomorphism, obtained by taking the inverse image functor $f^*(G_\mathbf{T})$ of the generic model $G_\mathbf{T}$ of **T** in **B**(**T**), i.e., $M = f^*(G_\mathbf{T})$. In particular, a set-theoretical model of the theory **T** is the inverse image of $G_\mathbf{T}$ in **Set**, the topos of sets, i.e., a geometric morphism $f\colon \mathbf{Set} \to \mathbf{B}(\mathbf{T})$. Furthermore, and this is a very important fact, the geometric sequents satisfied by $G_\mathbf{T}$ are *precisely* those provable in **T**.

We could reformulate the foregoing situation by saying that all the properties of the models of **T** *as models of* **T** are already in $G_\mathbf{T}$. Furthermore, it could be said that a property P is a property of a model *as a model of* **T** *in as much as* it is a property of the generic model $G_\mathbf{T}$. In practice, one wants to determine properties of a theory by considering properties of its classifying topos and generic model (and vice-versa). Once more, this is fundamentally geometric: the generic model is generic in the same sense that a triangle or a line is generic in geometry. Its properties are the properties it has *qua* triangle and any other property is simply irrelevant. Epistemologically, the classifying topos and the generic model constitute another way in which the essential features of a situation are revealed. Thus, in the same way that a transformation group can be said to point to what is essential and

7.3 Invariants Under Geometric Transformations

what is inessential in geometry, the classifying topos and the generic model point to what is essential and inessential in a logical theory.

This line of argument can be pushed a little further. It can be shown that, on the one hand, any geometric theory **T** has a classifying topos **B(T)** and a generic model $G_\mathbf{T}$ and, on the other hand, any Grothendieck topos $\mathscr{E}$ is equivalent to the classifying topos of some geometric theory **T**. (See [208], chap. 9 or [122], D3.) In the words of Johnstone:

> (...) we may interpret this result as saying that a Grothendieck topos is 'the same thing' as a Morita equivalence class of geometric theories – or perhaps more accurately, that the notion of Grothendieck topos represents the 'extensional essence' of the intuitive (intensional) notion of a geometric theory 'up to Morita equivalence'. ([122], D3.1, 897)

Recall that two (geometric) theories **T** and **T**′ are Morita equivalent if their corresponding conceptual categories are equivalent as categories, which, in the cases we are interested in, amounts to saying that their categories of models are equivalent. In other words, since a Grothendieck topos $\mathscr{E}$ *is* equivalent to the classifying topos **B(T)** of some geometric theory **T** and the latter constitutes, as I have just argued, the extensional essence of the theory, it follows that, up to Morita equivalence, $\mathscr{E}$ is the extensional essence of the geometric theory.

To get a better understanding of the classifying topos, it is perhaps better to consider an analogy between topos theory and ring theory that was systematically exploited by Joyal in the early seventies.[6]

Let $\mathbb{Z}$ denote the usual ring of integers and $\mathbb{Z}[X]$ denote the polynomial ring in the indeterminate X. Consider now a polynomial in X, for instance $p(X) = X^2 - X - 1$. The latter has no integer solution. There is a standard way to force the roots of this polynomial to exist in an extension of the ring $\mathbb{Z}[X]$, namely by taking the quotient ring $\mathbb{Z}[X]/(p(X))$, where $(p(X))$ denotes the ideal generated by the polynomial. The ring $\mathbb{Z}[X]/(p(X))$ contains the *generic* zero or root $G = X + (p(X))$ and it is a universal solution in the sense that there is an isomorphism $G^*\colon \mathrm{Hom}_{\mathbf{Ring}}(\mathbb{Z}[X]/(p(X)), R) \simeq \mathrm{Zero}_R(p(X))$, where, on the left-hand side, we are in the category of rings and, on the right hand side, we have the roots of $p(X)$ in R. Given these facts, the analogy becomes:

Ring Theory	Categorical Logic
Ring	Topos
Finitely presented ring	Coherent topos
$\mathbb{Z}$	**Set**
Ideal	Theory
Zero	Model
Proper ideal	Consistent theory
Generic zero	Generic model

From this point of view, the classifying topos of a theory **T** is the space **B(T)** which results from the category **Set** by forcing the existence of a generic model of **T** in the same way that the "generic" ring $\mathbb{Z}[X]/(p(X))$ results from $\mathbb{Z}[X]$ by forcing

---

[6] This was communicated to me by Gonzalo Reyes, who worked with Joyal during that period.

the existence of a generic zero of $p(X)$. The choice of terminology is not innocent: it is in the context of toposes of sheaves over a site, i.e., Grothendieck toposes, that the most general methods of forcing, in the standard logical sense of that expression, can be formulated. (See, for instance, [246].)

I will mention only one important example of a classifying topos. We have seen in Chapter 5 the category of simplicial sets, $\mathbf{Set}^{\Delta^\circ}$, which can now seen to be a Grothendieck topos. It was shown by André Joyal that it is in fact a classifying topos. It classifies the theory of linear orders with bottom and top elements, which I have described in the previous chapter. (For a proof, see [246], 21–26 or [200], 450–466.)

The classifying topos $\mathbf{B}(\mathbf{T})$ can be used to obtain various completeness theorems for theories, in particular coherent and geometric theories. Furthermore, these theorems have interesting, surprising and unexpected methodological and philosophical consequences.

Consider, for instance, the following "classical" completeness theorem for geometric theories. Let $\mathbf{T}$ be a geometric theory and α a geometric sequent (over the signature of $\mathbf{T}$). If α is satisfied in all $\mathbf{T}$-models in *Boolean* toposes, then it is provable in $\mathbf{T}$. The proof, taken from [122], D3.1, 899, follows from simple facts: it can be shown that there is, in this situation, a surjective geometric morphism $f\colon \mathscr{B} \to \mathbf{B}(\mathbf{T})$, where $\mathscr{B}$ is a Boolean topos. By assumption, the sequent α is satisfied in the $\mathbf{T}$-model $f^*(G_\mathbf{T})$. It can be shown that in this case, α is satisfied in $G_\mathbf{T}$ itself. But since sequents satisfied in $G_\mathbf{T}$ are provable in $\mathbf{T}$, the result follows at once.

One interesting consequence of the theorem is that if a geometric sequent α is derivable by classical means, then it has a *constructive* proof. In fact, by slightly strengthening the hypothesis, it is possible to conclude that if α is provable with the help of the axiom of choice, then there is a proof of α that avoids it.

Results for *coherent* theories are even more striking. Recall that a coherent language contains only symbols for finite conjunctions, finite disjunctions, the "truth", the "false" and the existential quantifier and the appropriate rules of deduction. A coherent theory $\mathbf{T}$ is given by a collection of axioms in a coherent language. I have already indicated that the conceptual category of a coherent theory is a coherent category. It can be shown that any topos is a coherent category and therefore coherent theories can be interpreted in toposes. This is a key link between first-order logic and toposes.

A Grothendieck topos equivalent to the classifying topos of a coherent theory $\mathbf{T}$ is called a *coherent topos*. (The terminology is certainly confusing. Grothendieck introduced coherent toposes in the sixties independently of any logical motivation. Joyal, while developing the categorical counterpart of first-order logic, then came up with the notion of coherent categories and coherent theories and proved, in collaboration with Reyes, that any coherent topos is equivalent to a classifying topos of a coherent theory.)

The "classical" completeness theorem for coherent logic then takes the following form. Let $\mathbf{T}$ be a coherent theory and α be a coherent sequent (over the signature of $\mathbf{T}$). If α is satisfied in all $\mathbf{T}$-models in $\mathbf{Set}$, then α is provable in $\mathbf{T}$. This is a striking result: if a sequent α is satisfied in all *set* models of the theory, then it is provable.

In other words, if α is satisfied or provable in **Set**, possibly by non-constructive means, then there is a *constructive* proof of α. It should be pointed out that the classical completeness theorem for coherent logic is equivalent, via classifying toposes, to a theorem in algebraic geometry which states that a coherent topos has enough points, sometimes known as Deligne's theorem for coherent toposes. (See [122], D3.3, 915.)

The preservation of geometric logic by geometric morphisms yields important results and opens up interesting avenues, for instance links between logic, algebraic geometry and algebraic topology. As far as mathematics is concerned, topos-theoretical ideas might provide the proper level of encoding of various important properties, just as groups encode various important properties of various algebraic, topological or geometric situations. Be this as it may, from a foundational point of view, one would also want to know whether other foundationally important structures are preserved by geometric morphisms or not. But here, the picture becomes somewhat more complicated—some would say, more interesting. For other kinds of morphisms between toposes have to be taken into account, namely logical morphisms. Once these have been defined, one can then elaborate a framework in which one can develop various parts of mathematics and investigate invariance from various angles. This setting provides a framework to find constructive proofs of (adequately translated) classical theorems of mathematics and even allows for a possible reinterpretation of Hilbert's program. One of the reasons underlying these possibilities is that the construction of the generic model does not depend on the use of the law of excluded middle nor the axiom of choice and thus one is able to prove the consistency of various theories in a relatively weak metalanguage, e.g., a constructive type theory.

7.4 Invariants Under Logical Transformations

As we have seen, an elementary topos contains all the algebraic structure encoding logical operations. In the early seventies, various mathematicians discovered that an elementary topos could be presented as a higher-order intuitionistic type theory. In fact, an elementary topos can be thought of as a higher-order intuitionistic type theory, once one knows certain facts about the theory, its interpretation and the associated conceptual category. I will first clarify these points.

Given the axioms of an elementary topos, the definition of a *logical morphism* is immediate. Let $\mathscr{E}$ and $\mathscr{F}$ be elementary toposes. A functor $F: \mathscr{E} \to \mathscr{F}$ is a *logical morphism* if it preserves all the elementary structure of $\mathscr{E}$, that is, it preserves all finite limits, the subobject classifier and the exponential, all up to isomorphism. Since all finite limits exist if and only if pullbacks and a terminal object exist, it is enough to require that F preserve these. Formally, this means that given, for instance, a product $X \times Y$ with the appropriate projections in $\mathscr{E}$, then $F(X \times Y)$ is a product in $\mathscr{F}$ and there is a unique isomorphism $F(X \times Y) = F(X) \times F(Y)$ in $\mathscr{F}$ preserving the projections. Similarly, preserving the subobject classifiers means that given the subobject classifier $\top\colon 1 \to \Omega_{\mathscr{E}}$ of $\mathscr{E}$, then $F(\top)\colon F(1) \to F(\Omega_{\mathscr{E}})$ is a subobject classifier of $\mathscr{F}$.

Thus, one can consider the (2-)category **Log** of toposes with *logical morphisms* between them.

It is natural to wonder at this stage whether the foregoing situation for geometric logic can be mimicked in the context of logical morphisms. In other words, is it possible to find an appropriate logical system L such that for any theory **T** in L, there is a topos $\mathscr{E}_{\mathbf{T}}$ and a canonical structure $M_{\mathbf{T}}$ in $\mathscr{E}_{\mathbf{T}}$ yielding an equivalence of categories $\mathbf{Log}(\mathscr{E}_{\mathbf{T}}, \mathscr{F}) \simeq \mathrm{Mod}_{\mathscr{F}}(\mathbf{T})$, where $\mathscr{F}$ is any topos and $\mathbf{Log}(\mathscr{E}_{\mathbf{T}}, \mathscr{F})$ denotes the category of *logical functors* and natural transformations between them? The answer is affirmative: there is such a logical framework.

However, before I introduce it, I have to admit that I have been sloppy in my presentation of the syntax of various logical systems so far. Because I want to interpret a formal system in a topos, I am forced to consider cases where models are empty, an assumption that is usually circumvented by fiat in standard accounts of semantics. One always *starts* with a non-empty domain of interpretation. However, if we want to interpret a formal system in a topos, we have to be careful. On the one hand, there is an "empty" object, namely the initial object 0, and, more troubling perhaps, there are toposes that have the unsettling property of having objects that are not empty without being fully inhabited. For instance, in the topos $\mathbf{Set}^2$ of pairs, the object $\langle 0, X \rangle$, where X is not empty, is not an initial object nor is it inhabited. Indeed, there is no morphism from the terminal object $\langle 1, 1 \rangle$ into it. On the other hand, empty domains of interpretation are usually rejected for good reasons. By allowing a predicate to be interpreted over an empty set, inferences that are clearly invalid become valid. Consider, for instance a unary predicate $P(x)$, vacuously interpreted. Then, clearly we have the inference $\forall x P(x) \vdash P(x)$ for it is impossible that $\forall x P(x)$ be true and $P(x)$ be false. The inference $P(x) \vdash \exists x P(x)$ is valid for similar reasons. It follows, by the transitivity of entailment, that $\forall x P(x) \vdash \exists x P(x)$, which is certainly not valid if the domain of interpretation is empty. Thus, something has to give: either we exclude empty domains, something that we do not want to do in a topos-theoretical context, or we drop the transitivity of entailment (a somewhat radical solution), or we introduce a way to keep track of empty terms in the formal system itself, so that the last entailment, for instance, will be rejected. The majority of categorical logicians have adopted the last solution.

There are numerous presentations of the type theory associated with toposes in the literature. (See [30], [160], [20], [258], [122].) I will roughly follow [160] and [122], with slight modifications in terminology.

Let us start with the syntactic framework. A *τ-signature* is given by a set S_{τ} of *sorts*. The set of types is closed under the following conditions:

1. *Basic types*: each sort is a type; the symbol Ω is a basic type;
2. *Product types*: If A and B are types, then $A \times B$ is a type; there is a distinguished type $\mathbf{1}$ (the empty product);
3. *Power types*: If A is a type, then PA is a type.

The *logical symbols* are $\top, \wedge, =, \in, \{|\}, \langle \rangle$.

7.4 Invariants Under Logical Transformations

Terms, together with their types, are defined recursively as usual, except that we keep track of free variables in them. To indicate that a term t has type A, we write $t\colon A$ and the finite set of free variables of a term t is denoted by $\mathrm{FV}(t)$.

1. For each type A, there is an infinite set of variables $x\colon A$; $\mathrm{FV}(x) = \{x\}$.
2. There is a distinguished term $*\colon 1$ with $\mathrm{FV}(*) = \emptyset$.
3. If $t_1\colon A$ and $t_2\colon B$, then $\langle t_1, t_2 \rangle \colon A \times B$; $\mathrm{FV}(\langle t_1, t_2 \rangle) = \mathrm{FV}(t_1) \cup \mathrm{FV}(t_2)$.
4. If $t\colon A \times B$, then $\mathrm{fst}(t)\colon A$ and $\mathrm{snd}(t)\colon B$; $\mathrm{FV}(\mathrm{fst}(t)) = \mathrm{FV}(\mathrm{snd}(t)) \cup \mathrm{FV}(t)$.
5. If φ is of type Ω and $x\colon A$, then $\{x \in A \mid \varphi(x)\}\colon PA$; $\mathrm{FV}(\{x \in A \mid \varphi(x)\}) = \mathrm{FV}(\varphi) \setminus \{x\}$.

Whenever the type of x is clear, we will write $\{x \mid \varphi\}$ instead.

There are two kinds of atomic formulas.

1. If t_1 and t_2 are of the same type A, then $t_1 =_A t_2$ is an atomic formula, thus of type Ω and $\mathrm{FV}(t_1 = t_2) = \mathrm{FV}(t_1) \cup \mathrm{FV}(t_2)$.
2. If $t_1\colon A$ and $\alpha\colon PA$, then $(t_1 \in \alpha)$ is an atomic formula; $\mathrm{FV}(t_1 \in \alpha) = \mathrm{FV}(t_1) \cup \mathrm{FV}(t_2)$.

Complex formulas are defined recursively in the usual manner.

1. If φ and ψ are formulas, then $(\varphi \wedge \psi)$ is a formula; $\mathrm{FV}(\varphi \wedge \psi) = \mathrm{FV}(\varphi) \cup \mathrm{FV}(\psi)$.
2. $\top$ is a formula; $\mathrm{FV}(\top) = \emptyset$.

As I have mentioned implicitly, formulas are in fact terms of type Ω. The reader will have noticed the absence of the remaining logical operators. In fact, they can all be defined from our basic set of connectives and terms. Although this is certainly not a reasonable way to proceed from a practical point of view, from a theoretical point of view, it shows that a purely algebraic presentation of a higher-order logic is feasible, an interesting result in its own right. Here are the definitions introducing the missing connectives:

D1 $\forall x \varphi \equiv [\{x \mid \varphi\} = \top]$.
D2 $\varphi \to \psi \equiv [\varphi \wedge \psi = \varphi]$.
D3 $\varphi \vee \psi \equiv \forall w[(((\varphi \to w) \wedge (\psi \to w)) \to w) = \top]$.
D4 $\bot \equiv \forall w (w = \top)$.
D5 $\exists x \varphi \equiv \forall w[(\forall x(\varphi \to w) \to w) = \top]$.
D6 $\neg \varphi \equiv \varphi \to \bot$.

Before I introduce the rules of the calculus in terms of sequents, I have to clarify the notion of a suitable context for a sequent. The latter notion is one way to preserve the transitivity of entailment and the possibility of empty types.

A *context* is a finite list $\mathbf{x} = x_1, \ldots, x_n$ of distinct variables. The *type* of a context $\mathbf{x}$ is the string of not necessarily distinct sorts of the variables occurring in it. The *empty context* is simply the case when there is no variable. If $\mathbf{x}$ is a context and y is a variable different from those occurring in $\mathbf{x}$, then $\mathbf{x}, y$ denotes the context obtained by appending y to the list $\mathbf{x}$. A context $\mathbf{x}$ is *suitable* for a formula φ if all free variables of φ occur in $\mathbf{x}$. Finally, a *sequent* over a τ–signature is an expression of the form $\varphi \vdash_\mathbf{x} \psi$, where φ and ψ are formulas over τ and **x** is a context suitable for both formulas. It is now easy to verify that in this way, the entailment relation is still transitive and that problems arising from empty types are avoided.

We can now present the rules of a τ-calculus.

1. Structural rules:

$$\varphi \vdash_\mathbf{x} \varphi \qquad \frac{\varphi \vdash_\mathbf{x} \psi \quad \psi \vdash_\mathbf{x} \chi}{\varphi \vdash_\mathbf{x} \chi}$$

$$\frac{\varphi \vdash_\mathbf{x} \psi}{\varphi \vdash_{\mathbf{x},y} \psi} \qquad \frac{\varphi(y) \vdash_{\mathbf{x},y} \psi(y)}{\varphi(b) \vdash_\mathbf{x} \psi(b)}$$

where, in the last rule, the substitution rule, y is a variable of type B and b is a term of type B with no free occurrences of variables other than those in the context. It is assumed that b is substitutable for x, with the usual proviso.

2. Logical rules

$$\varphi \vdash_\mathbf{x} \top \qquad \frac{\varphi \vdash_\mathbf{x} \psi \quad \varphi \vdash_\mathbf{x} \chi}{\varphi \vdash_\mathbf{x} \psi \wedge \chi}$$

$$\frac{\varphi \vdash_\mathbf{x} \psi \wedge \chi}{\varphi \vdash_{\mathbf{x},y} \psi} \qquad \frac{\varphi \vdash_\mathbf{x} \psi \wedge \chi}{\varphi \vdash_\mathbf{x} \chi}$$

3. Equality rules:

$$\top \vdash_\mathbf{x} x = x \qquad \varphi \wedge (x = t) \vdash_\mathbf{y} \varphi(t/x)$$

where, in the last rule, x and t are of the same type, **y** is any context containing x and the free variables of t, and t is substitutable for x in φ.

4. Other axioms

$$\top \vdash_w (w = \{x : A \mid x \in w\}), \text{ where } w : PA;$$

$$z \in \{y : A \mid \varphi\} \vdash_{\mathbf{x},z} \varphi(z/y) \qquad \varphi(z/y) \vdash_{\mathbf{x},z} z \in \{y : A \mid \varphi\}$$

where φ is a formula with free variables in the context **x**, *y*.

$$\top \vdash_\mathbf{x} x =_1 *, \text{ where } x : 1;$$

$$\top \vdash_{x,y} \text{fst}(\langle x, y \rangle) =_A x \qquad \top \vdash_{x,y} \text{snd}(\langle x, y \rangle) =_B x)$$

$$\top \vdash_z \langle \text{fst}(z), \text{snd}(z) \rangle =_{A \times B} z,$$

This completes the description of the τ-calculus. A τ-theory **T** over a τ-signature is given by a set of sequents over τ and these constitute the axioms of **T**. It should also be clear that the notion of a model M of a τ-theory **T** in a topos $\mathscr{E}$ can be defined. The soundness of the foregoing system is then easily proved.

It is also possible to construct the category of concepts $\mathscr{E}_\mathbf{T}$ from a τ-theory **T** and show that it is an elementary topos. And again, it is possible to define a canonical model $M_\mathbf{T}$ in $\mathscr{E}_\mathbf{T}$ with the property that a sequent $\varphi \vdash_\mathbf{x} \psi$ in the τ-signature is satisfied in $M_\mathbf{T}$ if and only if it is provable in **T**. Completeness follows: given a τ-theory **T**

7.4 Invariants Under Logical Transformations

and a sequent $\varphi \vdash_x \psi$ of **T**, then if $\varphi \vdash_x \psi$ is satisfied in all models of **T** in toposes, it is provable in **T**.

Let $\mathscr{F}$ be an elementary topos. We can consider the category of models $\mathrm{Mod}_{\mathscr{F}}(\mathbf{T})$ of **T** in $\mathscr{F}$ but with *isomorphisms* between them (thus, it is a groupoid). Morphisms are restricted to isomorphisms, since even if we are given morphisms between interpreted sorts, they will not give rise to morphisms between all interpreted *types*, in particular power types. Thus, suppose we have for each sort A and two interpretations of the sorts, MA and NA in a topos $\mathscr{F}$, morphisms $MA \to NA$. Consider now the passage from the pair (A,B) to the power type B^A. The latter construction is *contravariant* in the first variable: thus a morphism $MA \to NA$ gives rise to a morphism $NB^{NA} \to MB^{MA}$ in a natural way, but the latter goes in the wrong direction. We therefore do not get morphisms between all interpreted types in the right way. (We are assuming that $MB^{MA} = M(B^A)$.) By restricting morphisms to isomorphisms, we get around that difficulty.

It can be shown that any *logical* functor $\mathscr{E} \to \mathscr{F}$ gives rise to a functor $\mathrm{Mod}_{\mathscr{E}}(\mathbf{T}) \to \mathrm{Mod}_{\mathscr{F}}(\mathbf{T})$. We now get to the main result: for any topos $\mathscr{F}$ and any τ-theory **T**, there is an equivalence of categories $\mathbf{Log}(\mathscr{E}_\mathbf{T}, \mathscr{F})$. Thus, an $\mathscr{F}$-model of **T** corresponds to a logical isomorphism from $\mathscr{E}_\mathbf{T}$ to $\mathscr{F}$ and conversely. (See [122], D4.3, 970.) Once again, we can go in the opposite direction, meaning that given a (small) topos $\mathscr{E}$, there exists a τ-theory $\mathbf{T}_{\mathscr{E}}$, called the *theory* of $\mathscr{E}$, such that for any topos $\mathscr{F}$, there is an equivalence $\mathbf{Log}(\mathscr{E}, \mathscr{F}) \simeq \mathrm{Mod}_{\mathscr{F}}(\mathbf{T}_{\mathscr{E}})$. In words, a logical isomorphism from $\mathscr{E}$ to $\mathscr{F}$ corresponds to an $\mathscr{F}$-model of $\mathbf{T}_{\mathscr{E}}$ and conversely. The theory $\mathbf{T}_{\mathscr{E}}$ is constructed in the obvious way. The τ-signature is defined as follows: the types are the objects of $\mathscr{E}$. In particular, the type Ω is the object $\Omega_{\mathscr{E}}$ of $\mathscr{E}$. We can therefore define a language $L_{\mathscr{E}}$, called the *internal language of the topos* $\mathscr{E}$, based on this signature. Furthermore, there is an obvious canonical interpretation of that language in $\mathscr{E}$. The theory $\mathbf{T}_{\mathscr{E}}$ is the set of sequents in $L_{\mathscr{E}}$ that are satisfied under the canonical interpretation. (See [122], D4.2, 947, 956 and 972 for details. See also [160], 189–205 for a slightly different presentation.)

Thus, starting with a (small) topos $\mathscr{E}$, it is possible to construct $\mathbf{T}_{\mathscr{E}}$ from $\mathscr{E}$ and to construct $\mathscr{E}_{\mathbf{T}_{\mathscr{E}}}$, the category of concepts of $\mathbf{T}_{\mathscr{E}}$. It can be shown that there is an *equivalence* of categories $\mathscr{E} \simeq \mathscr{E}_{\mathbf{T}_{\mathscr{E}}}$. Starting instead with a τ-theory **T**, moving to its category of concepts $\mathscr{E}_\mathbf{T}$ and finally to its internal theory $\mathbf{T}_{\mathscr{E}_\mathbf{T}}$, one can show that **T** and $\mathbf{T}_{\mathscr{E}_\mathbf{T}}$ are Morita-equivalent, that is there are natural bijections between the isomorphisms classes of models of the theories in arbitrary toposes.

The foregoing results can be interpreted as saying that a (small) topos is essentially the same thing as a higher-order intuitionistic type theory. The logical and the geometrical sides are in this sense two faces of the same coin. How can we not transpose once more Cartan's claim linking the structure of a group to the underlying logic of a geometry? A paraphrase would give something like this: it is the whole logical structure of a higher-order type theory which is contained in the structure of the associated topos and even, in a more precise manner, in the law according to which morphisms of that topos compose with each other, *independently of the nature of the objects on which these operations act*. This resembles the case of geometry again where one can associate a certain group of transformations to a geometric theory, e.g., Euclidean geometry.

It should be emphasized that the internal language of a topos is not merely an alternative presentation of a topos; it is very often useful and yields simpler, more direct proofs of various results. (See, for various examples, [160], 148–160, [223], 126–132, and, for examples also using classifying toposes, [122], D4.1, 948–951.)

However, for this language to be considered adequate as a foundational framework, it is unquestionable that it ought to contain a type for natural numbers, or, equivalently, a type for an infinite list. Thus, we add to a τ-signature another basic type, denoted by $\mathbb{N}$, a new term, $0\colon \mathbb{N}$, a term-forming operation (namely if n is a term of type $\mathbb{N}$, then Sn is also a term of type $\mathbb{N}$), together with the following axioms (we now use the full language of the theory). The resulting signature will be called a τN-signature.

$$\top \vdash_x \neg(Sx = 0)$$
$$Sx = Sy \vdash_{x,y} x = y$$
$$(0 \in z) \wedge (\forall x)(x \in z \to Sx \in z) \vdash_z (\forall y)(y \in z)$$

where $x\colon \mathbb{N}$, $y\colon \mathbb{N}$ and $z\colon P\mathbb{N}$.

Now that there is a type of natural numbers, it is possible to construct other number systems in the standard algebraic fashion, for instance integers and rationals. The case of real numbers is subtler since we are in a constructive framework. Be that as it may, it seems reasonable to think of a τN-theory as a constructive foundational formal system.

All the foregoing concepts and results for τ-theories can be extended to τN-theories. τN-theories are sound with respect to toposes with natural number objects; given a τN-theory **T**, the category of concepts $\mathscr{E}_\mathbf{T}$ is a topos with a natural number objects containing a canonical structure $M_\mathbf{T}$ and completeness follows as in the previous cases; logical functors preserve natural number objects and therefore we have, as before, that for any τN-theory and any topos $\mathscr{F}$ with a natural number object, there is an equivalence $\mathbf{Log}(\mathscr{E}_\mathbf{T}, \mathscr{F})$; finally, given a small topos $\mathscr{E}$ with a natural number object, there is τN-theory $\mathbf{T}_\mathscr{E}$ such that every model of $\mathbf{T}_\mathscr{E}$ in a topos $\mathscr{F}$ with a natural number object corresponds to a logical functor $\mathscr{E}_\mathbf{T} \to \mathscr{F}$.

7.5 Invariant Foundational Frameworks

Consider the topos $\mathscr{E}_\mathbf{T}$ in the case when **T** is the empty or "pure type" theory over the "empty signature", that is the set $S_{\tau N}$ of sorts is empty. However, although the set of sorts is empty, there are still basic types, namely $1, \Omega$ and $\mathbb{N}$. Thus **T** nevertheless has many types: $1, P1, PP1, \ldots, P1 \times P1, \ldots, \mathbb{N}, P\mathbb{N}, PP\mathbb{N}, \mathbb{N} \times P\mathbb{N}, \ldots$ The topos $\mathscr{E}_\mathbf{T}$ is said to be freely-generated by the **T**-model $M_\mathbf{T}$ and it is called the *free topos with natural number object*. Interestingly enough, the free topos $\mathscr{E}_\mathbf{T}$ is the *initial object* in the (2-)category **Log**. This means that there is a *unique logical*

7.5 Invariant Foundational Frameworks

functor from $\mathscr{E}_\mathbf{T}$ to any topos $\mathscr{F}$ in **Log**. Like any initial object, $\mathscr{E}_\mathbf{T}$ is unique (up to a unique equivalence). Going back to the analogy with ring theory, the topos $\mathscr{E}_\mathbf{T}$ is to the category **Log** what the ring $\mathbb{Z}$ of integers is to the category of rings (with unit). Thus, in this sense, $\mathscr{E}_\mathbf{T}$ is the invariant logical content of all higher-order intuitionistic theories. This has been used by Lambek to argue that $\mathscr{E}_\mathbf{T}$ could be considered the appropriate foundational framework for the constructive part of arithmetic and analysis at least. He also argued that it could satisfy the moderate Platonist, the moderate formalist and the moderate logicist, thus providing a common ground for what he sees as constituting the standard positions in the foundations of mathematics. (See [53], [158], [159], [161].)

In fact, the free topos satisfies additional properties that an intuitionist should find appealing. First, Lambek argues that any topos $\mathscr{E}$ worthy of the name "universe of mathematics" should satisfy three properties, namely:

1. (*Consistency*) The topos $\mathscr{E}$ is not degenerate, i.e., the atomic formula $0 = 1$ is not true in $\mathscr{E}$, equivalently $\bot$ is not true in $\mathscr{E}$;
2. (*Disjunction property*) If $p \vee q$ is true then either p is true or q is true;
3. (*Existence property*) If $\exists x \varphi(x)$ is true and $x: A$, then $\varphi(a)$ is true for some object a of type A.

Lambek calls any topos satisfying these conditions a *local topos*. It can be shown that the free topos $\mathscr{E}_\mathbf{T}$ is a local topos. Recall that since provability and truth coincide in $\mathscr{E}_\mathbf{T}$, conditions 2 and 3 then become:

2. If $\vdash p \vee q$ then either $\vdash p$ or $\vdash q$;
3. If $\vdash \exists x \varphi(x)$ and $x: A$, then there is a term $a: A$ such that $\vdash \varphi(a)$.

Both properties are demonstrable properties of $\mathscr{E}_\mathbf{T}$, and since $\mathscr{E}_\mathbf{T}$ is non-degenerate by construction, it satisfies (1). These properties were shown to hold in an intuitionistic framework, the first one by Gödel and the second one by Kleene. Needless to say, these logical properties correspond to purely categorical properties of the topos, which are, informally, that the terminal object 1 is not initial, that 1 is indecomposable and, finally, that 1 is projective. (See [160], 229–230 for proofs.)

Furthermore, the free topos has another relevant property, namely that all numerals are *standard*, in the sense that all arrows $1 \to \mathbb{N}$ have the form $S^n 0$ for some natural number n. (This is not necessarily true in an arbitrary topos.)

It can also be shown that other well-known constructivist principles, which an intuitionist might not find desirable, are provable in the free topos. I will mention only two.

4. (Markov's principle) If $\vdash \forall x(\varphi(x) \vee \neg \varphi(x))$ and $\vdash \neg \forall x \neg \varphi(x)$, then $\vdash \exists x \varphi(x)$, where $x: A$;
5. (Independance of premises) If $\vdash \neg p \to \exists x \varphi(x)$, then $\vdash \exists x(\neg p \to \varphi(x))$, where $x: A$.

Thus, according to Lambek, some constructivists might consider the foregoing pure type theory **T** as being an adequate formal system for mathematics and $\mathscr{E}_\mathbf{T}$ a proper universe of mathematics, thus as constituting together an appropriate constructive foundations for (elementary) mathematics. In his own words:

> (...) the free topos is a suitable candidate for *the* world of mathematics acceptable to members of different philosophical schools, who do not insist on the principle of excluded middle and who are willing to compromise: (a) to moderate Platonists, because it is an initial object in the category of all toposes; (b) to moderate formalists or even nominalists, because it may be constructed from words (...); (c) to moderate intuitionists, because (...) [it] is an intuitionistic type theory (...) ([159], 153–154)

However, as Lambek is himself aware, Platonists, formalists, and intuitionists all have to compromise.

> Needless to say, such a compromise will be rejected by extreme Platonists, who may believe that mathematical entities are thoughts in the mind of a demiurge, by extreme nominalists, who may believe that only words are real, but that equivalence classes of synonymous words are not, and by extreme intuitionists, who may believe that infinite sets do not exist or that truth varies with historical time. ([159], 154)

Lambek mentions the assumption of the existence of the type $\mathbb{N}$ as being objectionable. More damaging to the constructivist, perhaps, is the presence of the power-type constructor PA for any type A, a questionable construction in and of itself that furthermore leads directly to impredicative definitions. A predicative type theory à la Martin-Löf, which has models in locally Cartesian closed categories, might constitute a more palatable framework. I will not present this framework here. (But see, for instance, [122], D4.4, or [251].)

Two important but seldom-mentioned aspects of the argument presented by Lambek have to be underlined. First, it is suggested that the pure type theory $\mathbf{T}$ could be an appropriate formal system for constructive mathematics. It would thus be the language of constructive mathematics. Its universe of interpretation would then be given by the free topos $\mathscr{E}_\mathbf{T}$. But the latter is *constructed* from $\mathbf{T}$ in an ambient theory. Thus, the construction of the universe depends on an underlying theory. Although the latter is very weak and constructive, it has to be kept in mind. The second point is more fundamental. The fact that $\mathscr{E}_\mathbf{T}$ is an *initial object* in the category **Log** is taken to be an important, if not crucial, property of the universe and, as a consequence, of $\mathbf{T}$ itself. However, that feature does not make sense without referring to the category **Log** and its properties. The main point I want to emphasize at this juncture is that Lambek sometimes argues as if $\mathbf{T}$ and $\mathscr{E}_\mathbf{T}$ can be taken together in an isolated fashion and considered as a foundational framework. But if being initial in **Log** is a crucial premiss of the argument in favor of $\mathbf{T}$ and $\mathscr{E}_\mathbf{T}$ as a foundational framework, then one cannot ignore the fact that the universe is unique up to a unique equivalence in a larger universe of universes. From our point of view, it is its *invariance* as an initial object that constitutes the key feature of $\mathscr{E}_\mathbf{T}$.

Of course, if classical logic is thought to be required in a foundational framework, then it is enough to add the axiom

$$\mathbf{T} \vdash \forall p\colon \Omega(\neg\neg p \to p)$$

or, equivalently, the law of excluded middle

$$\mathbf{T} \vdash \forall p\colon \Omega(p \vee \neg p)$$

7.5 Invariant Foundational Frameworks

to the foregoing τN-calculus. We now get *classical* τN-theories and given such a classical theory **T**, the topos $\mathscr{E}_T$ can be constructed in exactly the same manner as before. However, the latter topos has new properties.

A topos $\mathscr{E}$ is said to be *Boolean* if its internal language is classical. Equivalently, a topos $\mathscr{E}$ is Boolean if and only if one of the following conditions hold:

1. Every subobject in $\mathscr{E}$ has a complement, that is for every subobject $A \rightarrowtail X$ in $\mathscr{E}$, $A \vee \neg A = X$;
2. For every object X of $\mathscr{E}$, $\text{Sub}(X)$ is a Boolean algebra;
3. The morphism $\neg\neg \colon \Omega \to \Omega$ is the identity $1_\Omega \colon \Omega \to \Omega$;
4. The morphisms $\top \colon 1 \to \Omega$ and $\neg \circ \top \colon 1 \to \Omega$ induce an isomorphism $\Omega \simeq 1 + 1$.

It can be shown that any topos $\mathscr{E}_T$ constructed from a classical τN-theory **T** is Boolean.

The (2-)category **BoolLog** of Boolean toposes and logical morphisms between them is defined in the obvious manner. Starting with the pure type theory $\mathbf{T}_0$, it is possible to construct the free Boolean topos $\mathscr{E}_{\mathbf{T}_0}$, which is initial in the category **BoolLog**.

Can we proceed to soundness, completeness and all other results as we did in the constructive case? At this stage, the situation becomes somewhat more complicated, for Gödel's first incompleteness theorem rears its head.[7]

Completeness results for classical type theories have been known since the basic work of Henkin. Moreover, Henkin assumed that a model of a type theory had to be local. (Since the free topos for a constructive type theory *is* always local, there was no need to impose this condition on it.) Lambek and Scott follow suit. They therefore consider *local Boolean toposes* as legitimate models of classical τN-theories. A topos is Boolean local if and only if the following properties hold:

1. (*Consistency*): $\bot$ is not true;
2. (*Universal property*): If $\varphi(x)$ in the τN-signature of **T** is such that $\vdash \varphi(a)$ for all closed terms $a \colon A$, then $\vdash \forall x \varphi(x)$, i.e., it is a theorem of **T**.

It follows that in a Boolean local topos, the logical connectives are interpreted in the usual informal manner. Thus, (1) and (2) imply the disjunction and the existential properties and, furthermore:

3. (*Negation*) $\neg p$ is true if and only if p is not true;
4. (*Implication*) $p \to q$ is true if and only if, if p is true, then q is true.

These properties translate accordingly into categorical properties. In fact, Boolean local toposes are equivalent to well-pointed toposes. Let us first unpack the definition of being well-pointed.

---

[7] This might be taken as being a good reason *not* to include natural number object as part of the definition of elementary topos: it excludes the topos $\mathscr{S}_f$ of (hereditarily) finite sets. The latter is also a free Boolean topos, but without a natural number object.

A topos $\mathscr{E}$ is *well-pointed* if and only if its terminal object 1 is a generator (sometimes called a *separator*); that is, given two parallel morphisms $X \underset{g}{\overset{f}{\rightrightarrows}} Y$ in $\mathscr{E}$, either $f = g$ or there is a morphism $x \colon 1 \to X$ such that $f \circ x \neq g \circ x$. This condition is in fact equivalent to the condition (2) of a Boolean local topos. What it says is that the "elements" of X allow us to distinguish maps that are different in the universe. Thus it is enough to know the elements of the sets to know the universe.

It can be shown that any well-pointed topos is Boolean and two-valued; that is, there are only two morphisms into the subobject classifier Ω, namely $\top \colon 1 \to \Omega$ and $\bot \colon 1 \to \Omega$, or equivalently the terminal object 1 has only two subobjects, namely $0 \rightarrowtail 1$ and $1 \rightarrowtail 1$. (See [223], 211, for proofs.) Notice that being Boolean is *not* the same as being two-valued: a two-valued topos need not be Boolean and conversely a Boolean topos need not be two-valued. (See [200], 274.)

The main point is that a well-pointed topos behaves very much like a category of sets, in the standard Zermelo conception of a set. As we have just seen, it is Boolean and two-valued, every object or "set" is either initial or has global elements (which is certainly not the case in an arbitrary topos), being monic and epic can be defined in the usual manner on global elements. Moreover, in any well-pointed topos, supports split, that is if $X \neq 0$, then the unique map $!_X \colon X \to 1$ splits, which in turn means that there is a morphism $s \colon 1 \to X$ such that $!_X \circ s = 1$. (This latter claim is a special case of the categorical version of the axiom of choice. The full version of the axiom of choice states that any epimorphism has a section.) In fact, a well-pointed topos can be thought of as a model of Zermelo set theory with bounded comprehension, sometimes called restricted Zermelo set theory, Z^- set theory or Mac Lane set theory. (Bounded comprehension simply means that the quantified variables are typed. See [200], 332–343 for the proof that the theory of restricted Zermelo with choice is equiconsistent with the theory of well-pointed toposes with natural number object and choice.) Thus, every Boolean local topos is very much like a restricted Zermelo set theory. The basic idea is that the sets of such toposes should in some sense be completely determined by their points, or elements. After all, this is the fundamental property of standard sets: a set in the traditional conception is entirely determined by its elements and two sets are identical if and only if they have the same elements. The set-theoretical criterion of identity is radically different from the criterion of identity with which we have been working so far.

The surprise here is that the classical free topos $\mathscr{E}_{\mathbf{T}_0}$ is *not* local. (However, given the foregoing connection with a restricted Zermelo set theory, this is not so surprising after all.) The reason is simple: Gödel's first incompleteness theorem creeps in. Indeed, it is possible to reproduce Gödel's argument within $\mathbf{T}_0$ and thus obtaining a sentence G such that $\vdash G \vee \neg G$. If $\mathscr{E}_{\mathbf{T}_0}$ were local, it would satisfy the disjunction property and, thus we would have $\vdash G$ or $\vdash \neg G$, which is impossible by Gödel's theorem. Needless to say, the argument works for any undecidable sentence p.

From Lambek's perspective, the fact that $\mathscr{E}_{\mathbf{T}_0}$ fails to be local prevents it, in contrast with the constructive case, from being considered *the* universe of classical mathematics. Indeed, Lambek *defines* a model of a τN-theory to be a local topos. It is important to note immediately that, in the classical case, Lambek supposes

7.5 Invariant Foundational Frameworks

that one would like to identify the topos $\mathscr{E}_{T_0}$ with *the* universe of sets, since he assumes that mathematicians endorsing classical set theory believe that there *is* such a *unique* set-theoretical universe. (This is not to say that Lambek himself endorses such a view.)

If the goal is to try to find a *unique* universe of classical mathematics, then it seems that the topos-theoretical framework fails. Lambek does consider an alternative:

> It is tempting to abandon the search for a distinguished Boolean local topos and be satisfied with the sheaf of Boolean local toposes instead. ([159], 156)

Why would it be an alternative? In short, because of the following representation theorem: any small Boolean topos is equivalent to the topos of global sections of a sheaf of Boolean local toposes.

And we are back to sheaves. In fact, we are back to the analogy with rings. Indeed, starting with Grothendieck's representation theorem that asserts that every commutative ring is the ring of continuous sections of a sheaf of local rings, it becomes reasonable to ask whether a similar result could not be proved for toposes. After promising but not quite satisfactory results by Lambek and then by Lambek and Moerdijk, Awodey proved the completely analogous result stated in the foregoing paragraph. ([6])

We will concentrate on the meaning of the result, without entering fully into all the technical details. Let $\mathscr{E}$ be a small topos. We first construct a functor[8] $\mathscr{E}/: \mathscr{E}^\circ \to$ **CAT** into the category **CAT** of (possibly large) categories as follows: for each object X of $\mathscr{E}$, $(\mathscr{E}/)(X) =_{df} \mathscr{E}/X$, the slice topos and for each morphism $f: X \to Y$ in $\mathscr{E}$, we obtain a functor by pulling back along f. Composition can also be defined.[9] Notice that the functor $\mathscr{E}/$ can be thought of as a presheaf of *categories*. So far, we had only considered presheaves and sheaves of sets. There is absolutely nothing in the definition of a presheaf or of a sheaf that fixes the domain category. In so far as the category **CAT** is legitimate, the whole construction makes perfect sense. What Awodey shows first is that the functor $\mathscr{E}/$ is in fact equivalent to a sheaf of categories on $\mathscr{E}$. Thus, the small topos $\mathscr{E}$ is now considered as a site. (I will not specify the Grothendieck topology. Suffice it to say that it is the so-called finite epimorphism topology.) The Grothedieck topos $\mathrm{Sh}(\mathscr{E})$ of sheaves over $\mathscr{E}$ can be constructed. It can be shown that the topos $\mathrm{Sh}(\mathscr{E})$ has enough points and, by a result of Butz and Moerdijk, that for any such Grothendieck topos $\mathscr{E}$, there is a topological space $X_\mathscr{E}$ and a *geometric* morphism $\phi: \mathrm{Sh}(X_\mathscr{E}) \to \mathscr{E}$ (with additional properties). Thus, in this particular case, it can be shown that one obtains a geometric morphism $\phi: \mathrm{Sh}(X_\mathscr{E}) \to \mathrm{Sh}(\mathscr{E})$ and that $\mathrm{Sh}(X_\mathscr{E})(1,\tilde{\mathscr{E}}) \simeq \mathscr{E}$, where $\mathrm{Sh}(X_\mathscr{E})(1,\tilde{\mathscr{E}})$ is the topos of global sections of $\tilde{\mathscr{E}}$ on $X_\mathscr{E}$ and $\tilde{\mathscr{E}}$ denotes the inverse image of the sheaf equivalent to $\mathscr{E}/$ in $\mathrm{Sh}(X_\mathscr{E})$. Thus, the representation theorem, in its general form, states that any small topos $\mathscr{E}$ is equivalent to the topos of global sections of a sheaf of

---

[8] In fact, it is a pseudofunctor. Part of the proof of Awodey's result consists in showing that it is equivalent to a functor.

[9] The same remark appplies here. One has to show that the result is indeed a genuine composition, that is, not only defined up to a natural isomorphism.

local toposes on a topological space. Notice that this is a particular case of Joyal and Tierney's result: in this case we are dealing with a standard topological space and not a locale. Whenever $\mathscr{E}$ is a small Boolean topos, then the representation theorem asserts that $\mathscr{E}$ is equivalent to the topos of global sections of a sheaf of well-pointed toposes on a topological space.

If we start with a classical τN-theory **T**, we can construct its corresponding Boolean topos $\mathscr{S}_\mathbf{T}$ and, by the representation theorem, fix a sheaf representation $\tilde{\mathscr{S}}_\mathbf{T}$ of $\mathscr{S}_\mathbf{T}$ on a space X. As Freyd had already shown in the early seventies, there is then a faithful logical morphism

$$\langle \pi_x \rangle_{x \in X} : \mathscr{S}_\mathbf{T} \to \prod_{x \in X} (\tilde{\mathscr{S}}_\mathbf{T})_x$$

from the given Boolean topos into a product of well-pointed toposes. Since we still have that

$$\mathbf{Log}(\mathscr{S}_\mathbf{T}, \mathscr{F}) \simeq \mathrm{Mod}_\mathscr{F}(\mathbf{T}),$$

to every logical morphism $\mathscr{S}_\mathbf{T} \to \mathscr{F}$ corresponds a model $M_\mathbf{T}$ of **T** in $\mathscr{F}$. In particular, for each point $x \in X$, there is a model $\pi_x(M_\mathbf{T}) \in \mathrm{Mod}_{(\tilde{\mathscr{S}}_\mathbf{T})_x}(\mathbf{T})$. Now, given a **T**-sentence φ such that φ is true in every Boolean topos $\mathscr{F}$, in particular φ is true in every $\pi_x(M_\mathbf{T})$, and since $\langle \pi_x \rangle_{x \in X}$ is faithful, φ is true in $\mathscr{S}_\mathbf{T}$ and therefore, it is provable in **T**.

Informally, the models of a classical theory are the points of a topological space and the truth of a sentence of the theory varies continuously from point to point. In other words, this means that a sentence in the language of a classical theory **T** is provable if it is true in every **T**-model in every model of bounded Zermelo set theory. We might not be able to get *the* universe of sets, that is a fixed topos of sets, but completeness is attainable as long as these categories of sets vary, as long as we have variation in a space of categories of sets.

Notice, once again, that the metamathematics is taking place in a category of categories, namely a category of toposes with logical and/or geometric morphisms. These categories are always lurking in the background and they seem to be inescapable. Furthermore, the whole notion of invariance is *relative* to the context in which these categories are defined.

7.6 Using Geometric and Logical Invariants

As early as 1975, André Joyal suggested that the methods based on the notion of the classifying topos are reminiscent of Hilbert's ideas on the introduction and the elimination of ideal elements in mathematical reasoning, usually known as Hilbert's program. The basic idea is to find a geometric or coherent formulation of a given mathematical theory **T**, move to the category of concepts of **T** and the classifying topos of the theory. Then, by the completeness results for geometric and coherent theories, if there is a proof using classical logic and/or the axiom of choice, there is a constructive proof of the same result. The construction of the classifying topos amounts to the introduction of ideal elements that are then eliminated via the completeness results. Although my presentation makes it sounds as if it does not in fact deliver a concrete constructive proof of the results, in practice, such proofs have been found. (For a survey of such results in algebra, see for instance [50].)

As I have just said, the various theorems have to be "translated" adequately; that is, an appropriate version of the theorem has to be found and then it is possible to obtain a constructive proof of the new version. In many cases, these new versions become the standard version when they are interpreted in the topos of sets, or, in the case of topological spaces, when they are restricted to the correct class of topological spaces. It should also be mentioned that the proofs depend on an interplay between logical tools, geometrical tools and finally, categorical tools. They constitute perfect illustrations of the multifaceted aspect of topos theory. Here is a sample of some of the results that now have a constructive proof:

1. Tychonoff's theorem asserts that a product of compact spaces is compact. The standard proof depends on the axiom of choice. There are proofs that avoid the use of the axiom of choice. (See, for instance, [118] and [123].)
2. Tychonoff's theorem in turn implies the existence of the so-called Stone-Cech compactification for certain spaces X, which, in a sense, is the maximal compactification of a space relative to a given embedding. Various constructive proofs of this theorem can also be given. (See [10], [11], [13].)
3. The Stone-Weierstrass theorem can be proved constructively. (See [12].)
4. Gelfand duality can be given a contructive treatment. (See [14].)
5. The Hahn-Banach theorem in functional analysis is also known to depend on the axiom of choice. Properly reformulated, a constructive version of the theorem can also be given. (See [231].)

This is but a sample of an active field of research. Notice that in this case, topos-theoretical methods are heuristic: they are used to find constructive proofs of results that are known to depend on various parts of mathematics that are contentious from a constructive point of view.

7.7 Summing Up

All the themes touched upon in previous chapters find an echo in topos theory. It is intrinsically geometric, topological, algebraic and logical. A topos can be conceived of as a universe of sets, as a higher-order type theory, as a space or as an algebraic structure. It provides the means to apply and develop various techniques and tools: cohomology theories, homotopy theory, etc. I believe that it is as fundamental to mathematics as groups are, and for similar reasons. In the same way that the group axioms, simple and elegant, capture algebraically deep and fundamental aspects of a mathematical context and are for that reason extremely powerful and useful, the topos axioms, also simple and elegant, capture algebraically, logically and geometrically deep and fundamental aspects of a mathematical context. Furthermore, one can investigate the invariant content of various notions and situations, the invariance being now of at least two flavors, logical and geometrical. It is hard not to see in topos theory another embodiment of Klein's program in the way envisioned by Eilenberg and Mac Lane.

Conclusion

> Comprehending under an all-embracing unitary idea the opposing
> views of different schools of thought.
> (Klein, quoted by [98], 615)

Once Eilenberg and Mac Lane understood that a category was a natural generalization of a group of transformations, it must have been irresistible to conjecture that the algebra of such transformations, of the morphisms of a category, could be used in a way similar to the way groups are used in geometry. But if the group law and structure arose in a context where they immediately had a role to play in the solution of important mathematical problems, the same cannot be said of categories. As we have seen, categories were introduced for the sake of conceptual clarity. As such, they did not have a clear mathematical function. At first they were seen as providing a convenient framework, a language for various fields. Groups, as far as I know, never had the status of being a mere language. To many mathematicians, groups were central, powerful and revealing. It is hard to see how Galois theory could have arisen without groups. Modern differential geometry is inconceivable without Lie groups and Lie algebras. The fact that categories are often, if not almost always, thought of as providing a convenient language, whereas groups are not thought of that way, seems to be a fundamental difference between groups and categories. Of course, there was the element of *unity* provided by categories, comparable to the unity provided by groups towards the end and the beginning of the 20th century. It was clear to Eilenberg and Mac Lane that any well defined mathematical system gives rise to a category once one has defined the appropriate notion of morphism accompanying these systems. But this is not much of a gain as it stands; one wonders what to *do* with these categories.

The collection of morphisms of a category is given as a whole, and it is hard to see the underlying algebraic structure at work in this context. The forest hides the trees. Eilenberg and Mac Lane understood that they were introducing an *algebraic* framework but could not see how this algebraic machinery could work. It is as if someone somehow were to build a powerful computer and power it up, but have no idea how to program it or even that it was programmable. The flexibility, suppleness, versatility, power and strength of category theory were clearly not appreciated by its inventors or even by the first generation of mathematicians who applied it to algebraic topology and homological algebra. Category theory was in the background: useful, amusing, awfully general and abstract—but at the same time, indisputably clarifying, simplifying and helpful in the way it allowed the organization of concepts, results and proofs. But in the end, despite its usefulness, it was not clear that it was *necessary* or *indispensable*.

The algebraic structure at work emerged progressively in the fifties. Mac Lane recognized the possibility of using the algebra to define various central concepts like product and coproduct in order to characterize a type of category, but in the end the categories he defined did not do the work expected. Buchsbaum and Heller defined, by categorical means, types of categories that carved parts of homological algebra at its joints, but it was left to Grothendieck and Kan to give categories unexpected new roles and to put genuinely new categorical concepts to work. From that point on, category *theory* could be seen as a genuine extension of group *theory* in mathematics in general, although it is not clear that it was explicitly seen that way. The formulation, solution and understanding of various mathematical problems depend essentially on categorical concepts. Not only can one look at invariant properties of various mathematical structures, but—and this is a facet that we have not emphasized sufficiently—one can also consider *covariant* properties between various mathematical structures. Categories, functors and natural transformations become indispensable in the same way that group-theoretical concepts became indispensable in Galois theory or geometry. To say that categories are merely an organizational tool is like saying that groups *merely* organize elementary geometry or Galois theory. To make that claim is to fail to see the real work done by both concepts.

As we have already mentioned, Eilenberg and Mac Lane immediately saw that categories were entirely general and that almost any mathematical system could give rise to categories. The abstract categorical algebra arising in the sixties led to the possibility of conceiving the totality of mathematical concepts as being governed by categorical principles. While set theorists were considering various exotic set-theoretical universes, Lawvere proposed that one could instead frame mathematics within categorical universes instead. At that point, set theory and category theory could be seen, each in its own way, to provide the language and the tools to define, develop and analyze all the basic concepts of mathematics. But again, it is the *algebra* of categories that sets it apart and this algebra has an intrinsically geometrical flavor in the same way that the algebra of groups has an intrinsically geometrical flavor. In comparison, set theory is intrinsically combinatorial in its spirit. It is probably worth expanding on this contrast.

Since about the middle of the 19th century, mathematicians have been used to considering number systems—usually the natural numbers—as fundamental or given. The basic entities of mathematics are numbers, usually thought of as individual objects with well-defined properties. This is simple enough for numbers like 2 or $10^{100^{1000}}$, but not so simple for real numbers or complex numbers, since the latter are derived from a geometrical given, namely the real line. The real numbers could be reconstructed from the natural numbers, and the resulting system was a good model of the real geometric line. This made a rigorous development of analysis possible, so conceiving a line as somehow made up of points seemed a small price to pay. Of course, lurking in the background, one could find certain dubitable set-theoretical axioms, like the power-set axiom and the axiom of choice, and problems, like the continuum hypothesis. But the usefulness of the set-theoretical language and the set-theoretical universe was too great to consider abandoning it. Hilbert talked about Cantor's paradise; perhaps his opponents should have compared the language of transfinite sets to the siren's song.

But what if the real line is first and foremost a *line*, that is, a geometric object? We have known since antiquity that rather than thinking of the line as a sum of individually given points that are supposed to adhere to one another by some kind of miracle glue, we can think of it as a type of entity, with its own criterion of identity, that *can* have points as components. It seems that we are falling back on the perennial distinction between discrete and continuous quantities.

We submit that category theory and set theory differ radically on how they represent the nature of mathematical objects. (See [213] for more on this particular point.) A set is entirely determined by its elements, at least in the conception following Zermelo's axiomatization. A set is nothing but a bunch of elements considered as a whole. This conception is captured by the axiom of extensionality: two sets X and Y are identical if and only if they have the same elements. The *universe* of sets is simply the collection or sum of such sets. It has, as such, no criterion of identity since it is usually assumed to be unique. However, it could be considered to be a *class* and classes have essentially the same criterion of identity that sets have. In category theory, objects are treated differently. *Morphisms can* be identical and are identified, e.g., in commutative diagrams. The identity criterion for morphisms is given as primitive. Objects are *isomorphic*. This is a *derived* criterion of identity in a category. As for categories themselves, as we have seen, they are *equivalent*, not isomorphic. And this is only the tip of the iceberg.

To get a better grip of the situation without going into the technical details, it might be judicious to recall some elementary aspects of homotopy theory. (In fact, there are profound formal and conceptual links between category theory and homotopy theory, but we won't delve into them at this point!) We will be entirely informal, since the introduction of the formal definitions would needlessly encumber our discussion. Recall that a homotopy H between two parallel continuous maps $X \underset{g}{\overset{f}{\rightrightarrows}} Y$

is a continuous deformation of f into g. We have seen that being homotopic is an equivalence relation and therefore could be used as a criterion of identity. How about the homotopies themselves? In other words, suppose we have two homotopies H_1 and H_2 from f to g. Since H_1 and H_2 are maps, we could stipulate that $H_1 = H_2$ like any other maps. However, since they are parallel continuous maps, it is also possible to require that there be homotopies $\mathcal{H}: H_1 \to H_2$ and $\mathcal{G}: H_2 \to H_1$ between them such that $\mathcal{H} \circ \mathcal{G} = 1_{H_2}$ and $\mathcal{G} \circ \mathcal{H} = 1_{H_1}$, the identity deformations. How about $\mathcal{H}$ and $\mathcal{G}$ themselves, in other words homotopies of homotopies? Clearly, we could go on and on in this manner and introduce a hierarchy of criteria of identities. The main point is that the latter hierarchy is faithful to the nature of the objects involved.

The link with categories becomes immediate when the parallel with natural transformations and homotopies is brought forward. In fact, a natural isomorphism can be defined as a homotopy between functors. (See [35], 228.) It is also possible to define natural transformations in a similar fashion. (See [35], 233.) But this is not what we are interested in here. In the same way that the criterion of identity between homotopies should be provided by a homotopy of "higher-dimension", it makes sense to require that identities in a category of categories be replaced by a hierarchy of natural equivalences. The overall general picture emerging from this perspective can be sketched as follows.

One starts with "points", i.e., 0-dimensional objects, also called 0-cells, representing categories or sets (at this stage both ways of thinking are entirely acceptable) with paths between them. These paths represent functors and can informally be thought of as processes. They are also called 1-*cells*. As usual, paths compose and there is an identity path for each 0-dimensional object. But here, we face the first difference with the usual definition of a category. In the classical definition, composition has to satisfy the associativity, i.e., there is an *identity* between certain paths, and identities on 0-cells also have to satisfy certain conditions; again *identities* between paths have to be satisfied. Since objects are defined up to isomorphism, it seems reasonable to require that morphisms between 0-cells be defined up to isomorphism. Thus, given three paths or 1-cells f, g and h, there should be *a natural equivalence* $\alpha: f(gh) = (fg)h$ between the different ways they compose, and not an *identity* between them. Such a natural equivalence is called a 2-*cell*. The same should apply to the conditions on identities. If one thinks of α as a homotopy between paths, the analogy with the case of identities of homotopies becomes immediate. How are we to identify two such 2-cells? We have to introduce natural equivalences between them, that is 3-cells. And so on and so forth. Although moving from identities to natural equivalences seems conceptually justified, it introduces a level of complexity that rapidly becomes daunting. Of course, one can stop at any given level and decide that at that level, identities will be used. Thus, one can start with 0-cells with no paths between them. This is a universe of sets or a 0-category. One can start with 0-cells, paths between 0-cells, that is 1-cells, and *identities* between 1-cells, and then one recovers the standard definition of a category, now called a 1-category. But we can go one level up, that is 0-cells, 1-cells between them, 2-cells between 1-cells and *identities* between 2-cells. We thus get a (weak) 2-category. And so on and so forth. Of course, the process can be iterated ad infinitum and we get what is called a ω-category. The study of what are now called *higher-dimensional* categories is presently developing at a quick pace. (See, for an introduction, [181] or [46].) The main philosophical point I want to bring forward concerns the hierarchy of criteria of identities inherent to this universe and how it is different from the one found in a universe of ZF-type sets. What it shows is that there is a conceptual alternative to the cumulative hierarchy that could be used to conceive a universe of mathematical entities. In this alternative, one deals with types as basic objects. Of course, we still do not have a clear picture of a single category of categories with specific axioms that might constitute a genuine alternative to a set-theoretical universe, but no matter what these axioms turn out to be, the underlying framework will have to be along the lines of an ω-category.

From an epistemological point of view, the differences between categories and sets are also striking and significant. Here the comparison with geometry is again fruitful. In a set-theoretical framework, the world is made up of atoms, which in turn, coalesce to form sets, which in their turn, form new sets when they are collected together. We are therefore have a bottom-up structure where in order to know a given set one has to know its elements and, perhaps, what principle or construction were used to collect these elements to form a set. It is truly an atomistic epistemology. In a categorical framework, one is given a space as a whole and to know an object is to know how it is connected to the others in the space. To use the analogy with geometry, one has to "carve-out" the objects in the space by analyzing their algebraic properties *in the whole space*. It is a top-down approach. (For more on this point, see [7].) Of course, in a categorical framework, one quickly gets the feeling one is working in nested universes. Properties of objects *in* a category are reflected or determined by properties of functors *between* the given categories and other categories. Thus, universal morphisms correspond to adjoint functors. This reflection does not hold for all constructions defined within a category, but what can be done within a category depends, directly or indirectly, on the structure provided by these adjoint situations.

If foundational studies of mathematics rest upon the introduction, development and analysis of formal systems and their properties, then there is no doubt that category theory has something to offer the field and that what it has to offer ought to be seen in a geometric light. For instance, as we have seen, it is possible to characterize a hierarchy of formal systems, Cartesian logic, regular logic, coherent logic, geometric logic, intuitionistic logic, pretoposes, classical logic, lambda calculi, various type theories, etc., and to associate to these logical systems categories, for instance Cartesian, regular, coherent, pretoposes, Heyting, Boolean, toposes, Cartesian closed, locally Cartesian closed, etc., in such a way that one can prove various results for these systems, for instance completeness, to mention the most obvious. In each case, one can define morphisms between theories of a given kind and investigate invariance and coviariance between them. Furthermore, the categories are related to one another and it is therefore also possible to investigate relationships between them. In the same way that Klein's program opened the door to a systematic comparison between elementary geometrical systems, category theory and categorical logic provide the proper algebraic tools to perform a similar task.

If, on the other hand, foundational studies of mathematics are taken to also include the analysis and classification of the elements, concepts, constructions, and operations of mathematics itself, then it is clear that category theory provides a tool of exceptional power and clarity for this purpose. Moreover, unlike the use of formal systems in foundations, the use of category theory has actually shaped the development of mathematics itself, thus extending the notion of foundations to include the structure of modern mathematics. (See also [225].)

Needless to say, we haven't covered all aspects of category theory and categorical logic. We have barely mentioned higher-dimensional categories, we haven't said a word about monoidal categories, braided categories, etc., we haven't related how quantum logic is being reformulated in a categorical framework in an original and promising way, we haven't shown how linear logic and its variants can be modeled by various categories, etc.: in all these cases, we would have found additional elements showing that, indeed, as Eilenberg and Mac Lane had glimpsed without seeing all the details, category theory does indeed constitute an extension, in spirit, in essence, and in the role it gives to algebra, of Klein's program.

References

1. J. Adamek, P. T. Johnstone, J. A. Makowsky, and J. Rosicky. Finitary sketches. *J. Symbolic Logic*, 62(3):699–707, 1997.
2. J. Adamek and J. Rosicky. *Locally Presentable and Accessible Categories*, volume 189. Cambridge University Press, Cambridge, 1994. London Mathematical Society Lecture Note Series.
3. J. Adamek and J. Rosicky. On geometric and finitary sketches. *Appl. Categ. Structures*, 4(2–3):227–240, 1996.
4. M. A. Akivis and B. A. Rosenfeld. *Élie Cartan (1869–1951)*, volume 123 of *Translations of Mathematical Monographs*. American Mathematical Society, Providence, 1993.
5. M. Artin, A. Propanedioec, and J.-L. Verdier, editors. *Théorie des topos et cohomologie étale des schémas (SGA 4)*, volume 305 of *Lecture Notes in Mathematics*. Springer-Verlag, Berlin, 1973.
6. S. Awodey. Sheaf representation for topoi. *J. Pure Appl. Algebra*, 145(2):107–121, 2000.
7. S. Awodey. An answer to G. Hellman's question: Does category theory provide a framework for mathematical structuralism?. *Philos. Math. (3)*, 12(1):54–64, 2004.
8. S. Awodey. *Category Theory*, volume 49 of *Oxford Logic Guides*. Clarendon Press, Oxford, 2007.
9. F. Bachmann. *Aufbau der Geometrie aus dem Spiegelunsbegriff*, volume XCVI of *Die Grundlehren des Mathematischen Wissenschaften*. Springer-Verlag, Berlin, 2nd edition, 1973.
10. B. Banaschewski and C. J. Mulvey. Stone-Čech compactification of locales. I. *Houston J. Math.*, 6(3):301–312, 1980.
11. B. Banaschewski and C. J. Mulvey. Stone-Čech compactification of locales. II. *J. Pure Appl. Algebra*, 33(2):107–122, 1984.
12. B. Banaschewski and C. J. Mulvey. A constructive proof of the Stone-Weierstrass theorem. *J. Pure Appl. Algebra*, 116(1–3):25–40, 1997.
13. B. Banaschewski and C. J. Mulvey. The Stone-Čech compactification of locales. III. *J. Pure Appl. Algebra*, 185(1–3):25–33, 2003.
14. B. Banaschewski and C. J. Mulvey. A globalisation of the Gelfand duality theorem. *Ann. Pure Appl. Logic*, 137(1–3):62–103, 2006.
15. M. Barr and C. Wells. *Toposes, Triples and Theories*, volume 278 of *Grundlehren der Mathematischen Wissenschaften*. Springer-Verlag, New York, 1985.
16. J. Barwise and S. Feferman, editors. *Model-theoretic Logics*. Perspectives in Mathematical Logic. Springer-Verlag, New York, 1985.
17. H. Bass. *Lectures on Topics in Algebraic K-Theory*. Tata Institute for Fundamental Research Lectures on Mathematics. Tata Institute of Fundamental Research, Bombay, 1967.
18. J. C. Becker and D. H. Gottlieb. A history of duality in algebraic topology. In I. M. James, editor, *History of Topology*, pages 725–745. North-Holland, Amsterdam, 1999.

19. J. L. Bell. Category theory and the foundations of mathematics. *British J. Philos. Sci.*, 32(4):349–358, 1981.
20. J. L. Bell. *Toposes and Local Set Theories: An introduction*, volume 14 of *Oxford Logic Guides*. The Clarendon Press Oxford University Press, New York, 1988.
21. J. Bénabou. Algèbre élémentaire dans les catégories avec multiplication. *C. R. Acad. Sci. Paris*, 258:771–774, 1964.
22. J. Bénabou. Introduction to bicategories. In *Reports of the Midwest Category Seminar*, pages 1–77. Springer, Berlin, 1967.
23. J. Bénabou. Fibered categories and the foundations of naive category theory. *J. Symbolic Logic*, 50(1):10–37, 1985.
24. P. Benacerraf. What numbers could not be. *Philos. Rev.*, 74:47–73, 1965.
25. G. Birkhoff. A set of postulates for plane geometry, based on scale and protractor. *Ann. of Math.*, 33(2):329–345, 1932.
26. G. Blanc and M. R. Donnadieu. Axiomatisation de la catégorie des catégories. *Cahiers topologie géom différentielle*, XVII(2):1–36, 1976.
27. G. Blanc and A. Preller. Lawvere's basic theory of the category of categories. *J. Symbolic Logic*, 40(1):14–18, 1975.
28. A. Blass. The interaction between category theory and set theory. In *Mathematical Applications of Category Theory (Denver, Col., 1983)*, volume 30 of *Contemp. Math.*, pages 5–29. Amer. Math. Soc., Providence, RI, 1984.
29. O. Blumenthal. Lebensgeschichte. In *Gesammelte Abhandlungen, D. Hilbert*, volume 3, pages 388–429. Springer, Berlin, 2nd edition, 1970.
30. A. Boileau and A. Joyal. La logique des topos. *J. Symbolic Logic*, 46(1):6–16, 1981.
31. N. Bourbaki. *Éléments de Mathématiques. Les structures fondamentales de l'analyse. Livre I. Théorie des ensembles (Fascicule de résultats)*. Hermann, Paris, 1939.
32. D. A. Brannan, M. F. Esplen, and J. J. Gray. *Geometry*. Cambridge University Press, Cambridge, 1999.
33. E. H. Jr. Brown. Cohomology theories. *Ann. of Math. (2)*, 75:467–484, 1962.
34. R. Brown. From groups to groupoids: A brief survey. *Bull. London Math. Soc.*, 19(2):113–134, 1987.
35. R. Brown. *Topology and Groupoids*. BookSurge Publishing, North Charteston SC, 2006.
36. D. A. Buchsbaum. Exact categories and duality. *Trans. Amer. Math. Soc.*, 80:1–34, 1955.
37. I. Bucur and A. Deleanu. *Introduction to the Theory of Categories and Functors*. John Wiley & Sons, London, 1968.
38. C. Butz and P. T. Johnstone. Classifying toposes for first-order theories. *Ann. Pure Appl. Logic*, 91(1):33–58, 1998.
39. É. Cartan. *La Méthode du repère mobile: la théorie des groupes continus et les espaces généralisés*. Hermann, Paris, 1935.
40. É. Cartan. *Leçons sur la géométrie projective complexe*. Gauthier-Villars, Paris, 2nd edition, 1950.
41. É. Cartan. *Leçons sur la géométrie des espaces de Riemann*. Gauthier-Villars, Paris, 2nd edition, 1951.
42. É Cartan. *Notice sur les travaux scientifiques*. Discours de la méthode. Gauthier-Villars, Paris, 1974.
43. H. P. Cartan and S. Eilenberg. *Homological Algebra*. Princeton University Press, Princeton, 1956.
44. P. Cartier. A mad day's work: from Grothendieck to Connes and Kontsevich. the evolution of concepts of space and symmetry. *Bull. Amer. Math. Soc. (N.S.)*, 38(4):389–408, 2001.
45. C. Centazzo and E. M. Vitale. Sheaf theory. In M. C. Pedicchio and W. Tholen, editors, *Categorical Foundations*, Encyclopedia of Mathematics and its Applications, pages 311–358. Cambridge University Press, Cambridge, 2004.
46. E. Cheng and A. Lauda. Higher-dimensional categories: An illustrated guide book, 2004.
47. S.-S. Chern and C. Chevalley. Obituary: Elie Cartan and his mathematical work. *Bull. Amer. Math. Soc.*, 58:217–250, 1952.

References

48. P. Cohen. The independence of the continuum hypothesis. *Proc. Nat. Acad. Sci. U.S.A.*, 50:1143–1148, 1963.
49. P. Cohen. Comments on the foundations of set theory. In D. S. Scott, editor, *Axiomatic Set Theory*, pages 9–15. American Mathematical Society, Providence, 1971.
50. T. Coquand and H. Lombardi. A logical approach to abstract algebra. *Math. Structures Comput. Sci.*, 16(5):885–900, 2006.
51. L. Corry. *Modern Algebra and the Rise of Mathematical Structures*. Science Networks. Historical Studies. Birkhäuser, Basel, 1996.
52. M. Coste. Localisation, spectra and sheaf representation. In M. Fourman, C. Mulvey, and D. S. Scott, editors, *Applications of Sheaves*, Lecture Notes in Mathematics, pages 212–238. Springer-Verlag, Berlin, 1979.
53. J. Couture and J. Lambek. Philosophical reflections on the foundations of mathematics. *Erkenntnis*, 34(2):187–209, 1991.
54. H. S. M. Coxeter. The inversive plane and hyperbolic geometry. *Abh. Math. Sem. Univ. Hamburg*, 29:217–242, 1966.
55. P. Deligne. Quelques idées maîtresses de l'oeuvre d'Alexandre Grothendieck. In *Matériaux pour l'histoire des mathématiques au 20ème siècle*, volume 3 of *Séminaires et Congrès*, pages 11–19. Société Mathématique de France, Paris, 1998.
56. M. Detlefsen. Poincaré against the logicians. *Synthese*, 90(3):349–378, 1992.
57. M. Detlefsen. Hilbert's formalism. *Revue internationale de philosophie*, 47(186):285–304, 1993.
58. J. Dieudonné. *Linear Algebra and Geometry*. Houghton Mifflin, Boston, 1969.
59. J. Dieudonné. *A History of Algebraic and Differential Topology. 1900–1960*. Birkhäuser, Boston, 1989.
60. J. Dieudonné and A. Grothendieck. *Éléments de géométrie algébrique*. Springer-Verlag, Berlin, 1971.
61. E. J. Dubuc. *Kan Extensions in Enriched Category Theory*. Lecture Notes in Mathematics. Springer-Verlag, Berlin, 1970.
62. C. Ehresmann. Sur la théorie des espaces fibrés. In A. C. Ehresmann, editor, *Charles Ehresmann: Oeuvres complètes et commentées*, pages 133–145. Evrard, Amiens, 1937.
63. C. Ehresmann. Catégories structurées. *Annales scientifiques de l'École Normal Supérieure (3)*, 80:349–426, 1963.
64. C. Ehresmann. *Catégories et structures*. Dunod, Paris, 1965.
65. C. Ehresmann. Sur les structures algébriques. *C. R. Acad. Sci. Paris Sér. A-B*, 264:A840–A843, 1967.
66. S. Eilenberg. Cohomology and continuous mappings. *Ann. of Math. (2)*, 41:231–251, 1940.
67. S. Eilenberg. Review of "Éléments de mathématique. part I. Les structures fondamentales de l'analyse. livre I. Théorie des ensembles (Fascicule de résultats)", 1943. MR0004746.
68. S. Eilenberg. Singular homology theory. *Ann. of Math. (2)*, 45:407–447, 1944.
69. S. Eilenberg. Homology of spaces with operators. *Trans. Amer. Math. Soc.*, 61:378–417, 1947.
70. S. Eilenberg. Witold Hurewicz—personal reminiscences. In K. Kuperberg, editor, *Collected Works of Witold Hurewicz*, pages xlv–xlvi. American Mathematical Society, Providence, 1995.
71. S. Eilenberg and S. Mac Lane. Group extensions and homology. *Ann. of Math.*, 43(4):757–831, 1942.
72. S. Eilenberg and S. Mac Lane. Natural isomorphisms in group theory. *Proc. Nat. Acad. Sci. U.S.A.*, 28(12):537–543, 1942.
73. S. Eilenberg and S. Mac Lane. Relations between homology and homotopy groups. *Proc. Nat. Acad. Sci. U.S.A.*, 29:155–158, 1943.
74. S. Eilenberg and S. Mac Lane. A general theory of natural equivalences. *Trans. Amer. Math. Soc.*, 58:231–294, 1945.
75. S. Eilenberg and S. Mac Lane. Relations between homology and homotopy groups of spaces. *Ann. of Math. (2)*, 46:480–509, 1945.

76. S. Eilenberg and S. Mac Lane. Homology of spaces with operators II. *Trans. Amer. Math. Soc.*, 65:49–99, 1949.
77. S. Eilenberg and S. Mac Lane. Relations between homology and homotopy groups of spaces II. *Ann. of Math. (2)*, 51:514–533, 1950.
78. S. Eilenberg and S. Mac Lane. Acyclic models. *Amer. J. Math.*, 75:189–199, 1953.
79. S. Eilenberg and N. E. Steenrod. Axiomatic approach to homology theory. *Proc. Nat. Acad. Sci. U.S.A.*, 31:117–120, 1945.
80. S. Eilenberg and N. E. Steenrod. *Foundations of Algebraic Topology*. Princeton University Press, Princeton, 1952.
81. S. Eilenberg and J. A. Zilber. Semi-simplicial complexes and singular homology. *Ann. of Math. (2)*, 51:499–513, 1950.
82. G. Ewald. *Geometry: An Introduction*. Wadsworth Pub., Belmont, Calif., 1971.
83. E. Fadell. The contributions of Witold Hurewicz to algebraic topology. In K. Kuperberg, editor, *Complete Works of Witold Hurewicz*, pages xxxiii–xl. American Mathematical Society, Providence, 1995.
84. S. Feferman. Set-theoretical foundations of category theory. In S. Mac Lane, editor, *Reports of the Midwest Category Seminar. III*, volume 106 of *Lecture Notes in Mathematics*, pages 201–247. Springer-Verlag, New York, 1969.
85. S. Feferman. Categorical foundations and foundations of category theory. In *Logic, foundations of mathematics and computability theory (Proc. Fifth Internat. Congr. Logic, Methodology and Philos. of Sci., Univ. Western Ontario, London, Ont., 1975), Part I*, volume 9 of *Univ. Western Ontario Ser. Philos. Sci.*, pages 149–169. Reidel, Dordrecht, 1977.
86. R. H. Fox. On topologies for function spaces. *Bull. Amer. Math. Soc.*, 51:429–432, 1945.
87. G. Frege. *Philosophical and Mathematical Correspondence*. University of Chicago Press, Chicago, Ill., 1980. Edited and with an introduction by Gottfried Gabriel, Hans Hermes, Friedrich Kambartel, Christian Thiel and Albert Veraart, Abridged from the German edition and with a preface by Brian McGuinness, Translated from the German by Hans Kaal, With an appendix by Philip E. B. Jourdain.
88. H. Freudenthal. Lie groups in the foundations of geometry. *Adv. Math.*, 1(2):145–190, 1964.
89. P. Freyd. *Abelian Categories. An Introduction to the Theory of Functors*. Harper's Series in Modern Mathematics. Harper & Row, New York, 1964.
90. P. Freyd. The theories of functors and models. In J. W. Addison, L. Henkin, and A. Tarski, editors, *Theory of Models*, pages 107–120. North Holland, Amsterdam, 1965.
91. P. Freyd. Representations in Abelian categories. In *Proceedings of the Conference on Categorical Algebra, La Jolla, Calif., 1965*, pages 95–120. Springer-Verlag, New York, 1966.
92. P. Freyd. Homotopy is not concrete. In *The Steenrod Algebra and its Applications (Proc. Conf. to Celebrate N. E. Steenrod's Sixtieth Birthday, Battelle Memorial Inst., Columbus, Ohio, 1970)*, volume 168 of *Lecture Notes in Mathematics*, pages 25–34. Springer, Berlin, 1970.
93. P. Freyd. Properties invariant within equivalence types of categories. In A. Heller and M. Tierney, editors, *Algebra, Topology, and Category Theory (a collection of papers in honor of Samuel Eilenberg)*, pages 55–61. Academic Press, New York, 1976.
94. P. Freyd. Cartesian logic. *Theoret. Comput. Sci.*, 278(1–2):3–21, 2002. Mathematical foundations of programming semantics (Boulder, CO, 1996).
95. M. Friedman. *Kant and the Exact Sciences*. Harvard University Press, Cambridge, 1992.
96. P. Gabriel. Des catégories abéliennes. *Bull. Soc. Math. France*, 90:323–448, 1962.
97. P. Gabriel and F. Ulmer. *Lokal präsentierbare Kategorien*. Number 221 in Lecture Notes in Mathematics. Springer-Verlag, Berlin, 1971.
98. E. Glas. From form to function. *Stud. Hist. Philos. Sci.*, 24(4):611–631, 1993.
99. B. Gray. *Homotopy Theory*, volume 64 of *Pure and Applied Mathematics*. Academic Press, New York, 1975.
100. J. W. Gray. Fibred and cofibred categories. In *Proc. Conf. Categorical Algebra (La Jolla, Calif., 1965)*, pages 21–83. Springer-Verlag, New York, 1966.

References

101. J. W. Gray. The categorical comprehension scheme. In *Category Theory, Homology Theory and their Applications, III (Battelle Institute Conference, Seattle, Wash., 1968, Vol. Three)*, pages 242–312. Springer, Berlin, 1969.
102. J. W. Gray. The 2-adjointness of the fibred category construction. In *Symposia Mathematica, Vol. IV (INDAM, Rome, 1968/69)*, pages 457–492. Academic Press, London, 1970.
103. A. Grothendieck. Sur quelques points d'algèbre homologique. *Tôhoku Math. J.*, 9(2):119–221, 1957.
104. A. Grothendieck. *Fondements de la géométrie algébrique. [Extraits du Séminaire Bourbaki, 1957–1962.]*. Secrétariat mathématique, Paris, 1962.
105. A. Grothendieck. *Revêtement étale et groupe fondamental (SGA 1)*, volume 224 of *Springer Lecture Notes in Mathematics*. Springer-Verlag, Berlin, 1971.
106. A. Grothendieck. Récoltes et semailles, 1986. http://www.grothendieckcircle.org/
107. H. W. Guggenheimer. *Plane Geometry and its Groups*. Holden-Day, San Francisco, 1967.
108. J. Hafner and P. Mancosu. The varieties of mathematical explanation. In P. Mancosu, K. F. Jorgensen, and S. A. Pedersen, editors, *Visualization, Explanation and Reasoning Styles in Mathematics*, pages 215–250. Springer, Dordrecht, 2005.
109. M. Hallett. *Cantorian Set Theory and Limitation of Size*, volume 10 of *Oxford Logic Guides*. The Clarendon Press, Oxford, 1984.
110. J. D. Halpern. Review of "Observations on popular discussions of foundations", 1973. MR0294123.
111. T. Hawkins. *Emergence of the Theory of Lie Groups: An Essay in the History of Mathematics 1869–1926*. Sources and Studies in the History of Mathematics and Physical Sciences. Springer-Verlag, New York, 2000.
112. P. J. Hilton. The fundamental group as a functor. *Bull. Soc. Math. Belg.*, 14:153–177, 1962.
113. S.-T. Hu. An exposition of the relative homotopy theory. *Duke Math. J.*, 14:991–1033, 1947.
114. J. R. Isbell. Review of "The category of categories as a foundation for mathematics", 1967. MR0207517.
115. N. Jacobson. *Basic Algebra I*. W. H. Freeman and Co., San Francisco, 1974.
116. I. M. James. On the suspension triad. *Ann. of Math.*, 63:191–247, 1956.
117. P. T. Johnstone. *Topos Theory*, volume 10 of *London Mathematical Society Monographs*. Academic Press [Harcourt Brace Jovanovich Publishers], London, 1977.
118. P. T. Johnstone. Tychonoff's theorem without the axiom of choice. *Fund. Math.*, 113(1):21–35, 1981.
119. P. T. Johnstone. How general is a generalized space? In *Aspects of topology*, volume 93 of *London Math. Soc. Lecture Note Ser.*, pages 77–111. Cambridge University Press, Cambridge, 1985.
120. P. T. Johnstone. *Stone Spaces*, volume 3 of *Cambridge Studies in Advanced Mathematics*. Cambridge University Press, Cambridge, 1986. Reprint of the 1982 edition.
121. P. T. Johnstone. *Sketches of an Elephant: A Topos Theory Compendium. Vol. 1*, volume 43 of *Oxford Logic Guides*. The Clarendon Press Oxford University Press, New York, 2002.
122. P. T. Johnstone. *Sketches of an Elephant: A Topos Theory Compendium. Vol. 2*, volume 44 of *Oxford Logic Guides*. The Clarendon Press Oxford University Press, Oxford, 2002.
123. P. T. Johnstone and S. Vickers. Preframe presentations present. In A. Carboni, M. C. Pedicchio, and G. Rosolini, editors, *Category Theory (Como 1990)*, volume 1488 of *Lecture Notes in Mathematics*, pages 193–212. Springer-Verlag, Berlin, 1991.
124. A. Joyal and I. Moerdijk. Toposes are cohomologically equivalent to spaces. *Amer. J. Math.*, 112(1):87–95, 1990.
125. A. Joyal and I. Moerdijk. Toposes as homotopy groupoids. *Adv. Math.*, 80(1):22–38, 1990.
126. A. Joyal and G. E. Reyes. Forcing and generic models in categorical logic, 1977.
127. A. Joyal and R. Street. An introduction to Tanaka duality and quantum groups. In *Category Theory (Como 90)*, Lecture Notes in Mathematics, pages 413–492. Springer Verlag, Berlin, 1991.
128. A. Joyal and M. Tierney. An extension of the Galois theory of Grothendieck. *Memoirs of the American Mathematical Society*, 51(309):R4–R71, 1984.

129. D. L. Kan. Abstract homotopy I. *Proc. Nat. Acad. Sci. U.S.A.*, 41:1092–1096, 1955.
130. D. L. Kan. Abstract homotopy II. *Proc. Nat. Acad. Sci. U.S.A.*, 42:255–258, 1956.
131. D. L. Kan. Abstract homotopy III. *Proc. Nat. Acad. Sci. U.S.A.*, 42:419–421, 1956.
132. D. L. Kan. Abstract homotopy IV. *Proc. Nat. Acad. Sci. U.S.A.*, 42:542–544, 1956.
133. D. L. Kan. On c.s.s complexes. *Amer. J. Math.*, 79:449–476, 1957.
134. D. L. Kan. Adjoint functors. *Trans. Amer. Math. Soc.*, 87:294–329, 1958.
135. D. L. Kan. Functors involving c.s.s. complexes. *Trans. Amer. Math. Soc.*, 87:330–346, 1958.
136. J. L. Kelley. *General Topology*. D. Van Nostrand, New York, 1955.
137. J. Kim. *Supervenience and Mind*. Cambridge University Press, Cambridge, 1993.
138. F. Klein. A comparative review of recent researches in geometry. *Bull. Amer. Math. Soc.*, 2(10):215–249, 1893.
139. F. Klein. *Elementary Mathematics from an Elementary Standpoint: Geometry*. Dover, Mineola, 1939.
140. M. Kline. *Mathematical Thought from Ancient to Modern Times*. Oxford University Press, New York, 1972.
141. S. Kobayashi and K. Nomizu. *Foundations of Differential Geometry. Vol. I*. Interscience, New York, 1963.
142. A. Kock. Ehresmann and the fundamental structures of differential geometry seen from a synthetic viewpoint. In A. C. Ehresmann, editor, *Charles Ehresmann: Oeuvres complètes et commentées*, pages 549–554. Evrard, Amiens, 1984.
143. A. Kock and G. E. Reyes. Doctrines in categorical logic. In J. Barwise, editor, *Handbook of Mathematical Logic*, pages 283–313. North Holland, Amsterdam, 1977.
144. E. E. Kramer. *The Nature and Growth of Modern Mathematics*. Princeton University Press, Princeton, NJ, 1981.
145. G. Kreisel. Appendix II. Category theory and the foundations of mathematics. In S. Mac Lane, editor, *Reports of the Midwest Category Seminar. III*, volume 106 of *Lecture Notes in Mathematics*, pages 233–247. Springer Verlag, New York, 1969.
146. G. Kreisel. Observations on popular discussions of foundations. In D. S. Scott, editor, *Axiomatic Set Theory*, pages 189–198. American Mathematical Society, Providence, 1971.
147. G. Kreisel. Review of "Categorical algebra and set-theoretic foundations", 1972. MR0282791.
148. G. Kreisel and J.-L. Krivine. *Elements of Mathematical Logic. Model Theory*. Studies in Logic and the Foundations of Mathematics. North-Holland, Amsterdam, 1967.
149. R. Krömer. *Tool and Object. A History and Philosophy of Category Theory*, volume 32 of *Science Networks. Historical Studies*. Birkhäuser Verlag, Basel, 2007.
150. K. Kuperberg, editor. *Collected Works of Witold Hurewicz*. American Mathematical Society, Providence, 1995.
151. C. Lair. Catégories qualifiables et catégories esquissables. *Diagrammes*, 17:1–153, 1987.
152. C. Lair. Éléments de théorie des esquisses. 1. Graphes à composition. *Diagrammes*, 45/46:3–33, 2001.
153. C. Lair. Éléments de théorie des esquisses. 2. Systèmes tensoriels et systèmes enrichis de graphes à composition. *Diagrammes*, 47/48:34, 2002.
154. C. Lair. Éléments de théorie des esquisses. 3. Esquisses. *Diagrammes*, 49/50:58, 2003.
155. J. Lambek. Deductive systems and categories. I. Syntactic calculus and residuated categories. *Math. Systems Theory*, 2:287–318, 1968.
156. J. Lambek. Deductive systems and categories II. Standard constructions and closed categories. In P. J. Hilton, editor, *Category Theory, Homology Theory, and their Applications, I*, volume 86 of *Lecture Notes in Mathematics*, pages 76–122, Springer-Verlag, Berlin, 1969.
157. J. Lambek. Deductive systems and categories III. Cartesian closed categories, intuitionistic propositional calculus, and combinatory logic. In F. W. Lawvere, editor, *Toposes, Algebraic Geometry and Logic*, volume 274 of *Lecture Notes in Mathematics*, pages 57–82, Springer-Verlag, Berlin, 1972.
158. J. Lambek. Are the traditional philosophies of mathematics really incompatible? *Math. Intelligencer*, 16(1):56–62, 1994.

159. J. Lambek. What is the world of mathematics? provinces of logic determined. *Ann. Pure Appl. Logic*, 126(1–3):149–158, 2004.
160. J. Lambek and P. J. Scott. *Introduction to Higher Order Categorical Logic*. Cambridge Studies in Advanced Mathematics. Cambridge University Press, Cambridge, 1986.
161. J. Lambek and P. J. Scott. Reflections on a categorical foundations of mathematics, 2007.
162. E. Landry and J.-P. Marquis. Categories in context: Historical, foundational and philosophical. *Philos. Math. (3)*, 13(1):1–43, 2005.
163. S. Lavine. *Understanding the Infinite*. Harvard University Press, Cambridge, 1994.
164. F. W. Lawvere. *Functorial Semantics of Algebraic Theories*. Ph.d. thesis, Columbia University, 1963.
165. F. W. Lawvere. An elementary theory of the category of sets. *Proc. Nat. Acad. Sci. U.S.A.*, 52:1506–1511, 1964.
166. F. W. Lawvere. Algebraic theories, algebraic categories, and algebraic functors. *Theory of Models (Proc. 1963 Internat. Sympos. Berkeley)*, pages 413–418, North-Holland, Amsterdam, 1965.
167. F. W. Lawvere. The category of categories as a foundation for mathematics. In *Proceedings of the Conference on Categorical Algebra (La Jolla, Calif. 1965)*, pages 1–20. Springer-Verlag, New York, 1966.
168. F. W. Lawvere. Functorial semantics of elementary theories. *Journal of Symbolic Logic*, 31:294–295, 1966.
169. F. W. Lawvere. Theories as categories and the completeness theorem. *Journal of Symbolic Logic*, 32:562, 1967.
170. F. W. Lawvere. Some algebraic problems in the context of functorial semantics of algebraic theories. *Reports of the Midwest Category Seminar. II*, pages 41–61, Springer, Berlin, 1968.
171. F. W. Lawvere. Adjointness in foundations. *Dialectica*, 23:281–296, 1969.
172. F. W. Lawvere. Diagonal arguments and cartesian closed categories. In P. Hilton, editor, *Category Theory, Homology Theory and their Applications II*, volume 92 of *Lecture Notes in Mathematics*, pages 134–145. Springer-Verlag, Berlin, 1969.
173. F. W. Lawvere. Equality in hyperdoctrines and comprehension schema as an adjoint functor. *Applications of Categorical Algebra (Proc. Sympos. Pure Math., Vol. XVII, New York, 1968)*, pages 1–14, 1970. Amer. Math. Soc. Providence, R.I.
174. F. W. Lawvere. Quantifiers and sheaves. In M. Berger, J. Dieudonné, J. Leray, J.-L. Lions, P. Malliavin, and J.-P. Serre, editors, *Actes du Congrès International des Mathématiciens*, pages 329–334, Nice, 1970. Gauthier-Villars.
175. F. W. Lawvere. Categorical dynamics. In A. Kock, editor, *Topos Theoretical Methods in Geometry*, volume 30, pages 1–28. Aarhus University, Aarhus, 1979.
176. F. W. Lawvere. Cohesive toposes and Cantor's "lauter Einsen". *Philos. Math. (3)*, 2(1):5–15, 1994. Categories in the Foundations of Mathematics and Language.
177. F. W. Lawvere. Comments on the development of topos theory. In J.-P. Pier, editor, *Development of Mathematics: 1950–2000*, volume II, pages 715–734. Birkhäuser, Basel, 2000.
178. F. W. Lawvere. Categories of spaces may not be generalized spaces as exemplified by directed graphs. *Repr. Theory Appl. Categ.*, (9):1–7 (electronic), 2005. Reprinted from Rev. Colombiana Mat. **20** (1986), no. 3-4, 179–185.
179. F. W. Lawvere and R. Rosebrugh. *Sets for Mathematics*. Cambridge University Press, Cambridge, 2003.
180. F. W. Lawvere and S. H. Schanuel. *Conceptual Mathematics. A First Introduction to Categories*. Cambridge University Press, Cambridge, revised edition of the 1991 original edition, 1997.
181. Tom Leinster. A survey of definitions of n-category. *Theory Appl. Categ.*, 10:1–70 (electronic), 2002.
182. F. E. J. Linton. Some aspects of equational categories. In *Proc. Conf. Categorical Algebra (La Jolla, Calif., 1965)*, pages 84–94. Springer, New York, 1966.
183. E. J. Lowe. The metaphysics of abstract objects. *J. Philos.*, 92(10):509–524, 1995.
184. S. Lubkin. Imbedding of Abelian categories. *Trans. Amer. Math. Soc.*, 97:410–417, 1960.

185. S. Mac Lane. Groups, categories and duality. *Proc. Nat. Acad. Sci. U.S.A.*, 34:263–267, 1948.
186. S. Mac Lane. Duality for groups. *Bull. Amer. Math. Soc.*, 56:485–516, 1950.
187. S. Mac Lane. Locally small categories and the foundations of set theory. In *Infinitistic Methods (Proc. Sympos. Foundations of Math., Warsaw, 1959)*, pages 25–43. Pergamon, Oxford, 1961.
188. S. Mac Lane. Natural associativity and commutativity. *Rice Univ. Studies*, 49(4):28–46, 1963.
189. S. Mac Lane. Categorical algebra. *Bull. Amer. Math. Soc.*, 71:40–106, 1965.
190. S. Mac Lane. Foundations of mathematics: Category theory. In R. Klibansky, editor, *Contemporary Philosophy*, volume I, pages 286–294. La Nuova Italia Editrice, Firenze, 1968.
191. S. Mac Lane. Foundations for categories and sets. In P. Hilton, editor, *Category Theory, Homology Theory and their Applications. II*, volume 92 of *Lecture Notes in Mathematics*, pages 146–164. Springer Verlag, New York, 1969.
192. S. Mac Lane. One universe as a foundation for category theory. In S. Mac Lane, editor, *Reports of the Midwest Category Seminar. III*, volume 106 of *Lecture Notes in Mathematics*, pages 192–200. Springer Verlag, New York, 1969.
193. S. Mac Lane. The influence of M. H. Stone on the origins of category theory. In F. E. Browder, editor, *Functional Analysis and Related Fields*, pages 229–235. Springer-Verlag, New York, 1970.
194. S. Mac Lane. Categorical algebra and set-theoretic foundations. In *Axiomatic Set Theory (Proc. Sympos. Pure Math., Vol. XIII, Part I, Univ. California, Los Angeles, Calif., 1967)*, pages 231–240. Amer. Math. Soc., Providence, R.I., 1971.
195. S. Mac Lane. *Categories for the Working Mathematician*, volume 5 of *Graduate Text in Mathematics*. Springer-Verlag, New York, 1971.
196. S. Mac Lane. Concepts and categories in perspective. In P. Duren, editor, *A Century of Mathematics in America, Part I*, pages 323–365. American Mathematical Society, Providence, 1988.
197. S. Mac Lane. The developments and prospects for category theory. *Appl. Categ. Structures*, 4(2–3):129–136, 1996.
198. S. Mac Lane. *Categories for the Working Mathematician*, volume 5 of *Graduate Texts in Mathematics*. Springer-Verlag, New York, 2nd edition, 1998.
199. S. Mac Lane and G. Birkhoff. *Algebra*. The Macmillan Co., New York, 1967.
200. S. Mac Lane and I. Moerdijk. *Sheaves in Geometry and Logic. A first introduction to topos theory*. Universitext. Springer-Verlag, New York, 1994.
201. M. Makkai. Strong conceptual completeness for first-order logic. *Ann. Pure Appl. Logic*, 40(2):167–215, 1988.
202. M. Makkai. Duality and definability in first order logic. *Mem. Amer. Math. Soc.*, 105(503):x+106, 1993.
203. M. Makkai. Generalized sketches as a framework for completeness theorems. I. *J. Pure Appl. Algebra*, 115(1):49–79, 1997.
204. M. Makkai. Generalized sketches as a framework for completeness theorems. II. *J. Pure Appl. Algebra*, 115(2):179–212, 1997.
205. M. Makkai. Generalized sketches as a framework for completeness theorems. III. *J. Pure Appl. Algebra*, 115(3):241–274, 1997.
206. M. Makkai. Towards a categorical foundation of mathematics. In *Logic Colloquium '95 (Haifa)*, volume 11 of *Lecture Notes Logic*, pages 153–190. Springer, Berlin, 1998.
207. M. Makkai and R. Paré. *Accessible Categories: The foundations of Categorical Model Theory*, volume 104 of *Contemporary Mathematics*. American Mathematical Society, Providence, RI, 1989.
208. M. Makkai and G. E. Reyes. *First Order Categorical Logic: Model-theoretical Methods in the Theory of Topoi and Related Categories*, volume 611 of *Lecture notes in mathematics*. Springer-Verlag, Berlin, New York, 1977.
209. Y. I. Manin. *Topics in Noncommutative Geometry*. M.B. Porter Lectures. Princeton University Press, Princeton, 1991.

References

210. Y. I. Manin. Interrelations between mathematics and physics. In *Matériaux pour l'Histoire des Mathématiques au 20ème siècle*, pages 157–168. Société Mathématique de France, Paris, 1998.
211. J.-P. Marquis. Abstract mathematical tools and machines for mathematics. *Philos. Math. (3)*, 5(3):250–272, 1997.
212. J.-P. Marquis. Mathematical engineering and mathematical change. *Interna. Stud. Philos. Sci.*, 13(3):245–259, 1999.
213. J.-P. Marquis. Categories, sets and the nature of mathematical entities. In J. Van Benthem, G. Heinsmann, M. Rebuschi, and H. Visser, editors, *The Age of Alternative Logics: assessing philosophy of logic and mathematics today*, Logic, Epistemology, and the Unity of Science, pages 181–192. Springer, Dordrecht, 2006.
214. J.-P. Marquis. A path to the epistemology of mathematics: Homotopy theory. In J. Ferreiros and J. J. Gray, editors, *The Architecture of Modern Mathematics*, pages 239–260. Oxford University Press, Oxford, 2006.
215. W. S. Massey. *Singular Homology Theory*, volume 70 of *Graduate Texts in Mathematics*. Springer-Verlag, New York, 1980.
216. A. R. D. Mathias. What is Mac Lane missing? In H. Judah, W. Just, and W. H. Woodin, editors, *Set Theory of the Continuum*, volume 26 of *Mathematical Sciences Research Institute Publications*, pages 113–118. Springer Verlag, New York, 1992.
217. A. R. D. Mathias. The strength of Mac Lane set theory. *Ann. Pure Appl. Logic*, 110(1–3):107–234, 2001.
218. J.-P. May. Stable algebraic topology. 1945–1966. In I. M. James, editor, *History of Topology*, pages 665–723. North-Holland, Amsterdam, 1999.
219. C. McLarty. Left exact logic. *J. Pure Appl. Algebra*, 41(1):63–66, 1986.
220. C. McLarty. Stable surjection logic. *Diagrammes*, 22:45–57, 1989. Journées E.L.I.T. (Esquisses, Logique et Informatique Théorique), Vol. 1 (Paris, 1988).
221. C. McLarty. The uses and abuses of the history of topos theory. *British J. Philos. Sci.*, 41(3):351–375, 1990.
222. C. McLarty. Axiomatizing a category of categories. *J. Symbolic Logic*, 56(4):1243–1260, 1991.
223. C. McLarty. *Elementary Categories, Elementary Toposes*, volume 21 of *Oxford Logic Guides*. The Clarendon Press, Oxford, 1995.
224. C. McLarty. Exploring categorical structuralism. *Philos. Math. (3)*, 12(1):37–53, 2004.
225. C. McLarty. Learning from questions on categorical foundations. *Philos. Math. (3)*, 13(1):44–60, 2005.
226. B. P. McLaughlin. Varieties of supervenience. In E. E. Savellos, editor, *Supervenience: New Essays*, pages 16–59. Needham Heights, Cambridge, 1995.
227. B. Mitchell. The full imbedding theorem. *Amer. J. Math.*, 86:619–637, 1964.
228. B. Mitchell. *Theory of Categories*. Academic Press, New York, 1965.
229. I. Moerdijk. Toposes and groupoids. In *Categorical algebra and its applications (Louvain-La-Neuve, 1987)*, volume 1348 of *Lecture Notes in Math.*, pages 280–298. Springer, Berlin, 1988.
230. I. Moerdijk. *Classifying Spaces and Classifying Topoi*, volume 1616 of *Lecture Notes in Mathematics*. Springer-Verlag, Berlin, 1995.
231. C. Mulvey and J. W. Pelletier. A globalization of the Hahn-Banach theorem. *Adv. Math.*, 89(1):1–59, 1991.
232. J. R. Munkres. *Topology: A First Course*. Prentice-Hall, Englewoods Cliffs, 1975.
233. M. A. Naimark and A. I. Stern. *Theory of Group Representations*, volume 246 of *Grundlehren der Mathematischen Wissenschaften*. Springer Verlag, New York, 1982.
234. V. V. Nikulin and I. R. Shafarevich. *Geometries and Groups*. Universitext: Springer Series in Soviet Mathematics. Springer-Verlag, Berlin, 1987.
235. B. Pareigis. *Categories and Functors*, volume 39 of *Pure and Applied Mathematics*. Academic Press, New York, 1970.
236. W. R. Parzynski and P. W. Zipse. *Introduction to Mathematical Analysis*. International Series in Pure and Applied Mathematics. McGraw-Hill, New York, 1982.

237. A. M. Pitts. Conceptual completeness for first-order intuitionistic logic: An application of categorical logic. *Ann. Pure Appl. Logic*, 41(1):33–81, 1989.
238. B. Plotkin. Algebra, categories and databases. In M. Hazewinkel, editor, *Handbook of Algebra*, volume 2, pages 79–148. North-Holland, Amsterdam, 2000.
239. H. Poincaré. Rapport sur les travaux de M. Cartan. *Acta Mathematica*, 38:137–145, 1914.
240. H. Poincaré. Les travaux de M. Cartan. In M. A. Akivis and B. A. Rosenfeld, editors, *Élie Cartan (1869–1951)*, volume 123 of *Translations of Mathematical Monographs*, pages 311–315. American Mathematical Society, Providence, 1993.
241. M. Resnik and D. Kushner. Explanation, independence and realism in mathematics. *British J. Philos. Sci.*, 38(2):141–158, 1987.
242. R. Rosebrugh and R. J. Wood. An adjoint characterization of the category of sets. *Proc. Amer. Math. Soc.*, 122(2):409–413, 1994.
243. J. J. Rotman. *An Introduction to Algebraic Topology*, volume 119 of *Graduate Texts in Mathematics*. Springer Verlag, New York, 1988.
244. D. E. Rowe. Klein, Lie and the Erlanger Programm. In L. Boi, D. Flament, and J.-M. Salanskis, editors, *1830–1930: A Century of Geometry*, pages 45–54. Springer-Verlag, Berlin, 1992.
245. B. Russell. *Introduction to Mathematical Philosophy*. G. Allen & Unwin, London, 1919.
246. A. Scedrov. *Forcing and Classifying Topoi*, volume 48 of *Memoirs of the American Mathematical Society*. American Mathematical Society, Providence, 1984.
247. D. I. Schlomiuk. An elementary theory of the category of topological spaces. *Trans. Amer. Math. Soc.*, 149:259–278, 1970.
248. J.-P. Serre. Homologie singulière des espaces fibrés. *Ann. of Math.*, 54(3):425–505, 1951.
249. E. Sharpe. *Differential Geometry*, volume 166 of *Graduate Texts in Mathematics*. Springer Verlag, New York, 1996.
250. S. Shelah. Logical dreams. *Bull. Amer. Math. Soc. (N.S.)*, 40(2):203–228, 2003.
251. G. Sommaruga. *History and Philosophy of Constructive Type Theory*, volume 290 of *Synthese Library*. Kluwer, Dordrecht, 2000.
252. E. H. Spanier. *Algebraic Topology*. McGraw-Hill, New York, 1966.
253. D. I. M. Spraggon. The influence of Klein's Erlangen Program. In M. Kinyon, editor, *Proceedings of the Canadian Society for the History and Philosophy of Mathematics*, 2001.
254. R. Stalnaker. Varieties of supervenience. *Noûs*, 30(10):221–241, 1996.
255. M. H. Stone. Applications of the theory of Boolean rings to general topology. *Trans. Amer. Math. Soc.*, 41(3):375–481, 1937.
256. M. H. Stone. Remarks of Professor Stone. In F. Browder, editor, *Functional Analysis and Related Fields*, pages 235–241. Springer-Verlag, New York, 1970.
257. P. F. Strawson. *Entity and identity: And Other Essays*. Clarendon Press, Oxford, 1997.
258. P. Taylor. *Practical Foundations of Mathematics*, volume 59 of *Cambridge Studies in Pure Mathematics*. Cambridge University Press, Cambridge, 1999.
259. P. Taylor. Geometric and higher order logic in terms of abstract Stone duality. *Theory Appl. Categ.*, 7(15):284–338 (electronic), 2000.
260. P. Taylor. Subspaces in abstract Stone duality. *Theory Appl. Categ.*, 10(13):301–368 (electronic), 2002.
261. J. Tits. Les groupes de Lie exceptionnels et leur interprétation géométrique. *Bull. Soc. Math. Belg.*, 8:48–81, 1956.
262. R. Vanden Eynde. Development of the concept of homotopy. In I. M. James, editor, *History of Topology*, pages 65–102. North-Holland, Amsterdam, 1999.
263. R. Vaught. *Set Theory: An Introduction*. Birkhäuser, Boston, 1995.
264. O. Veblen and J. H. C. Whitehead. *The Foundations of Differential Geometry*. Cambridge University Press, Cambridge, 1932.
265. H. Wallman. Lattices and topological spaces. *Ann. of Math. (2)*, 39(1):112–126, 1938.
266. C. Weibel. History of homological algebra. In I. M. James, editor, *History of Topology*, pages 797–836. North-Holland, Amsterdam, 1999.

267. C. Wells. A generalization of the concept of sketch. *Theoret. Comput. Sci.*, 70(1):159–178, 1990. Fourth Workshop on Mathematical Foundations of Programming Semantics (Boulder, CO, 1988).
268. H. Weyl. *The Classical Groups. Their Invariants and Representations*. Princeton University Press, Princeton, 1939.
269. J. B. Wilker. Inversive geometry. In C. Davis, B. Grünbaum, and F. A. Sherk, editors, *The Geometric Vein*, pages 379–442. Springer Verlag, New York, 1981.
270. E. Zermelo. Über Grenzzahlen und Mengenbereiche: neue Untersuchungen über die Grundlagender Mengenlehre. *Fundamenta Mathematicae*, 14:339–344, 1930.
271. M. Zisman. Fibre bundles, fibre maps. In I. M. James, editor, *History of Topology*, pages 605–629. North-Holland, Amsterdam, 1999.

Index

A

$\mathscr{A}$-algebra 195
Abelian category *see* Category, Abelian
Action of a category on a set 51
Additive category *see* Category, additive
Adjoint functor 65, 110, 132, 148
   As criterion of identity 113
   As criterion of meaningfulness 113
Adjointness 65, 159
Adjunction 94, 127, 153
   Counit of an — 143
   Unit of an — 143
Algebra of type $\mathscr{A}$ 195
Algebraic semantics 196
Algebraic structure 196
Algebraic theory *see* Theory, algebraic
Alphabet *see* Type, similarity
Arrow
   Epi — 95
   Monic — 94
Awodey, S. 281
Axiom of extensionality 87
Axiomatization of structure reduction to set theory 188

B

Bénabou, J. 197, 268
Beck-Chevalley conditions 225
Bell, J. 57–59, 205
Benacerraf, P. 28
Bernays, P. and P. Levy 183
Bicategory *see* Category, bicategory
Bimorphism 95
Blass, A. 52
Borsuk, K. and S. Eilenberg 41, 42

Bourbaki, N. 40, 76, 77, 110, 111, 176, 197, 198
Brouwer, L. E. J. 114
Brouwer, L. E. J. 114
Brown, E. H. Jr. 262
Buchsbaum, D. A. 91, 100
Buchsbaum, D. A. 72, 84–86, 90–94, 100, 101, 133, 286

C

c.s.s. complex 127, 128
c.s.s. map 128
Canonical language of a category 240
   Extended — 240
Cantor, G. 182, 185, 189, 212
Carnap, R. 49
Cartan, É. 9, 10, 19, 21, 22, 34, 35, 56, 167, 170, 174, 193, 245, 249, 275
Cartan, H. 100, 128, 139, 186
Cartan, H. and S. Eilenberg 67, 71–74, 80, 85, 90–92, 100, 106, 122, 129, 133
Categorical doctrines 241
Categorical property 88, 172
Category 72, 92, 102
   — of concepts 235
   1 61
   2 62
   3 62
   ω-category 288
   Abelian — 84, 86, 93, 98
   Abstract — 166, 177
   Additive — 96
   Algebraic — 195
   As a space 105
   As criterion of identity 167
   As generalization of monoids 48

303

As generalization of partial orders 48
Bicategory 83, 84
Boolean — 241
Cartesian — 241
— of (set-)models 242
— of étales bundles 162
— of functors 61
— of presheaves 163
— of sheaves 163
Coherent — 241
Concrete — 177
Discrete — 205
Dual — 84, 93
Eilenberg and Mac Lane's definition 45
Equivalent — 94, 161
Finite — 87
Grothendieck's definition 93
Heyting — 241
Higher-dimensional — 288
Large — 53, 177
Left-Exact *see* Category, Cartesian
Locally small — 180
Mac Lane's definition 83
Product — 50
Representation of a — in **Set** 49
Simplicial — 128
Sketchable — 230
Small — 53, 177
U-category 178
U-small — 178
Category description theorem 207
Category of categories 51
   Axiomatization of the — 59, 201
Category theory 41, 92, 112, 147, 175, 185, 212, 287
   As generalization of Erlangen Program 43, 59
   Foundations of — 52, 53, 63, 101, 176
   Set-theoretical foundations of — 177
Cayley's representation theorem 48
   Generalization to categories 48, 49
Cayley, A. 15
CDT *see* Category description theorem
Čech, E. 115
Chern, S.-S. and C. Chevalley 34
Class 53, 180
   Proper — 53
Codiagonal 152
Coequalizer 89
Cohen, P. 176, 187, 210
Cohomology theory 68
Cokernel 98
Colimit 145
Commutative diagram 69

Compactification 75
   One-point — 75
   Stone-Čech — 75
Complete semi-simplicial complexes *see* c.s.s. complexes
Completeness
   Conceptual — *see* Theory, conceptually complete
   Strong conceptual — *see* Theory, strongly conceptually complete
Concept
   Extension of a — 33
   Geometric — 23
   Self-dual — 88
Conceptual frame 216
Conceptual level 234
Conceptual, the 212
Cone 227
Congruence
   Affine-congruence 18
   Euclidean-congruence 17
Context 273
   Empty — 273
   Suitable — 273
Context principle 171
Coproduct 88
Corry, L. 214
Cover 250
Covering system 251
Criterion of identity
   — for categories 87, 94, 162
   — for geometries 32
   External — 171
   Internal — 171
Criterion of meaningfulness 172

D

Darboux, J.-G. 23, 24
Dedekind, R. 185, 189
Dense below subset 255
Descartes, R. 70
Diagonal 152
Diagram over a category 144
Dieudonné, J. 69, 103, 254
Differential geometry
   As generalization of Erlangen PRogram 170
Direct limit 144
Doctrines *see* Categorical doctrines
Dual statement 84
Duality 78, 83, 158
   — theorems 163
   Gefland — 248

Index

Pontrjagin — 78, 110, 158
Principle of — *see* Principle of Duality
Stone — 158

E

Eckmann, B. 141
Ehresmann, C. 147, 167, 168, 170, 174, 175, 197, 202, 225, 233
Eilenberg, S. 3, 41, 42, 76, 100, 110–113, 116, 119, 121, 122, 125–128, 139, 145, 158, 177, 186, 189, 191, 208
Eilenberg, S. and J. A. Zilber 121
Eilenberg, S. and J. A. Zilber 122, 124, 128
Eilenberg, S. and N. E. Steenrod 119, 139
Eilenberg, S. and N. E. Steenrod 67–73, 77–80, 91, 93, 100, 105, 115, 122–126, 129, 133, 174
Eilenberg, S. and S. Mac Lane 9, 41, 42, 44, 47, 67, 73, 119, 123, 132, 147, 168, 179, 191
Eilenberg, S. and S. Mac Lane 1, 3–5, 9–11, 40, 42–45, 47–55, 59–61, 63–66, 71, 73, 79, 85, 94, 97, 100, 101, 105, 108–113, 120–126, 129, 132–136, 138, 139, 143, 145, 152, 157, 159, 164, 165, 167, 175, 186, 189, 192, 201, 202, 208, 210, 247, 249, 284–286, 290
Element 209
Elementary theory of abstract categories 202
  Axiomatization of — 202
Elementary theory of the category of sets 208
Elementary topos *see* Topos, elementary
Entailment 216
Epimorphism *see* Arrow, epi
Equalizer 88
Equational — 231
Equivalence of categories 92, 94, 161
Equivalence of completeness and representation theorems 217, 245
Equivalence of sketches 230
Erlangen Program 3, 4, 7, 9–12, 15, 25, 34, 35, 37–41, 43, 46, 59, 63–66, 74, 93, 97, 102, 106, 113, 132, 148, 158, 164, 167, 170, 174, 208, 210, 214, 215, 246, 247, 284, 289, 290
ETAC *see* Elementary theory of abstract category
Étales bundles 162
ETCS *see* Elementary theory of the category of sets
(Euclidean) n-simplex 129
Evaluation map 119

Exact sequence 80
Exponential law 119

F

Feferman, S. 6, 55, 57–59, 186
Fiber 169
Fibered product 222
Finite limit sketch 227
Formal system 236
Formal, the 212
Formula
  Coherent — 236
    Implication of — 237
  Geometric — 237
    Implication of — 237
Foundations of Algebraic Topology 67, 72, 73, 124
Foundations of mathematics 184, 289
  Categorical — 55, 57, 191
    strong sense 57
    weak sense 58
Frame 256
Frege, G. 33, 182
Frege-diagrammatic sentence *see* Freyd-diagrammatic sentence
Freyd adjoint functor theorem 137
Freyd, P. 6, 98, 99, 139, 147, 172, 173, 175, 191, 197, 202, 213, 225, 282
  Axiomatization of Abelian categories 99
Freyd-diagrammatic sentence 173
Function
  Monotone — 128
Functor 49
  Additive — 98
  Algebraic — 195
  As basic object of mathematics 106
  As object of a category 105
  Composition of — 49
  Conservative — 245
  Contravariant — 49
  Covariant — 49
  Diagonal — 144
  Direct image — 259
  Faithful — 49, 104
  Forgetful — 50
  Homotopy invariant — 115
  Inverse image — 259
  Lifted — 130
  Quotient — 64
  Representable — 65, 102, 106
  Representable — and universal morphism 107
  Representation — 149

Representation of a — 106
Semantic — 196
Subobject — 262
Functor category   *see* Category of functors
Functorial construction   152
Fundamental group   *see* Group, fundamental

G

G-space
   étale —   258
G-space over a locale   257
G-space over a space   255
   étale —   256
Gödel, K.   212, 217, 277, 280
Gabriel, P. and F. Ulmer   197, 233
Galois connections   157
Geometric morphisms   *see* Morphism, geometric
Geometric property   21, 22
   Euclidean property   17
Geometric transformations
   Affine transformations   18
   Collineation   18
      Algebraic representation   18
   Complex linear transformations   14
   Inversion   13
   Isometry   12, 17
      Algebraic representation   17
Geometry
   Affine plane —   18
   Algebraic form of a —   39, 40
   Equivalence of —   26
   Generating element of a —   28
   — of the complex projective line   14
   Inversive —   12, 15
   Möbius —   *see* Inversive geometry
   Plane Euclidean —   17
   Spherical —   13
   Subgeometry of a —   31
Germ   169
Giraud's theorem   253
Giraud, J.   253
Grothendieck topology   *see* Topology, Grothendieck
Grothendieck Topos   *see* Topos, Grothendieck
Grothendieck universe   178
Grothendieck, A.   1, 2, 5, 7, 73, 74, 84–86, 90–94, 96, 98–109, 111, 119, 133, 147, 157, 164, 176, 178–180, 182, 186, 189, 192, 194, 198, 201, 246–249, 252–254, 256, 259, 261, 266, 268, 270, 286
   Foundations of algebraic geometry   102

Group
   Center of a —   64
   Fundamental group   117
   — in a category   226
   — of affine transformations   18
   — of isometries   17
   Transformation —   16, 20
      As identity criterion   20
Group action   24
   Effective —   25
   Transitive —   24
Group representation   25
Groupoid   168
   As generalization of group and pseudogroup   169
   Continuous —   *see* Localic
   Fine —   170

H

Hakim, M.   268
Halmos, P. R.   215
Halpern, J. D.   187
Heller, A.   100, 101, 133, 286
Henkin, L.   279
Hesse, O.   25
Hilbert, D.   3, 26, 76, 282, 287
Hilton, P.   123, 141, 142
Hom-functor relative to a functor   140
Hom-set   45
Homological Algebra   67, 71–74, 91
Homological algebra   71
Homology theory   67, 68
   Axiomatization of —   67
   Equivalence   68
Homomorphism
   — of models   242
   Injective —   80
   Surjective —   80
Homotopic relative mappings   114
Homotopy   114
   Free —   114
Homotopy class   115
Homotopy equivalence   115
Homotopy equivalent   161
Homotopy theory   114, 115
Homotopy type   115
Hopf, H.   115
Hu, S.-T.   123
Hurewicz isomorphism theorem   121
Hurewicz, W.   3, 80, 114–116, 118, 120, 121, 123, 141

Index

I

Initial object 88
Internal theory of an algebra 220
Interpretation 217, 241
Isbell, J. R. 197, 207
Isomorphic objects 47
Isomorphism of categories 94
Isomorphism of objects 47

J

j-closure 266
j-sheaf 266
James, I. M. 133, 141, 143
Johnstone, P. T. 210, 261, 265, 269
Joyal and Tierney's representation theorem 258
Joyal, A. 246, 268–270, 282
Joyal, A. and G. Reyes 215
Joyal, A. and M. Tierney 256, 257, 259, 282

K

Kan extension 130
Kan, D. 5, 42, 73, 74, 109–115, 118, 121, 122, 124–145, 152, 153, 155, 164, 180, 186, 189, 191, 192, 286
Kant, I. 24, 49
Kelley, J. L. 119
Kernel 79, 98
Killing, W. 34, 249
Kleene, S. 277
Klein's classification of geometries 31, 32, 64
Klein's program *see* Erlangen Program
Klein, F. 3, 4, 7, 10, 15, 16, 19, 23–28, 30, 31, 34–37, 39, 41, 44, 45, 56, 64, 65, 76, 106, 157, 166, 167, 174, 190, 247, 249
Kline, M. 31
Kock, A. and G. Reyes 214
Krömer, R. 176
Kreisel G. and J.-L. Krivine 187
Kreisel, G. 6, 182, 184–189, 199–201
Kuhn, T. 112

L

$\mathbb{L}$-algebra 217
Lagrange, J.-L. 114
Lambek, J. 197, 203, 212, 277, 278, 280, 281
Lambek, J. and I. Moerdijk 281
Lambek, J. and P. J. Scott 279

Lawvere, F. W. 111, 119, 139, 177, 180, 182, 183, 194, 195, 199, 200, 202, 208, 210, 213, 214
Lawvere, F. W. and M. Tierney 247
Lawvere, F. W. and S. Schanuel 213
Lawvere, F. W. 6, 59, 63, 86, 105, 106, 147, 176, 177, 179, 181, 190–215, 220, 224, 235, 242, 246, 259
Lawvere, F. W. and M. Tierney 177, 247, 261, 264
Lawvere, F. W. and R. Rosebrugh 199
Lawvere-Tiernay topology *see* Topology, Lawvere-Tierney
LE-sketch *see* Finite limit sketch
Lie, S. 23, 24, 34, 36, 166, 167, 190, 247, 249
Lindenbaum-Tarski algebra 218
Linton, F. E. J. 196, 197
Local homeomorphism 162
Local operators 265
Locale 257
Localic 256, 257
Localization 250
Localization system 251
Locke, J. 166
Logical axiom scheme 216
Lubkin, S. 99

M

Mac Lane, S. 10, 42, 53, 77, 83, 91, 111, 119, 137, 177, 179, 180, 182, 198, 208
Mac Lane, S. 3, 5, 6, 11, 44, 48, 77–86, 90–94, 100, 101, 109–113, 122, 125, 132–134, 137, 139, 147, 148, 176, 177, 179–182, 184, 189, 192, 198, 201, 286
Makkai, M. 234
Manifold
 Connected — 24
Map
 Admissible — 170
 Degeneracy — 129
 Face — 129
Massey, W. S. 124
McLarty, C. 84, 86
Mendelson, E. 198
Metacategory 181
Metagraph 181
Metric space
 Complete — 74
 Completion of a — 75
Mitchell, B. 99
Model 213, 217
 Generic — 219

— of an algebraic theory   195
(Set-)model   242
Universal   *see* Model, generic
Module
  Infinitely divisible —   81
  Injective —   80
  Projective —   80
Monomorphism   *see* Arrow, monic
  Dense —   266
Moore, J. C.   125
Morita equivalence   243
Morita invariance   243
Morita, K.   243
Morphism
  — of étale G-spaces   258
  Coimage of a —   99
  Geometric —   260, 267
  Image of a —   99
  Logical —   271
  — of G-spaces   258

N

Natural isomorphism   60
Natural transformation   60
Noether, E.   3
Noll, W.   198
Normal subschool   180

O

Organization of mathematics   184
Orthogonal matrix   12

P

Pareigis, B.   195
Particular   20
Path   116
  Closed —   116
  Constant —   117
  End of a —   116
  Inverse —   117
  Origin of a —   116
Path class   117
Peano, G.   33
Peano-Lawvere characterization of the natural numbers   209
Plücker, J.   23
Poincaré, H.   55, 56, 116
Point
  Tangent —   13
Pointed space   *see* Space, pointed
Power object   262

Presheaf   252
Pretopos   241
Principle of duality   81, 84, 93
Principle of transference   4, 25
Product   149
  Cartesian —   82, 87
  Direct —   81
  Free —   82
  — of objects   86
  Wedge —   142
Pseudogroup   168
Pullback   221
Puppe, D.   125

Q

Quantifiers
  Stable under substitution   224
Quotient-objet   85

R

Reduced suspension   141
Reflection principle   182
Representation functor   *see* Functor, representation
Retraction   96
Reyes, G.   268–270
Robinson, A.   187
Rowe, D. E.   34
Rule of inference   216
  Conclusion of a —   216
  Premises of a —   216
Russell, B.   186, 187, 212

S

Samuel, P.   76, 77
Schlomiuk, D.   210
Section   96
  Global —   260
Semantic consequence   217
Semantics   234
Sequent over a *tau*-signature   273
Serre, J.-P.   121, 133, 141, 143
Set   *see* Category, discrete
  Abstract —   261
  Formal —   237
  — of components   206
  — of morphisms   206
  — of objects   206
  Small —   179
Sets   53
Sheaf   163, 252, 266

Index

Sheaf theory 86
Sieve 250
Simplicial object 129
Simplicial set 128
Site 163, 251
Space
  Loop— 142
  Path component of a— 116
  Path connected— 116
  Path in a— *see* Path
  Pointed— 118
Spanier, E. H. 69
Stabilizer subgroup of a generating element 29
Steenrod, N. E. 42
Steenrod, N. E. 41, 42, 123, 125, 139, 186
Stone representation theorem 110
Stone, M. 111, 112, 119, 137, 158, 159
Subfunctor 64
Subobject 85, 231
  Dense— 266
Subobject classifier 262
Subobject functor *see* Functor, subobject
Subtransformation 64
Supervenience 37
  Formal— 36, 38, 69
  — of group theory over geometry 37, 38
Sur Quelques Points d'algèbre homologique 90, 92
Syntax 234

T

T-algebra 231
Tarski, A. 198, 203, 212, 215
τ-signature 272
Taylor, P. 66, 199
Terminal object 88
Theory
  Algebraic— 195
  Coherent— 237
  Conceptually complete— 242
  Geometric— 237
  Logical form of a— 39
  Strongly conceptually complete— 244
Theory of schools 180
Tierney, M. 6, 193, 262, 268
Tohoku *see Sur Quelques Points d'algèbre homologique*
Token 20, 82, 230
Topological space
  Compact— 75
  Open cover of a— 75
Topology
  Grothendieck— 251
  Lawvere-Tierney 264

Topos
  Boolean— 279
  Classifying— 267
  Cocomplete— 268
  Coherent— 270
  Elementary— 261, 262
  Free—with natural number object 276
    As foundational framework 277
  Grothendieck— 249, 253
  Local— 277
  — of G-equivariant sheaves 256
  Well-pointed— 280
Topos theory 247
Truesdell, C. 139, 198
Type 20, 82, 87, 230
  Similarity— 236
  — of a context 273

U

Uniqueness of the completion property 75
Unit sphere S^2 13
Universal algebra 85
Universal mapping 76
Universal morphism 65, 88
Universal property 85, 87
Universe *see* Grothendieck universe, 179

V

Variable models of set-theory 180
Vaught, R. L. 119
Veblen, O. and J. H. C. Whitehead 168
Verdier, J.-L. 178
Vietoris, L. 121

W

Wallman, H. 257
Weibel, C. 71
Weil conjectures 74, 192
Weil, A. 40, 157
Weyl, H. 34, 157, 170
World War II 112

Y

Yoneda lemma 105
Yoneda, N 105

Z

Zermelo, E. 182
Zero object 88
Zilber, J. A. 125